Steve O'Meara's
Herschel 400 Observing Guide

The Herschel 400 is a list of 400 galaxies, nebulae, and star
clusters, picked from over 2500 deep-sky objects discovered
and cataloged by the great eighteenth-century astronomer Sir
William Herschel and his sister Caroline. It comprises

- 231 galaxies
- 107 open clusters
- 33 globular clusters
- 20 planetary nebulae
- 2 halves of a single planetary nebula
- 7 bright nebulae

In this guide Steve O'Meara takes the observer through
the list, season by season, month by month, night by night,
object by object. He works through the objects in a carefully
planned and methodical way, taking in some of the most
dramatic non-Messier galaxies, nebulae, and star clusters
in the night sky.

 Ideal for astronomers who have tackled the Messier objects,
this richly illustrated guide will help the amateur astronomer
hone their observing skills.

STEVE O'MEARA earned a Bachelor of Science degree from
Northeastern University and has spent much of his career on
the editorial staff of Sky & Telescope magazine. The Texas Star
Party gave him its Omega Centauri Award for "advancing
astronomy through observation, writing, and promotion, and
for his love of the sky," and the International Astronomical
Union named asteroid 3637 O'Meara in his honor.

Steve O'Meara's
Herschel 400
Observing Guide

How to find and explore 400 star clusters,
nebulae, and galaxies discovered by
William and Caroline Herschel

CAMBRIDGE
UNIVERSITY PRESS

Shaftesbury Road, Cambridge CB2 8EA, United Kingdom

One Liberty Plaza, 20th Floor, New York, NY 10006, USA

477 Williamstown Road, Port Melbourne, VIC 3207, Australia

314–321, 3rd Floor, Plot 3, Splendor Forum, Jasola District Centre, New Delhi – 110025, India

103 Penang Road, #05–06/07, Visioncrest Commercial, Singapore 238467

Cambridge University Press is part of Cambridge University Press & Assessment, a department of the University of Cambridge.

We share the University's mission to contribute to society through the pursuit of education, learning and research at the highest international levels of excellence.

www.cambridge.org
Information on this title: www.cambridge.org/9780521858939

First published 2007 (version 9, April 2024)

Printed in Great Britain by Ashford Colour Press Ltd., April 2024

A catalogue record for this publication is available from the British Library

ISBN 978-0-521-85893-9 Hardback
ISBN 978-1-107-63200-4 Paperback

To Donna, my North Star. Whenever I look into the face of the night, I am reminded of your beauty.

To Milky Way, Miranda Piewacket, and Pele, three angels living in my heart.

And to Daisy Duke Such a Joy, my angel here on Earth.

Contents

Preface

Congratulations, you are about to begin a remarkable visual journey, and this book is to be your faithful guide. The Herschel 400 is a list of 400 galaxies, nebulae, and star clusters, culled from the more than 2,500 deep-sky objects discovered and cataloged by the great eighteenth-century astronomer, Sir William Herschel and his sister Caroline. The list comprises 231 galaxies, 107 open star clusters, 33 globular star clusters, 20 planetary nebulae, 2 halves of a single planetary nebula, and 7 bright nebulae. It contains some of the most dramatic non-Messier galaxies, nebulae, and star clusters in the night sky; it also includes some objects at the very limits of detection in modest-sized telescopes.

Members of the Ancient City Astronomy Club (ACAC) of St Augustine, Florida, created the list in response to a letter published in the April, 1976 issue of Sky & Telescope (page 235). In that letter James Mullaney (Pittsburgh, Pennsylvania) suggested that amateurs set up an informal "Herschel Club" with the goal of observing the brightest of the more than 2,500 objects discovered and cataloged by William Herschel. "The total count of the Herschel objects is inconveniently large," Mullaney explained, "but can be reduced to about 615 by excluding his classes II and III (see page 3), which are largely made up of more difficult and less interesting specimens."

While ACAC members agreed with Mullaney's contention that most of the objects in Herschel's catalog were too faint for amateur astronomers to detect with modest-sized telescopes from the suburbs, they decided not to discount the class II and III objects; after careful consideration, they came up with a list of 400 Herschel objects that they said would challenge observers with 6-inch or larger telescopes under skies that were affected somewhat by light pollution (no pain, no gain). So, the Herschel 400 is not a list of Herschel's "brightest and best" deep-sky objects. It is a list designed to hone (or test) the observing skills of amateur astronomers living in the Northern Hemisphere, and there is much to be said for that.

Since its creation, the Herschel 400 list has become very popular. A Herschel 400 Club is now part of the Astronomical League (AL) – an organization composed of more than 240 amateur astronomical societies across the United States – which awards certificates to members who complete the Herschel 400 list. The AL promotes these certificates as "prized possessions among serious amateurs as an indication of the advanced level of their amateur capabilities." The key word here is "advanced." If you complete the Herschel 400 list, you will be considered an advanced observer. The craze is catching on. Other astronomy clubs and institutions now routinely challenge members to observe the Herschel 400 objects as well, and some offer their own awards. But, until now, no book has been available to help observers find these objects in any systematic or detailed way.

This Herschel 400 observing guide, then, will be a valuable asset to the growing masses of visual observers who desire a long-term project and challenge. Its purpose is mainly to guide you, star by star, to each of the 400 objects in the Herschel 400 list, but it will also help you to grow as an observer. Observers usually take interest in the Herschel 400 list after they succeed in finding all of the Messier objects – the 109 "nebulae" and star clusters cataloged by the eighteenth-century French comet hunter Charles Messier* – and want to continue searching for more rewards. Finding the Messier objects has long been regarded as an important first step for beginners wanting to learn the deep sky. Doing so not only helps them to become familiar with the varied appearances of deep-sky objects, but it also helps them become more adept at finding their way around the starry heavens.

But completing the Messier list is not a requirement to completing the Herschel 400 list. In fact, this book is designed to help observers of all levels find these objects in a carefully planned and methodical way – season by season, month by month, night by night, object by object.

Four things are required: a general knowledge of the night sky, an understanding of how to use binoculars and a telescope, a strong sense of commitment, and the will to succeed. By sticking to the strategy outlined in this book, you will not only master the Herschel 400 list but prove to all that you have mastered the deep sky.

This book is divided into four seasons: winter, spring, summer, and fall. Each season is further broken down by month (three months per season). Each month is subdivided into seven nights – the minimum number of nights I expect you will need to observe the target objects for each month. The number of objects I have selected for each night's observing varies; it depends largely on how difficult I believe each object will be to find or see. Note that the number of target objects on many spring evenings is high; that's because most of the objects are galaxies in the rich Virgo Cluster of galaxies, and the targets lie so close to one another that several can be seen in the same field of view.

* Deep-Sky Companions: The Messier Objects, Stephen James O'Meara (Cambridge, Cambridge University Press; Cambridge, MA, Sky Publishing, 2000).

I have taken great care to offer you a logical plan of attack for each night, taking into consideration the object's declination and when it is reasonably high enough in the sky for you to see well. I have considered how troublesome an object will be to find – whether it is near a bright guide star or isolated in a vast or seemingly empty field of dim suns. And I have considered how difficult it will be to see each object in a modest-sized telescope.

With these points in mind, each target opens with a general description that details the object being sought, where it is located in the sky, and, if applicable, its appearance to the unaided eye or binoculars. Many times, the general descriptions provide you with some words of warning or advice, which should prepare you for the search, or tell you what you should expect to see or how best to see it. Each evening's hunt is accompanied by a general sky chart that plots the target objects as well as a series of detailed star charts that not only zoom in on the object's position but also show the proposed plan of attack, which is detailed in the text. The object's general description is further complemented with a general view – a more detailed description of the object as seen at various magnifications in a 4-inch telescope under a dark sky.

It's important for me to stress that the main purpose of this guide is to help you find and see (not study) each object. My goal is to help you achieve success in the hunt – in the most swift and efficient manner; if you feel you'd like to study a particular object in more detail at a later date, I have provided a place for you in Appendix B to check off the object for further study – preferably after you've completed the Herschel 400 list.

Appendix B is, in fact, a checklist for you to keep track of when and where you observed each object. It also includes spaces for you to write down important information, such as telescope size, magnification, atmospheric seeing and transparency, and any other special notes you want to record. It is a personal log that you can return to weeks, months, or years later, to see how you are progressing as an observer. Two other appendices complete the work. Appendix A tabulates each object's type, constellation, position, magnitude, angular size, visibility rating, and Herschel catalog number. And Appendix C lists the book's photo credits.

I would like to thank Simon Mitton for encouraging me to pursue the book, and Vince Higgs, Lindsay Barnes, and the editorial staff at Cambridge University Press for helping me take the book through to completion. I sincerely thank Sue Tritton at the Plate Library of the Royal Observatory Edinburgh for granting me permission to use the Digitized Sky Survey images taken with the UK Schmidt Telescope; her contribution was invaluable. I give a tip of the hat to Al and David Nagler of Tele Vue Optics in Sufferin, New York, for making such a superb telescope that lets me reach the limits of vision in a small telescope. And I thank my friends and colleagues Larry Mitchell and Barbara Wilson of Houston, Texas, for their help in the histories of William and Caroline Herschel. I would also like to thank Michael Tabb of the William Herschel Society for his time and patience with me in Bath, England; the Society's venture is a most honorable and worthy one.

Finally, I would like to hug my beautiful wife, Donna, and Daisy Duke, our loving papillon, for tolerating my long absences at night as I flirted with the heavens, I thank them for their love, support, and understanding. Of course, any errors that might have materialized in this work belong solely to yours truly.

Stephen James O'Meara
Volcano, Hawaii
June 2006

Introduction

Before you set off to observe the Herschel 400 objects, you should know a little about the man who discovered the vast majority of them, and his equally remarkable sister.

Friedrich Wilhelm (later William) Herschel (1738–1822) was the pre-eminent astronomer of his time. Born on November 15, 1738, in Hanover, Germany, Herschel began his career as a musician and composer. In the fall of 1757, the 18-year-young man moved to England. After 10 years of a weathery existence, he settled down in Bath, where he made a comfortable living as an organist, teacher, and concert director. In 1772, Herschel's beloved sister, Caroline (1750–1848), joined him in Bath, where she helped keep her brother's house while enjoying a brief career as a vocalist. She also arrived in Bath at a most auspicious time: when William's passion for music was being eclipsed by his passion for making telescopes and observing the night sky.

William's career changed dramatically after March 13, 1781, when he discovered a new planet (Uranus) with a 6.2-inch reflector set up in his backyard. The following year, King George III offered him the position of court astronomer with an annual salary (of £200) for life; the reward allowed Herschel to give up music and make astronomy his career. The King subsequently made an allowance of £50 a year to Caroline as her brother's assistant, making her the first professional female astronomer. In 1772, William and Caroline moved to Dachet near Windsor Castle, where William began making increasingly larger reflecting telescopes. By the fall of 1783, he was sweeping the heavens with his large 20-foot reflecting telescope, which had an aperture of 18.7 inches. Mainly he used this telescope to survey double stars; little was known about the nature of these stars and their motions, and Herschel's examination of them – he often observed 400 a night – was the first systematic survey.

Meanwhile, Caroline's astronomical career was also blossoming. In August 1783, William surprised her with a gift of a homemade telescope, which she used to sweep the heavens for new comets; her success in this new venture led to her own fame and glory. While sweeping the heavens, she also discovered several deep-sky objects not on Messier's list. Her success impressed William, who suddenly became curious about the abundance of these new, unseen wonders. As Barbara Wilson notes in her biography of Caroline,* it was Caroline's "rash of deep-sky discoveries that prompted William to turn his attention away from double-star observing and to start his greatest endeavor – a systematic search for nebulae with his large 20-foot telescope, which began shortly after Caroline made her twelfth deep-sky discovery on October 30, 1783."

After some experimental attempts, William began a new systematic survey of the heavens with his large 20-foot reflector. After 20 years of review, William had discovered and cataloged (with the help of his sister) no fewer than 2,508 new "nebulae" and "clusters." His first official series of sweeps commenced on December 19, 1783, and the fruits of his labors led to the creation of his first *Catalogue of One Thousand New Nebulae and Clusters of Stars*, which was published on April 27, 1786. He published his *Catalogue of a Second Thousand of New Nebulae and Clusters of Stars* on June 11, 1789, and a final *Catalogue of 500 New Nebulae, Nebulous Stars, Planetary Nebulae, and Clusters of Stars*, was published on July 1, 1802.

I say "nebulae" and "clusters" because in Herschel's day no one knew the true nature of these objects. As William peered into the eyepiece of his large telescope, he could only imagine what wonders were before him. Yet the man displayed remarkable aptitude and deductive reasoning as he contemplated their natures. As Larry Mitchell explains, in his biography of William,** "Herschel carefully analyzed everything he saw in the night sky, and he tried to understand what all astronomical objects were composed of, and how and why they acquired their diverse forms."

To make better sense of the varied objects he was seeing, Herschel created a system to classify them. The code he created is the letter H, followed by a Roman numeral and an Arabic number – H I-11, for instance. The H stands for

---

* "Caroline Herschel: no ordinary eighteenth-century woman." Barbara Wilson, in *Deep-Sky Companions: Hidden Treasures*, Stephen James O'Meara (Cambridge: Cambridge University Press, 2007).

** "William Herschel: the greatest visual observer of all time." Larry Mitchell, in *Deep-Sky Companions: The Caldwell Objects*, Stephen James O'Meara (Cambridge, MA: Sky Publishing and Cambridge: Cambridge University Press, 2002).

Had William and Caroline been alive today, they would have marveled at the images of their discoveries taken with our great telescopes. Today, we now know that many of the Herschel nebulae are actually galaxies – vast citadels of stars, dust, and gas, held together by gravity; these island universes range from tiny dwarfs measuring a few hundred light-years across and containing a few million stars, to spectacular systems spanning over hundreds of thousands of light-years and containing several trillion stars (below, left). Some of the nebulae are planetary nebulae – luminous shells of matter spewed forth by sun-like stars as they near the end of their lives (below, middle). Still other nebulae turned out to be very distant globular clusters, orbs of ancient starlight, some 10 to 14 billion years old that reside in the Milky Way's halo tens of thousands of light-years distant (below right).

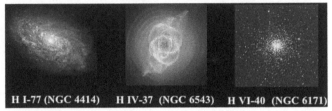

H I-77 (NGC 4414)   H IV-37 (NGC 6543)   H VI-40 (NGC 6171)

William Herschel, the Roman numeral identifies the class into which Herschel placed each object:

   I (bright nebulae)
  II (faint nebulae)
 III (very faint nebulae)
  IV (planetary nebulae: stars with burs, with milky chevelure, with short rays, remarkable shapes, etc.)
   V (very large nebulae)
  VI (very compressed and rich clusters of stars)
 VII (pretty much compressed clusters of large or small stars)
VIII (coarsely scattered clusters of stars).

The Arabic numeral that follows is simply the order in which that object appears in that class. So H I-11 is the 11th object in Herschel Class I (bright nebulae).

With this and other data before him, Herschel tried to fathom the construction of the universe. At first, he believed that all nebulae could be resolved into stars, given sufficient aperture. But by the turn of the nineteenth century, he had changed his mind, believing that some nebulae were indeed composed of some form of luminous matter. In William's new cosmology, the universe was in a state of flux.

As Mitchell explains, Herschel believed that "[n]ebulae and star systems slowly developed over time under the constant action of gravity, and the source of their luminosity was unknown. A nebula that was a little brighter in the middle than along its periphery had not undergone much central attraction and therefore was not very advanced. A nebula 'gradually brighter in the middle' was in a more advanced evolutionary state, while one that appeared 'gradually much brighter in the middle' was even more evolved." To Herschel, planetary nebulae were highly evolved nebulae, and globular star clusters (which he found lying near dark vacancies in the heavens) were objects that had somehow congealed to form these dark voids.

How lucky we are today. When we look through our telescopes at the faint, fuzzy glows that Herschel tried so desperately to understand, we see them with the added dimension of knowledge. So no matter how dim and faint a Herschel 400 object appears in your telescope, be thankful that, at least, as the poet Robert Frost had penned, we are, "Acquainted with the Night". Besides, to quote Frost again, "Do we know any better where we are[?]"

My observing site and telescope

The observations for this book were made from Volcano, Hawaii, where I live. Volcano is on the island of Hawaii – the youngest, largest, and most southerly of the Hawaiian islands. Also called the Big Island, Hawaii comprises five coalescing shield volcanoes of various ages. The tallest, Mauna Kea, is nearly 14,000-feet high and is occasionally snowcapped. It is also home to many of the world's largest and most technologically advanced telescopes, including the twin 10-meter Keck telescopes, the 8.2-meter Subaru telescope, and the 8-meter Gemini North telescope.

I'm often asked why I don't observe from the summit of Mauna Kea. While viewing the heavens from this lofty peak is an experience almost beyond imagining, conditions at the summit can be severe. (Remember, any professional astronomer observing atop Mauna Kea does so inside an enclosed and heated structure and is most likely monitoring a computer

screen, not standing outside with a telescope enduring the elements.) Outside the mighty observatories, the air is crisp and dry, with temperatures that typically hover near, or fall below, freezing. Winds can be strong; during severe conditions, winds can whip up to over 100 miles per hour. Also, the atmospheric pressure at the summit is 40 percent less than at sea level, so less oxygen is available to the lungs, and acute mountain sickness is common (though I have yet to suffer such conditions). But what concerns me, a visual observer, the most is that less oxygen is available to my eyes and mind, putting me at a disadvantage fully to appreciate and see the wonders above, unless I'm sucking on bottled oxygen.

Besides, there's no need for me to travel to Mauna Kea, because my home and its surroundings have world-class skies. I live at an altitude of 3600 feet, just a few miles from the 4200-foot-high summit of Kilauea, a gently sloping shield volcano that has been in near-continuous eruption since January 1983. Most of the observations for this book were made in the summit area of Kilauea Volcano (shown below). The large circular depression near the middle of this frame is the volcano's caldera, which measures about 2.5 miles long and 2 miles wide. The smaller pit is Halemaumau crater, which spans about 1000 feet (305 meters). Some of the observations were made from my front yard (see above right).

I observed all of the Herschel 400 objects with an old Tele Vue 4-inch f/5 Genesis refractor. I generally used only three eyepieces (also made by Tele Vue): a 22-mm Panoptic, a 7-mm Nagler, and a 4.8-mm Nagler, which provided magnifications of 23×, 72×, and 105×, respectively. Some of the planetary nebulae, especially, required higher magnifications to see well. In these cases, I employed either a 1.8× or a 3× Barlow lens in combination with the eyepieces listed above. As a finder I use a Tele Vue Qwik Point (it's like a laser pointer). The telescope sits in the cradle of a sturdy Gibraltar altazimuth mount; the entire set-up can be broken down in two minutes in case I need to be mobile.

It may surprise some that all 400 Herschel objects can be seen in a 4-inch telescope. But there's no need to be surprised. As I mentioned in the Preface, the members of the Ancient City Astronomy Club created the Herschel 400 list to challenge observers using 6-inch or larger telescopes under skies that were affected somewhat by *light pollution*. Tests with the 4-inch Genesis under dark Hawaiian skies at altitude prove that it can perform as well as an 8- to 10-inch Schmidt-Cassegrain telescope from a suburban site.

Now consider that although William's 18.7-inch telescope was large by the standards of his day, the mirror in that telescope was of inferior quality by the standards of today. Herschel's mirrors (see photo below) were not finely polished silvered glass. They were made of low-reflectivity speculum metal – a copper–tin alloy containing 45 percent tin. A speculum-metal mirror tarnishes quickly and loses reflectivity. Experiments have shown that reflectivity of speculum-metal mirrors varies from 63 percent at 4500 Angstroms to 75 percent at 6500 Å. In a study published in a 1947 *Journal of Scientific Instruments*, Tolansky and Donaldson (University of Manchester) reported that after keeping speculum mirrors for six months in a damp environment, their reflectivity decreased by 10 percent in the red region and by 2 percent in the blue region. So Herschel had, in essence, *discovered* the faintest objects in the Herschel 400 list with a telescope equivalent to a modern 10-inch reflector with excellent coatings.

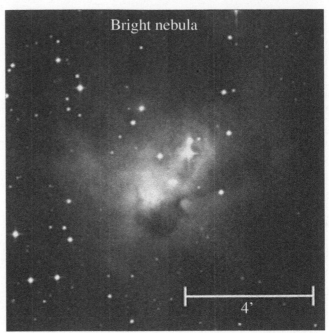

Bright nebula

4'

off its own light (emission) or shines by reflecting the light of nearby stars (reflection).

A *planetary nebula* is a luminous shell of gas cast off and caused to fluoresce by an evolved star of less than about four solar masses. William Herschel coined the term "planetary nebula" because, through his telescope, these objects appeared round in form and resembled the green, gas-giant planet Uranus, which he discovered. Note that not all planetary nebulae listed by Herschel are, in fact, true planetary nebulae in the modern sense; again, Herschel's

How to use this book

To find a Herschel object, first turn to the season and month you want to begin and review the table of essential data for the first target of the first night. The table includes the following data: NGC number, object type; constellation; equinox 2000.0 coordinates; apparent magnitude; angular size or dimensions; and object rating.

The **NGC** refers to the object's *New General Catalogue* (NGC) number, published in 1888 by Johann Louis Emil Dreyer. The NGC is an enlarged version of the *General Catalogue of Nebulae and Clusters* (GC) published in 1864 by William Herschel's son, John. All of the "nebulae" and clusters discovered by William, then, can be identified either by William's original code, its GC number, or an NGC number. Today, deep-sky objects are widely identified only by their NGC numbers; for instance, H VIII-8 is commonly referred to today as NGC 1647, an open star cluster in Taurus.

Object **type** identifies the object's class: a *bright nebula* is a luminous, interstellar cloud of dust and gas that either gives

Planetary nebula

2'

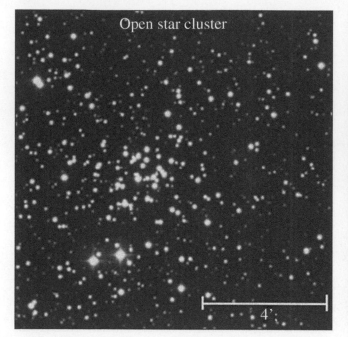

Open star cluster

4'

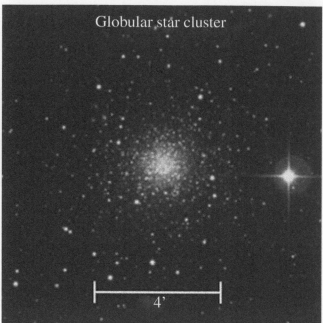

Globular star cluster

4'

classification refers to the object's visual appearance through a telescope, not to its astrophysical nature.

Open star clusters are loose and irregularly shaped collections of dozens or hundreds of young stars that travel in the thin disk of stars, dust, and gas comprising the plane of our galaxy. They occupy a volume of space typically less than 50 light-years across, are loosely held together by gravity, and are fated to disperse over a period of several hundred million years.

Globular star clusters are spherically symmetric collections of old stars that share a common origin. They lie far above or below the plane of our galaxy (in its halo), contain from tens of thousands to millions of stars, and measure from 100 to 300 light-years across.

Galaxies, also called island universes, are giant assemblies of stars, gas, and dust into which most of the visible matter of the universe is concentrated. They range in size from the smallest dwarf galaxies only a few hundred light-years across with just a few million suns, through normal galaxies like our own Milky Way, with a few hundred billion stars, to giant ellipticals spanning over hundreds of thousands of light-years and containing several trillion stars. Galaxies are of three basic types: spirals (normal, barred, or mixed), ellipticals, and irregular. Lenticular galaxies are those midway in form between a spiral and an elliptical. We see the various types of galaxies in a variety of orientations — from face-on to edge-on, to, in the case of some ellipticals, end-on.

Constellation refers to the grouping of stars within a region of sky that has been divided by international agreement. There are 88 official constellations.

The object's **coordinates** are given in *right ascension* (RA) and *declination* (Dec). Think of these latter terms as celestial longitude and latitude. Note: if you plan to star-hop to the

Normal spiral   Barred spiral   Mixed spiral

Elliptical   Irregular   Lenticular

Edge-on

Abbrev.	Constellation	Latin genitive	Abbrev.	Constellation	Latin genitive
And	Andromeda	Andromedae	CMa	Canis Major	Canis Majoris
Ant	Antlia	Antliae	CMi	Canis Minor	Canis Minoris
Aps	Apus	Apodis	Cap	Capricornus	Capricorni
Aqr	Aquarius	Aquarii	Car	Carina	Carinae
Aql	Aquila	Aquilae	Cas	Cassiopeia	Cassiopeiae
Ara	Ara	Arae	Cen	Centaurus	Centauri
Ari	Aries	Arietis	Cep	Cepheus	Cephei
Aur	Auriga	Aurigae	Cet	Cetus	Ceti
Boo	Bootes	Bootis	Cha	Chamaeleon	Chamaeleontis
Cae	Caelum	Caeli	Cir	Circinus	Circini
Cam	Camelopardalis	Camelopardalis	Col	Columba	Columbae
Cnc	Cancer	Cancri	Com	Coma Berenices	Comae Berenices
CVn	Canes Venatici	Canum Venaticorum	CrA	Corona Australis	Coronae Australis

Abbrev.	Constellation	Latin genitive	Abbrev.	Constellation	Latin genitive
CrB	Corona Borealis	Coronae Borealis	Oct	Octans	Octantis
Crv	Corvus	Corvi	Oph	Ophiuchus	Ophiuchi
Crt	Crater	Crateris	Ori	Orion	Orionis
Cru	Crux	Crucis	Pav	Pavo	Pavonis
Cyg	Cygnus	Cygni	Peg	Pegasus	Pegasi
Del	Delphinus	Delphini	Per	Perseus	Persei
Dor	Dorado	Doradus	Phe	Phoenix	Phoenicis
Dra	Draco	Draconis	Pic	Pictor	Pictoris
Equ	Equuleus	Equulei	Psc	Pisces	Piscium
Eri	Eridanus	Eridani	PsA	Pisces Austrinus	Piscis Austrini
For	Fornax	Fornacis	Pup	Puppis	Puppis
Gem	Gemini	Geminorum	Pyx	Pyxis	Pyxidis
Gru	Grus	Gruis	Ret	Reticulum	Reticuli
Her	Hercules	Herculis	Sge	Sagitta	Sagittae
Hor	Horologium	Horologii	Sgr	Sagittarius	Sagittarii
Hya	Hydra	Hydrae	Sco	Scorpius	Scorpii
Hyi	Hydrus	Hydri	Scl	Sculptor	Sculptoris
Ind	Indus	Indi	Sct	Scutum	Scuti
Lac	Lacerta	Lacertae	Ser	Serpens	Serpentis
Leo	Leo	Leonis	Sex	Sextans	Sextantis
LMi	Leo Minor	Leo Minoris	Tau	Taurus	Tauri
Lep	Lepus	Leporis	Tel	Telescopium	Telescopii
Lib	Libra	Librae	Tri	Triangulum	Trianguli
Lup	Lupus	Lupi	TrA	Triangulum Australe	Triangulum Australis
Lyn	Lynx	Lyncis	Tuc	Tucana	Tucanae
Lyr	Lyra	Lyrae	UMa	Ursa Major	Ursae Majoris
Men	Mensa	Mensae	UMi	Ursa Minor	Ursae Minoris
Mic	Microscopium	Microscopii	Vel	Vela	Velorum
Mon	Monoceros	Monocerotis	Vir	Virgo	Virginis
Mus	Musca	Muscae	Vol	Volans	Volantis
Nor	Norma	Normae	Vul	Vulpecula	Vulpeculae

object using the method in this book, you do not need to use these coordinates. They are here for the benefit of those who will be using Go To telescopes, or for those who want to find an object on their own star atlases or charts that have right ascension and declination grids (the star charts in this book do not have these grids). The coordinates are precise for "equinox 2000.0". The coordinate system is in constant change. Gravitational tugs by the Sun, Moon, and planets, cause the Earth's axis to wobble like a top. It takes about 26,000 years for the axis to complete a wobble. Although this is a long time, the gradual shift adds up, so every 50 years or so star charts are revised to incorporate this shift, or precession, of the coordinate system against the backdrop of stars. For this book, the coordinates given correspond exactly to the year 2000, hence equinox 2000.0.

Magnitude refers to an object's apparent brightness. The brighter an object appears, the smaller the numerical value of its apparent magnitude. On the brighter side of the magnitude scale, the values soar into the negative numbers. Sirius, the brightest star in the night sky, for instance, shines at magnitude -1.6. As a general rule, the faintest stars visible at a glance to the unaided eye hover at around 6th magnitude. Mathematically, a 1st-magnitude star is 2.512 times brighter than a 2nd-magnitude star, which is 2.512 times brighter than a 3rd-magnitude star, and so on. The math works out nicely so that a star of 1st magnitude is exactly 100 times brighter than a star of 6th magnitude. The faintest star visible *at a glance* in 7×50 binoculars is of about 9th magnitude, and 12th magnitude is the faintest star visible in a 4-inch telescope without effort. But these numbers are *very* conservative; the limit you see will vary wildly depending on your location, the clarity of the atmosphere, the degree of light pollution, your visual acuity, the time you spend looking behind the eyepiece, and your expertise.

When it comes to observing deep-sky objects, "magnitude" is also deceiving. In most cases, the Herschel 400 objects do not appear as point sources – although the nuclei of some galaxies, and the view of some planetary nebulae at low power, do appear starlike (which is important to keep in mind). For the most part, the light of these deep-sky objects is spread across a specific area of sky. A 10th-magnitude galaxy, then, will *appear* dimmer than a 10th-magnitude star, because the light is no longer concentrated but diffused over a greater area of sky. Imagine how the concentration of light differs when you use the different settings of a flashlight with an adjustable beam. The wider the beam, the less intense the beam appears. This dimming effect is intensified under less than perfect sky conditions. It is easier to see the flashlight beam in a lighted room, for instance, when the beam is concentrated. For the same reason, that's why you can see a 4th-magnitude star in the daytime through your telescope but not a 4th-magnitude nebula.

You can get a sense of how difficult the object will be to see by comparing its magnitude with the object's apparent

size – given in the table as its diameter (Diam) or dimensions (Dim). The apparent size of a deep-sky object is an angular measure of its dimensions against the celestial sphere. The units of angular measure are degrees ($°$), arc minutes ($'$), and arc seconds ($''$): $1°$ is $1/360$ of a circle; $1'$ is $1/60$ of a degree; and $1''$ is $1/60$ of an arc minute. The larger a diffuse object appears against the night sky, the more difficult it will be to pick out from the sky background, and vice versa.

The tabular data listed above were drawn from a variety of modern sources: three primary sources were the books in my *Deep-Sky Companions* series: *Deep-Sky Companions: The Messier Objects*, *Deep-Sky Companions: The Caldwell Objects*, and *Deep-Sky Companions: Hidden Treasures*. Otherwise, the data came from the following sources:

Stellar magnitudes	Alan Hirshfeld, Roger W. Sinnott, and Francois Ochsenbein, eds. *Sky Catalogue 2000.0*, 2nd edn., vol. 1 (Cambridge: Cambridge University Press and Cambridge, MA: Sky Publishing, 1991).
Open star clusters	Brent A. Archinal, and Steven J. Hynes. *Star Clusters* (Richmond, VA: Willmann-Bell, Inc, 2003).
Globular star clusters	Brent A. Archinal, and Steven J. Hynes. *Star Clusters* (Richmond, VA: Willmann-Bell, Inc, 2003.
Planetary nebulae	Brian A. Skiff, "Precise positions for the NGC/IC planetary nebulae." Webb Society Quarterly Journal 105:15, 1996. (Position.) Christian B. Luginbuhl and Brian A. Skiff. *Observing Handbook and Catalogue of Deep-Sky Objects* (Cambridge: Cambridge University Press, 1998). (Dimensions and central star magnitudes.) Murray Cragin, James Lucyk, and Barry Rappaport. *The Deep-Sky Field Guide to Uranometria 2000.0*, 1st edn (Richmond, VA: Willmann-Bell, Inc., 1993).
Diffuse nebulae	*The Deep-Sky Field Guide to Uranometria 2000.0*, 1st edn. Alan Hirshfeld and Roger Sinnott, *Sky Catalogue 2000.0*, vol. 2 (Cambridge: Cambridge University Press and Cambridge, MA: Sky Publishing, 1993).
Galaxies	*The Deep-Sky Field Guide to Uranometria 2000.0*, 1st edn. Alan Hirshfeld and Roger Sinnott, *Sky Catalogue 2000.0*, vol. 2 (Cambridge: Cambridge University Press and Cambridge, MA: Sky Publishing, 1993).

The table ends with an object **rating**, which is a five-point scale I created for this book. The number reflects how easy or difficult an object is to see through a 1 inch telescope under dark skies (or an 8- to 10-inch telescope under slightly light-polluted skies). A rating of 1 means that the object is very difficult to see; a rating of 5 means the object is easy to see. Some objects have an intermediate rating: 1.5, or 2.5, for instance.

	H400 Rating scale
1	Very difficult
2	Difficult
3	Somewhat difficult
4	Fairly easy
5	Easy

A black-and-white photograph of the Herschel object being reviewed accompanies each data table. The photograph shows the object in rich detail with north up and west to the right. All the images are reproduced digitized photographs taken by enormous Schmidt telescopes in both hemispheres. These photos have been made available to astronomers and scientists worldwide by the visionary architects of the Digitized Sky Survey (DSS), which can be perused on the World Wide Web at http://archive. stsci.edu/dss/. (The copyright for the DSS photos of objects used in this book rests with the Anglo-Australian Observatory Board, the United Kingdom Particle Physics and Astronomy Research Council, the California Institute of Technology, and the Associated Universities for Research in Astronomy; they are used here with permission.) Detailed credits appear in Appendix C.

Why use a photograph instead of a drawing? I used a 4-inch telescope to observe the Herschel objects, which is considered small by today's standards. Most observers will be using larger (much larger?) telescopes that will show more detail. So a drawing of a 12th-magnitude galaxy as seen through a 4-inch (essentially a ghost mote of light) does little to help someone using, say, a decent 12-inch reflector under a dark sky, which might show the galaxy as a beautiful little spiral system (arms and all!). The photograph, on the other hand, demonstrates a more "perfect" view; it shows you the true glory of the object. How much detail you see in that object will depend, again, on a number of variables. I must also stress again that the principal purpose of this book is to help you *find* each Herschel object.

To help give you an idea of what the Herschel 400 objects looked like through my 4-inch refractor after some study, I include on the next few pages some drawings I made of various Herschel objects in various classes at different ratings. I suggest you tab these drawings and refer to them if you ever have trouble seeing an object, because they should help you get an idea of what to look for or expect to see.

Each drawing has a rating above it, and its NGC number below it. A scale bar also accompanies each drawing. As you review these drawings, think, "variations on a theme." As you start to observe some open star clusters, for instance, you'll discover that the vast majority of the brightest ones are well-resolved splashes of stars of mixed magnitudes; what

will differ is their shape, size, and degree of concentration. Note too how a globular star cluster with a rating of 1.5 can look like an oblique or face-on galaxy with a rating of 2. (This also demonstrates how difficult it was for Herschel to fathom what it was he was seeing.) I've broken down the galaxy drawings into three general subcategories: the top row shows three edge-on systems at different ratings, the middle row shows three oblique galaxies at different ratings, and the last row shows three face-on galaxies at different ratings. So, if you are searching for an edge-on lenticular galaxy in Virgo that has a rating of 2, it might look something like NGC 4419.

Once you have reviewed an object's data table and photograph, read the **General description** that follows. It tells you what the object is, where it can be found in the sky, what's the brightest star near it, and, if it is visible to the naked eye or binoculars, what it looks like. It also prepares you for the telescopic view.

The next section, **Directions**, tells you, step by step, how to go about finding the Herschel 400 object using the star charts that open each night's observing plan. Use these charts in concert with your own detailed star charts. To find a given Herschel object, I will direct you to a specific wide-field star chart. The purpose of this chart is to show brightest stars in a constellation near the Herschel objects of interest. Each wide-field chart shows stars roughly to 6th magnitude, but generally only in the region around the Herschel objects. The brightest stars in each constellation have been labeled with either a Bayer (Greek) letter or a Flamsteed number. The Greek letters belong to a nomenclature system introduced in 1603 by the Bavarian astronomer Johann Bayer, who labeled stars in each constellation according to their brightness. The most prominent star was given the letter alpha (α); the faintest became omega (ω). The brightest star near NGC 1647, for example, is Alpha Tauri; note that the Greek letter is followed by the Latin genitive of the constellation. There are exceptions, however, such as with the stars in the Big Dipper, which are labeled in order of right ascension, from west to east, not by brightness.

The Greek alphabet (lower case)					
α	Alpha	ι	Iota	ρ	Rho
β	Beta	κ	Kappa	σ	Sigma
γ	Gamma	λ	Lambda	τ	Tau
δ	Delta	μ	Mu	υ	Upsilon
ε	Epsilon	ν	Nu	ϕ	Phi
ζ	Zeta	ξ	Xi	χ	Chi
η	Eta	o	Omicron	ψ	Psi
θ	Theta	π	Pi	ω	Omega

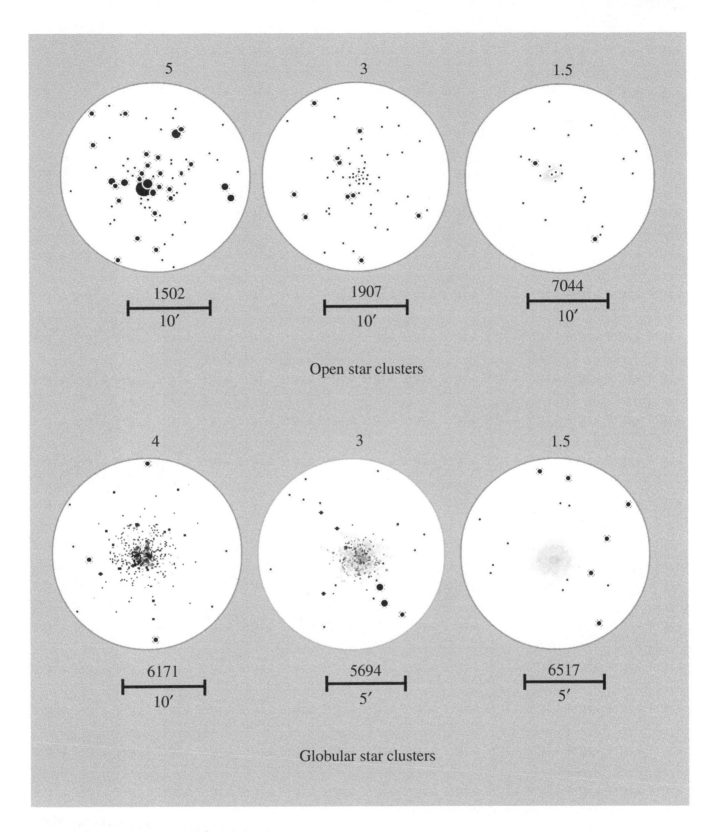

Open star clusters

Globular star clusters

The Flamsteed number identifications are Roman numerals that precede the Latin genitive of the constellation: NGC 7662, for example, is just 25′ southwest of 5th-magnitude 13 Andromedae (the Flamsteed star).

Some stars may have an italicized lower-case letter, like *a* or *b*; these are additional guide stars, which you'll find mentioned in the text as Star *a* or Star *b*, etc. One symbol, a circle, is used to mark the location of each Herschel object in the wide-field star charts. A label with the object's NGC number accompanies each circle unless the box is too populated with circles; note that the NGC prefix does not appear with the numbers. Most Herschel objects and the brightest star near them appear in

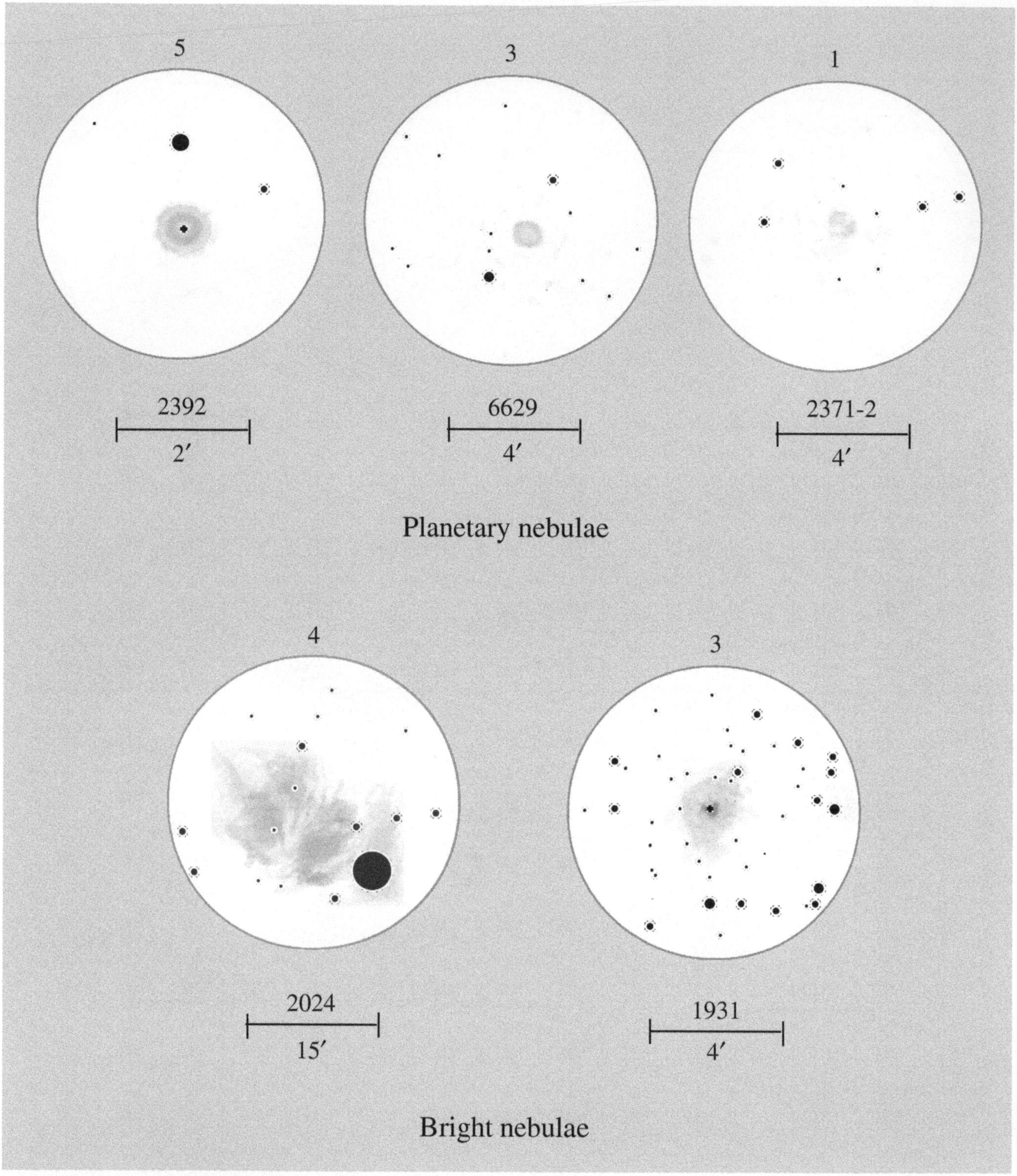

5
3
1
2392
2′
6629
4′
2371-2
4′
Planetary nebulae
4
3
2024
15′
1931
4′
Bright nebulae

boxes labeled with lower-case Roman letters. The box shows the areas covered in the accompanying detailed star charts.

The detailed finder charts have the same orientations as the wide-field charts, and they work on the same principles, only on a smaller scale; they show stars to about magnitude 11, and sometimes to magnitude 12. Any stellar magnitudes mentioned in the text are rough, and are rounded off in most cases to the nearest half magnitude (6th magnitude, 6.5 magnitude, 7th magnitude, 7.5 magnitude, and so on). A scale bar appears at the bottom of each detailed star chart. Note too that in the detailed finder the italicized letters may also refer to asterisms described in the text. In creating these

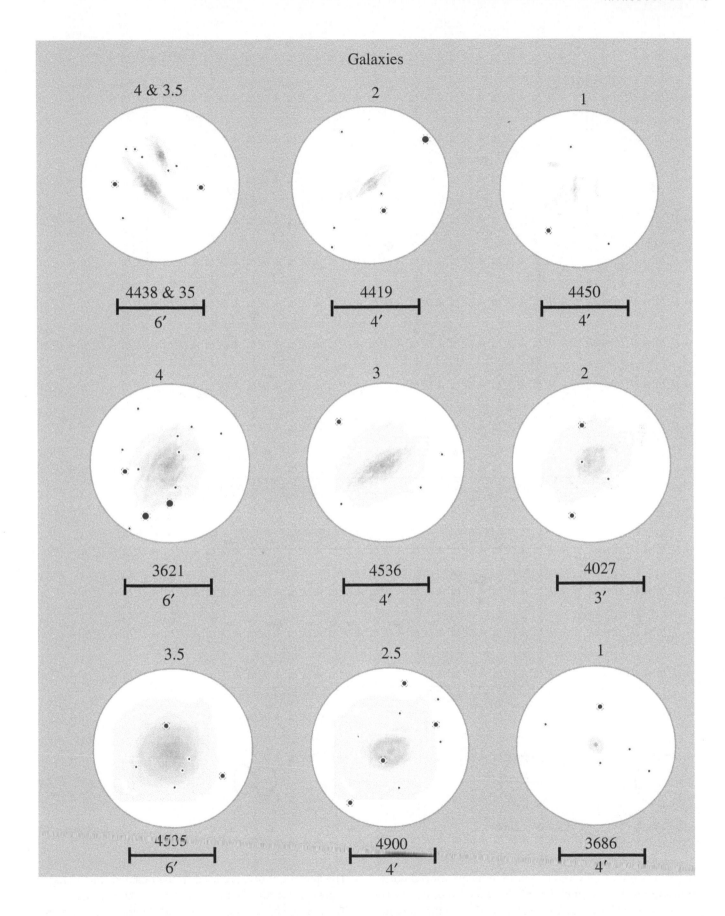

Galaxies

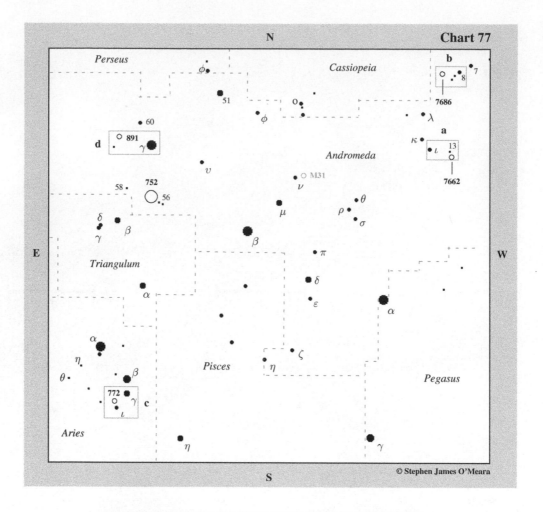

Chart 77

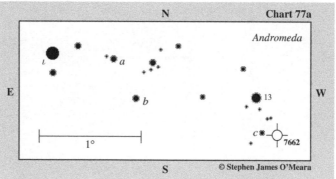

Chart 77a

Open cluster	Globular cluster	Planetary nebula	Bright nebulae	galaxies

charts, my philosophy was to get rid of the peripheral noise. Why clutter the view with lots of dim stars and other objects when all you want to do is see where in the sky the Herschel object lies and which bright stars are nearest to it?

The above symbols are used in the detailed finder charts to represent the different classes of deep-sky objects:

To find an object, first use the wide-field star chart mentioned in the text to locate the bright Bayer or Flamsteed star

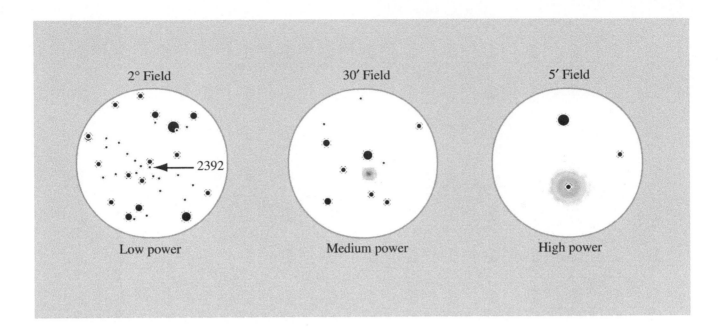

near your Herschel target; in some cases, you might need to use binoculars, which is an essential tool for this project! Next, center that star in your telescope, then switch to the detailed finder chart. Now find the part of the text that describes how to locate the object and simply follow the directions. For instance, after directing you to a bright star – Iota (ι) Andromedae – near NGC 7662, a planetary nebula in Andromeda, I ask you to switch to Chart 77a (see opposite) and move 35′ west to 7.5-magnitude Star a. Now dip nearly 30′ southwest to similarly bright Star b. Now make a slow and careful sweep $1\frac{1}{4}°$ due west to 13 Andromedae . . . your target is only 25′ southwest of 13 Andromedae.

Note that not all non-Herschel 400 deep-sky objects appear on the charts. If you come across a faint and fuzzy object near your target, do not assume you have discovered a comet. I have included only some of the bright non-Herschel 400 objects near your target, or those you might bump into or notice along the way.

Once you locate your target, read **The quick view**, which describes how an object appears in my 4-inch telescope under dark skies. Note that the visual dimensions in the text may vary from the photographic dimensions in the table at different magnifications. For some of the more glorious objects, though, I do go into a little more detail.

To help you understand how the appearance of a deep-sky object appears at different magnifications, I have prepared the above sketch, which shows the planetary nebula NGC 2392 at low, medium, and very high power.

You'll find that, on occasion, I will say that the object looks like the head of a comet. If you've never seen a dim comet without a tail through a telescope, they generally appear as circular, or nearly circular, nebulous glows with various degrees of central condensation. The following drawings will help you to imagine three basic types of appearances.

The grand mechanism of the night sky

As you venture out into the night and turn your telescope to the stars, try to imagine how frustrating it must have been for William Herschel to have to try to discern the structure of the universe through his visual scans. He was like a prisoner chained to a tree on a deserted island, trying to fathom what wonders lie beyond his watery horizons – his knowledge coming from only his immediate surroundings. But one thing was certain, while many of the nebulous glows that Herschel saw were feeble and faint, their magnificence was nonetheless awesome. His discoveries inspired his son John to pick up the torch and set down the road his father had paved. The spirit of William lives on in these words that reflect John's passion as he pondered the grand mechanism of the night sky, as relayed to us by the nineteenth-century observer William Henry Smyth in his "*A Cycle of Celestial Objects*" (republished in *The Bedford Catalogue*, William Bell, Inc., 1986):

How much, how much is escaping us! How unworthy is it in them who call themselves philosophers, to let these great phenomena of nature, these slow but majestic manifestations of the power and the glory of God, glide by unnoticed, and drop out of memory beyond the reach of recovery, because we will not take the pains to note them in their unobtrusive and furtive passage, because we see them in their every-day dress, and mark

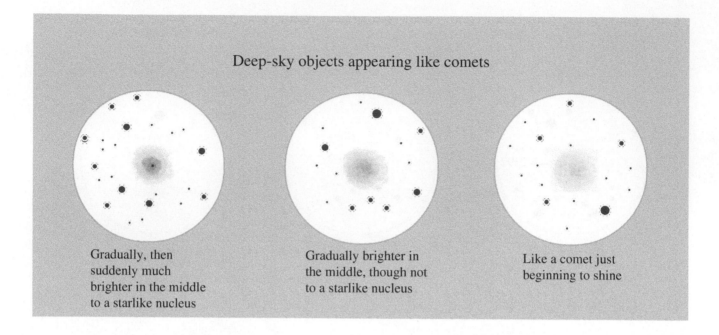

Deep-sky objects appearing like comets

Gradually, then
suddenly much
brighter in the middle
to a starlike nucleus

Gradually brighter in
the middle, though not
to a starlike nucleus

Like a comet just
beginning to shine

no sudden change, and conclude that all is dead, because we will not look for signs of life; and that all is uninteresting, because we are not impressed and dazzled.

Now it is your turn to pick up the Herschel torch. Let it light your way across the heavens as you seek out the wondrous "worlds" discovered by William and his sister. It is your turn to "take the pains to note them." And see, if by the grace of God, you cannot be dazzled and impressed by their visual dimness and ultimate obscurity.

FURTHER READING

The Herschel Objects and How to Observe Them: Exploring Sir William Herschel's Star Clusters, Nebulae, and Galaxies, James Mullaney (New York, Springer, 2007).

Deep-Sky Companions: The Messier Objects, Stephen James O'Meara (Cambridge, Cambridge University Press; Cambridge, MA, Sky Publishing, 2000).

Deep-Sky Companions: The Caldwell Objects, Stephen James O'Meara (Cambridge, Cambridge University Press; Cambridge, MA, Sky Publishing).

Winter

1 · January

Star charts for first night

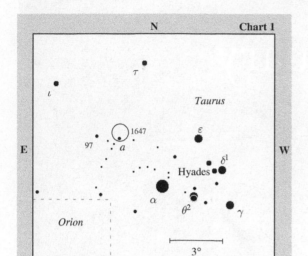

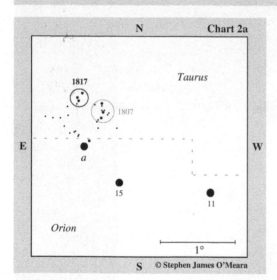

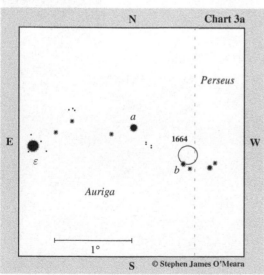

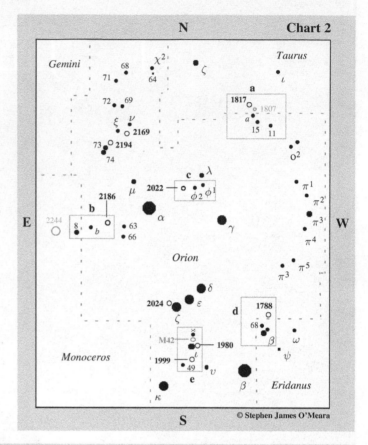

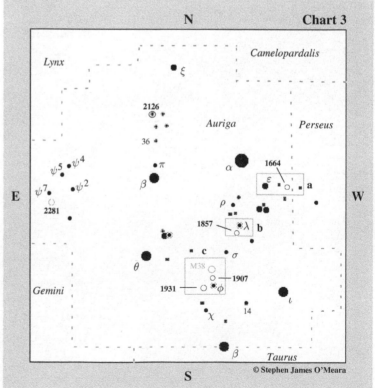

FIRST NIGHT

1. NGC 1647 (H VIII-8)

Type	Con	RA	Dec	Mag	Size	Rating
Open cluster	Taurus	04^h 45.7^m	+19° 07′	6.2	40.0′	4

General description

NGC 1647 is a beautiful open cluster $3\frac{1}{2}°$ northeast of Alpha (α) Tauri (Aldebaran) near the V-shaped face of Taurus, the Bull. It can be be seen with the unaided eye from a dark-sky site. Through 7×50 binoculars NGC 1647 is a lovely cluster, appearing as a round ghostly glow with an apparent size larger than that of the full Moon. It is a fine view in a telescope, especially at low power. Look for a triangular-shaped core with loose arms of stars emanating from it.

Directions

Use Chart 1 to locate 1st-magnitude Alpha (α) Tauri (Aldebaran). Now look a little more than 8° northeast for 4.6-magnitude Iota (ι) Tauri. NGC 1647 lies halfway between Iota Tauri and Aldebaran, just $1\frac{1}{4}°$ due west of 5th-magnitude 97 Tauri. A 7th-magnitude star (a) abuts the cluster to the south. If you place Aldebaran in the southwest edge of your binocular field, NGC 1647 will appear near the field's center.

The quick view

At 23×, the cluster is very round, with most of its stars packed within $\frac{1}{2}°$ of the cluster's core. A pair of 9th-magnitude

double stars dominate the view. These stars, and a similarly bright companion equidistant to the southeast, give the cluster its distinct, triangular-shaped heart. Other bright stars spiral out from this triangular core in crooked or disjointed arms. The overall impression is one of a face-on spiral galaxy whose dashing arms are resolved into streams of individual suns. Return to this object later if you want to study it with higher powers.

2. NGC 1817 (H VII-4)

Type	Con	RA	Dec	Mag	Size	Rating
Open cluster	Taurus	05^h 12.4^m	+16° 41′	7.7	20.0′	3

General description

NGC 1817 is a moderately dim and neglected open star cluster in Taurus, at the northern tip of Orion's shield. From a dark-sky site, it can be seen in 7×50 binoculars as a distinct fuzzy glow next to open cluster NGC 1807. Together, they form a dim double cluster that will be difficult to appreciate in small telescopes under bright skies.

Directions

Use Chart 2 to follow Orion's Shield northward until you identify the 5th-magnitude stars 11 and 15 Orionis. Another 5th-magnitude star, Star a, is 40′ northeast of 15 Orionis. These three stars form a graceful 2° curve. Under a dark sky, try to spy NGC 1817 in binoculars 40′ northeast of Star a. Otherwise, center Star a in your telescope and use Chart 2a to star-hop to the cluster. Small-telescope users should prepare themselves for an underwhelming sight.

The quick view

At 23×, both NGC 1817 and 1807 appear; at first, simply as two linear groups of stars separated by 20′. NGC 1817, our target, is the more dynamic and irregular group to the northeast. It contains four bright stars that shine between 9th- and 10th-magnitude and form a short lightning-bolt pattern oriented northwest–southeast. (Neighboring NGC 1807 looks more like a long branch of stars.) With averted vision, the east side of NGC 1817's lightning bolt swells into a ball of noisy starlight. At 72×, NGC 1817 loses its luster. Some two-dozen suns are dimly visible in this fuzzy ball in my 4-inch, but actually it contains more than 280 members (most being fainter than 12th magnitude) spread across 20′ of sky. Clearly this object is better suited for larger telescopes.

Field note

Paradoxically, NGC 1807 is 0.7 magnitude brighter than NGC 1817, but it is the less visually obtrusive, containing only 37 members in an area 12′ across.

3. NGC 1664 (H VIII-59)

Type	Con	RA	Dec	Mag	Size	Rating
Open cluster	Auriga	04^h 51.1^m	+43° 41′	7.6	18.0′	3

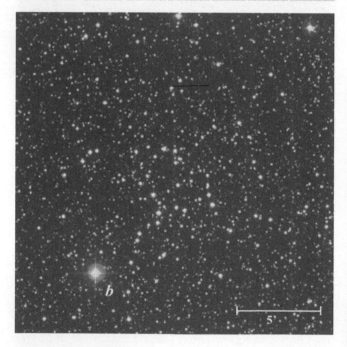

General description

NGC 1664 is a moderately dim open cluster near Epsilon (ε) Aurigae, one of the Kids. The cluster is barely visible from a dark-sky site in 7×50 binoculars, appearing as a breath of light next to a magnitude 7.5 star. In small telescopes the cluster is best appreciated at low power where it mingles with the surrounding Auriga Milky Way.

Directions

Use Chart 3 to locate 3rd-magnitude Epsilon (ε) Aurigae, center it in your telescope, then switch to Chart 3a. From Epsilon, sweep 1$\frac{1}{4}$° west–northwest to 6th-magnitude Star *a*. NGC 1664 lies 45′ to the southwest; 7.5-magnitude Star *b* abuts the cluster to the southeast.

The quick view

At 23×, NGC 1664 is a fine little cluster, appearing as a roughly 20′-wide asymmetrical haze that blooms from Star *b*. With imagination, Star *b* looks like the flash from a meteorite strike, while the cluster is the debris cloud splashing obliquely away from it to the northwest. A smaller and tighter triangular grouping of stars lies immediately southeast of Star *b*; seen together, this grouping and the cluster look like an uneven bow tie of stars, or a butterfly with a damaged wing. A magnification of 72× reveals the northwestern group to be composed of about two-dozen suns that comprise a short string of stars connected to a oval loop. At higher powers the cluster looks like an earring dangling in the moonlight. Larger scopes will show more than 100 stars here.

4. NGC 2126 (H VIII-68)

Type	Con	RA	Dec	Mag	Size	Rating
Open cluster	Auriga	06^h 02.6^m	+49° 52′	10.2	6.0′	2

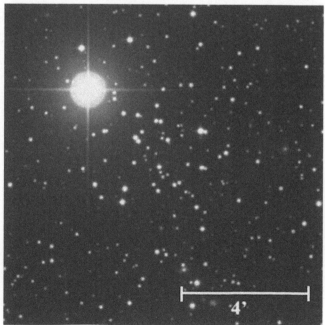

General description

NGC 2126 is a dim and difficult open cluster in remote northern Auriga, about half way between Beta (β) and Xi (ξ) Aurigae. Fortunately, a 6th-magnitude star is superimposed on this dim haze, making it easy to locate. In small telescopes the cluster is visible at low power under a very dark sky. Think, "averted vision"; higher magnifications in the 4-inch are almost useless on this ill-defined fog of faint light.

Directions

Use Chart 3 to locate 5th-magnitude Xi Aurigae, which forms the northern apex of a near equilateral triangle with 2nd-magnitude Beta Aurigae and 1st-magnitude Alpha (α) Aurigae. The 6th-magnitude star superimposed on NGC 2626 marks the northeast corner of a 1°-wide trapezoid of 6th-magnitude suns. Use binoculars to confirm this trapezoid, then use low power in your telescope to center the cluster.

You could also star hop from Beta Aurigae. From Beta Aurigae, move about 1° north to magnitude 4.5 Pi (π) Aurigae, which is part of a roughly 20'-wide near-equilateral triangle with two 6th-magnitude suns. Center Pi Aurigae,

then hop 45' due north to a 20-wide right-triangle of roughly 8th-magnitude suns. A 1° sweep north–northeast will bring you to 6th-magnitude 36 Aur, which is part of yet another 20'-wide acute triangle. Another 1° sweep will bring you to the southeast corner of the 1°-wide trapezoid mentioned above. You want to make one last 1° sweep north–northeast to the 6th-magnitude sun superimposed on NGC 2126.

The quick view

At 23×, NGC 2126 is a very faint halo of light with a 6th-magnitude star at its northeastern flank. With averted vision, the 6'-wide cluster appears elliptical, with its major axis oriented northeast–southwest. It also appears very granular. The cluster contains 40 stars of 13th magnitude and fainter, so it is a bear to resolve in a small telescope. Unless you have sufficient aperture, expect to see just a smidgen of diffuse light kissing a bright star. (Don't breathe on your eyepiece.)

Star charts for second night

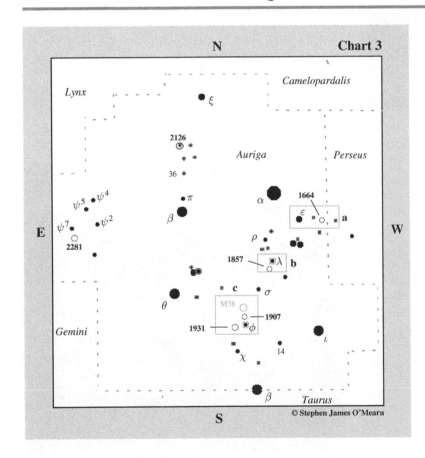

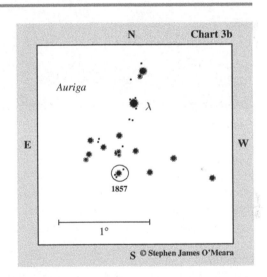

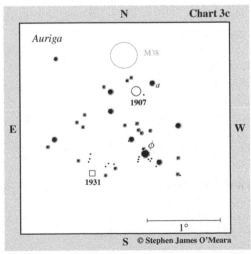

Second night

1. NGC 1857 (H VII-33)

Type	Con	RA	Dec	Mag	Size	Rating
Open cluster	Auriga	05ʰ 20.1ᵐ	+39° 21′	7.0	10.0′	2

General description

NGC 1857 is a dim open cluster near 5th-magnitude Lambda (λ) Aurigae. The cluster's brightest star (magnitude 7.5) is visible from a dark-sky site in 7×50 binoculars; with effort the cluster itself can be imagined as a dim and insignificant shimmer of light around it. In small telescopes, NGC 1857 is best viewed at low power.

Directions

Use Chart 3 to locate Lambda Aurigae, which is about 6° south–southeast of Alpha (α) Aurigae. It forms the southern end of a line of three similarly bright stars, with Rho (ρ) Aurigae being at the northern end. Use binoculars to identify the field. Center Lambda Aurigae in your telescope at low power, then switch to Chart 3b. NGC 1857 will be a tiny aggregation of suns about 45′ south–southeast of Lambda Aurigae.

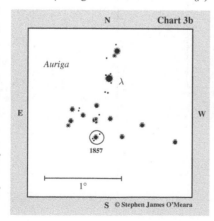

Field note

Do not be fooled by the roughly 1°-wide splash of 7th- and 8th-magnitude suns immediately to the cluster's north, which can easily be mistaken for the cluster at low power. To spy the cluster, use averted vision and think, "tiny!"

The quick view

At 23×, measuring a mere 10′ across, NGC 1857 is a tough catch, appearing as a faint fog of light surrounding a 7.5-magnitude star. With concentration, about a half dozen suns crowd around that bright central star, whose light over-powers the dimmer members. A pair of roughly 11th-magnitude stars can be seen immediately to the southeast and a solitary 9.5-magnitude star shines immediately to the northwest. About two dozen dim suns pop in and out of view with averted vision, forming an irregular skirt of inconceivably dim light. A magnification of 72× brings out some of the brighter members from the background, making them easier to detect. The cluster's southeastern edge appears to broaden before it abruptly ends. The notion that this is a cluster is all but lost at high magnification in a small telescope. The cluster has only about 40 members, so larger telescopes will glorify the view, mainly by bringing out faint background stars in the Milky Way.

2. NGC 1907 (H VII-39)

Type	Con	RA	Dec	Mag	Size	Rating
Open cluster	Auriga	05ʰ 28.1ᵐ	+35° 19.5′	8.2	5.0′	3

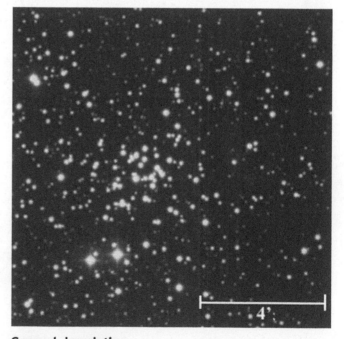

General description

NGC 1907 is another small and dim open star cluster. But it is also compact and rich in stars, so it is not difficult to see under a dark sky. The cluster is also very accessible, being within 30′ of M38, one of the finest open star clusters in the Messier catalog.

Directions

Use Chart 3 to locate M38, which is almost 2° southeast of 5th-magnitude Sigma (σ) Aurigae or a little more than 1° north of 5th-magnitude Phi (φ) Aurigae. Confirm the position of the cluster relative to these stars in your binoculars. Center M38 in your telescope, then use Chart 3c to confirm NGC 1907's position 30′ to the south–southwest.

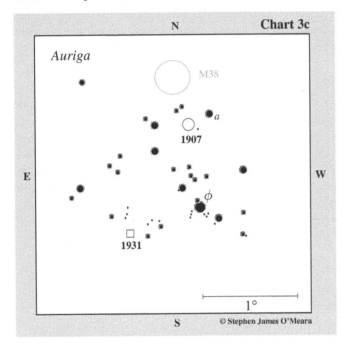

General description

NGC 1931 is a small but bright emission and reflection nebula near 5th-magnitude Phi (φ) Aurigae. It is best seen with moderate to high magnifications.

Directions

Using Chart 3c, from 1907, swing 50′ south to 5th-magnitude Phi Aurigae. Phi Aurigae is easily identifiable, since it is flanked by a 6th-magnitude star about 10′ to the southwest and a 7th-magnitude sun less than 20′ to the northeast. Phi Aurigae is itself like a mini-cluster, being surrounded by about a half-dozen dim suns. NGC 1931 lies 50′ southeast of Phi Aurigae. Using low power, look for a 20′-long, upward-curving arc of three 7.5-magnitude suns, oriented east to west. NGC 1931 is the easternmost "star" in that arc.

The quick view

At 23×, NGC 1931 appears stellar with a direct glance. But if you use averted vision and concentrate, you'll see that it has a dense nebulous quality. It begs for magnification. A magnification of 72×, shows NGC 1931 to be a beautiful 4′-wide stellarlike knot of nebulosity centered on a roughly 12th-magnitude sun. Other dim companions can be glimpsed nearby, but they are distracting. The nebula takes magnification well, and at 105×, it appears more intense and obvious, with a smooth milky texture that gradually and briefly fades as you look outward from the bright core.

The quick view

At 23× in the 4-inch, NGC 1907 is just a little puff of "smoke" near M38, like a 5′-wide comet just beginning to shine. With some effort, the cluster appears partially resolved, or "noisy," like a tight ball of stellar static. Because it is so compact, the cluster shows up better at 72×. Use of increasingly higher powers tends to peel away the outer layers bringing you closer and closer to resolving its tight circular core. Dark veins of broken starlight mar its ancient-looking face, as if it were some mummified remains about to fall apart.

Stop. Do not move the telescope. Your next target is nearby!

3. NGC 1931 (H I-261)

Type	Con	RA	Dec	Mag	Diam	Rating
Emission and reflection nebula	Auriga	05ʰ 31.4ᵐ	+34° 15′	–	4.0′	3

4. NGC 2281 (H VIII-71)

Type	Con	RA	Dec	Mag	Diam	Rating
Open cluster	Auriga	06ʰ 48.3ᵐ	+41° 05′	5.4	25.0′	4

General description

NGC 2281 is a bright and neglected open cluster in the remote eastern corridor of Auriga. In 7 × 50 binoculars, the cluster appears as a large diffuse glow nestled between two 7th- and 8th-magnitude stars; both of these stars have dramatic golden hues. The cluster is very pretty and bright in a small telescope.

Directions

Use Chart 3 to locate 2nd-magnitude Beta (β) Aurigae and 2.6-magnitude Theta (θ) Aurigae. NGC 2281 forms a near equilateral triangle with these stars to the east. You'll find the cluster about 45′ south–southwest of 5th-magnitude Psi^7 (ψ^7) Aurigae, which is part of a pretty, 3°-wide, Y-shaped asterism of similarly bright stars – the others being Psi^2 (ψ^2), Psi^4 (ψ^4), and Psi^5 (ψ^5) Aurigae. Confirm these stars in your binoculars, then center Psi^7 (the southernmost one) in your telescope at low power. Again, NGC 2281 should be easily identified some 45′ to the south–southwest.

The quick view

At 23× in the 4-inch, NGC 2281 is a beautiful cluster with some two dozen suns visible around a tight, diamond-shaped asterism at the cluster's very core. The cluster is very loose, very pretty, and bright. The surrounding Milky Way makes it difficult to judge the cluster's true extent. At 72×, the central diamond, which is composed of four equally bright stars, is quite brilliant and mesmerizing. The diamond is surrounded by a loose and coarse gathering of mixed suns that form various geometrical patterns.

Star charts for third night

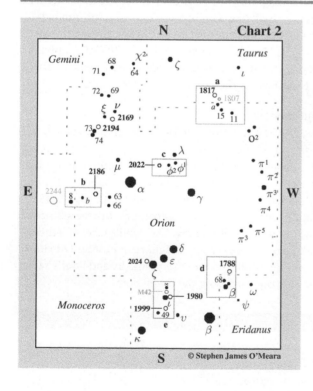

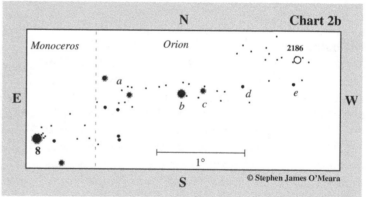

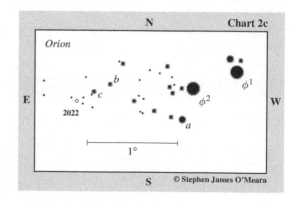

THIRD NIGHT

1. NGC 2169 (H VIII-24)

Type	Con	RA	Dec	Mag	Diam	Rating
Open cluster	Orion	06ʰ 08.4ᵐ	+13° 58′	5.9	6.0′	4

General description
NGC 2169 is a beautiful and bright open star cluster in the northern reaches of Orion. It is just visible to the unaided eye under very dark skies and is a cinch to see in binoculars. Small telescopes will show it as a trapezoid of tightly knit suns that looks like a magnified view of M42's Trapezium cluster without the nebulosity. Commonly called the Shopping Cart Cluster, the group can also be viewed as a Little Pleiades or the number "37."

Directions
Shining at 6th-magnitude, NGC 2169 is easy to find. Use Chart 2 to find the 4th-magnitude stars Xi (ξ) and Nu (ν) Orionis (which comprise Orion's right hand). Now just raise your binoculars about 35′ southwest of the midpoint between those two stars. You should immediately see a tight double star centered in a 6′-wide "mist" of light. Look no further, you have spotted your target!

The quick view
At 23× in the 4-inch, the cluster's 30-odd members are separated into two distinct groups of bright stars – one to the northwest, the other to the southeast – each of which is shaped like the lower-case Greek letter Lambda (λ). The

northwestern Lambda is short and squat; the southeastern Lambda is long and lean. These stellar groupings are bright and should be a welcome sight even under city skies. At 72×, NGC 2169 is equally striking. A quick count shows 14 obvious suns; two dozen stars are visible with concentration. The cluster's brightest member shines at 7th magnitude and is a double star (Struve 848).

Stop. Do not move the telescope. Your next target is nearby!

2. NGC 2194 (H VI-5)

Type	Con	RA	Dec	Mag	Diam	Rating
Open cluster	Orion	06ʰ 13.8ᵐ	+12° 48′	8.5	9.0′	2.5

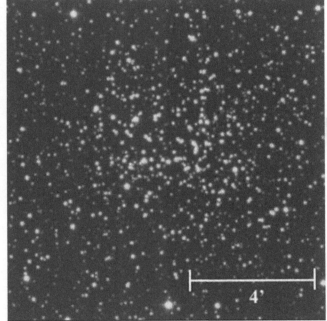

General description
Shining at magnitude 8.5, NGC 2194 may be a challenge for small-telescope users in light-polluted areas. Under dark skies, it is a dim glow in a 4-inch. Surprisingly, this cluster is better known than nearby NGC 2169, but that's because NGC 2194 is quite impressive in a large telescope.

Directions
Use Chart 2 to locate the two 4th-magnitude stars Xi (ξ) and Nu (ν) Orionis – the same stars you used to locate NGC 2169. Now look a little more than 2° southeast of Xi Orionis, where you should see a close pair of 5th-magnitude stars – 73 and 74 Orionis. Center 73 Orionis in your scope. NGC 2194 is only 30′ to the northwest. The key to finding this object in a small telescope is to go slow. Sweep gently yet determinedly with averted vision. When you reach the field, stop, breathe, and tap the tube gently. NGC 2169 should emerge from the background.

The quick view

At 23× in the 4-inch, NGC 2194 is a dim round glow. It measures 9′-wide and looks like a faint comet with no central condensation or a tail. At 72×, the cluster appears very mottled and has fuzzy and spiked edges – a comet having a "bad-hair day." With averted vision, an elongated E-shaped core appears weakly branded on a dim background haze. Large telescopes will see this tight fuzzy ball resolve into nearly 200 stars shining at 13th magnitude and fainter. You might want to check this object off and return to it later for a more detailed look.

3. NGC 2186 (H VII-25)

Type	Con	RA	Dec	Mag	Diam	Rating
Open cluster	Orion	06h 12.1m	+05° 27.5′	8.7	5.0′	2

General description

NGC 2186 is a small and very dim open cluster in eastern Orion, near Monoceros. It is a challenge to find in a small telescope and appears only as a thin, round haze. The directions below give the most direct route for star-hopping.

Directions

Use Chart 2 to find 1st-magnitude Alpha (α) Orionis (Betelgeuse). Next look about 7° to the southeast for 4th-magnitude 8 Monocerotis, which is less than 2° west–southwest of NGC 2244 – the open star cluster at the heart of the wreathlike Rosette Nebula. Use binoculars to confirm the location of 8 Monocerotis, then center it in your telescope at low power. Now switch to Chart 2b. Using low power, move 1° northwest of 8 Monocerotis to a pair of 8th-magnitude

stars (a), oriented northeast to southwest and separated by 20′. Just 45′ west of the southernmost star in that pair is 6th-magnitude Star b with a 7th-magnitude companion (c) 15′ to the west–northwest. Less than 30′ northwest of Star c is a solitary 8th-magnitude sun (d). Equidistant to the west is similarly bright Star e. NGC 2186 lies about 30′ north of Star e. It's nestled in a group of roughly 10th-magnitude suns.

The quick view

At 23× in the 4-inch, NGC 2186 is a small and difficult haze. Averted vision is required to pick it out from the Milky Way background. At 72×, the cluster's heart is marked by a pair of 12th-magnitude suns, which forms the nucleus of a tiny 5′-wide ellipse of dimmer stars whose major axis is oriented northeast to southwest. The cluster contains only 30 stars, which are bordered to the southeast by a 10th-magnitude star and to the northwest by an 11th-magnitude star.

4. NGC 2022 (H IV-34)

Type	Con	RA	Dec	Mag	Dim	Rating
Planetary nebula	Orion	05h 42.1m	+09° 05′	11.9	28″ × 27″	2

General description

NGC 2022 is a fine planetary nebula for 8-inch and larger telescopes, but it's a bit of a challenge for a 4-inch telescope under dark skies. The nebula appears as a small gray annulus under high magnification. Otherwise it is a slightly swollen "star."

Directions

Use Chart 2 to locate 3rd-magnitude Lambda (λ) Orionis and its two 4th-magnitude attendants – Phi¹ (φ¹) and Phi² (φ²)

Orionis. Center Phi^2 Orionis in your telescope, then switch to Chart 2c. From Phi^2 Orionis move 20′ south–southeast to 6th-magnitude Star *a*. Moving northeast, follow the roughly 50′-long chain of 8th- to 9th-magnitude stars that ends at Star *b*. Just 10′ southeast of Star *b* you'll find a nice pair of stars: 7.5-magnitude Star *c* and a 9th-magnitude companion to the southeast. NGC 2022 lies about 10′ further to the southeast.

The quick view

At 23× in the 4-inch, NGC 2022 requires averted vision and a dedicated search to see it as an ill-defined puff of dim gray light that's just brighter than the background sky. At 72×, the planetary nebula is visible as a diffuse orb in a line of 12th- and 13th-magnitude stars. Higher powers reveal the annulus, whose inner edge is sharp. The hole might not be obvious, though, until you use averted vision.

Star charts for fourth night

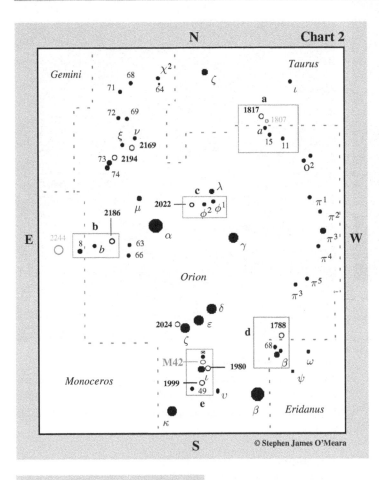

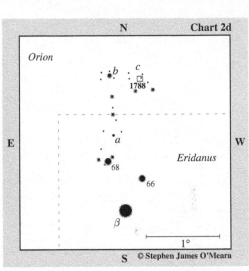

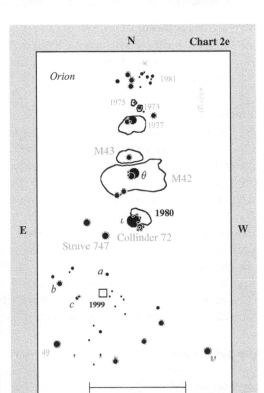

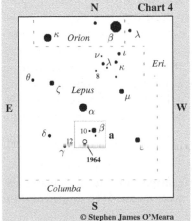

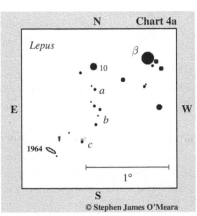

FOURTH NIGHT

1. NGC 1964 (H IV-21)

Type	Con	RA	Dec	Mag	Dim	Rating
Spiral galaxy	Lepus	05^h 33.4^m	−21° 57′	10.7	5.0′ × 2.1′	3

4′

General description

NGC 1964 is a small but condensed galaxy that is not too difficult to see in a small telescope at moderate magnification. But you need to know exactly where to look.

Directions

Use Chart 4 to locate 3rd-magnitude Beta (β) Leporis. The 5th-magnitude star 10 Leporis lies 40′ to the east–southeast. Center 10 Leporis in your telescope, then switch to Chart 4a. Using low power, dip 15′ south of 10 Leporis to 9th-magnitude Star a, which has a 10th-magnitude companion to the northeast. Now move 20′ due south to a very acute triangle (b) of 8th- and 9th-magnitude stars with a north–south orientation. Center the southernmost star in that triangle, then move 15′ east–southeast to a pair of 10th-magnitude suns (c). NGC 1964 lies a little more than 20′ east–southeast of Pair c and about 15′ south of a 9th-magnitude star with an 11th-magnitude companion to the south.

The quick view

At 23× in the 4-inch, NGC 1964 has a moderately condensed core surrounded by a fuzzy 5′-wide disk. The galaxy is not readily visible at this power; it takes concentration. I suggest changing to at least 40× once you get the field. At 72×, the galaxy's bright, highly concentrated core is revealed, which is slightly egg-shaped. The core sits in a mottled oval disk oriented northeast to southwest. Larger telescopes will show that the mottling seen in small telescopes is due to at least three roughly 14th-magnitude stars in the galaxy's halo.

2. NGC 1788 (H V-32)

Type	Con	RA	Dec	Mag	Dim	Rating
Reflection nebula	Orion	05^h 06.9^m	−03° 21′	−	5′ × 3′	3

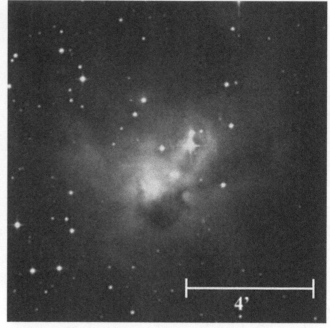

4′

General description

NGC 1788 is a small but surprisingly bright reflection nebula next to a 10th-magnitude star. In small telescopes it appears more extensive at lower power than at high power.

Directions

Use Chart 2 to locate 0-magnitude Beta (β) Orionis (Rigel), the left knee of Orion, then 3rd-magnitude Beta (β) Eridani 3½° further to the northwest. Beta Eridani marks the southern apex of a 40′-wide isosceles triangle with the 5th-magnitude stars 66 and 68 Eridani. Center 68 Eridani in your telescope then switch to Chart 2d. Using low power, you want to follow the roughly 1°-long line of 7th- to 9th-magnitude stars that flows northward from 68 Eridani. Begin by moving 20′ north–northwest to 9th-magnitude Star a and stop when you get to 8th-magnitude Star b. NGC 1788 lies only about 25′ to the west–southwest of Star b. It lies southeast of a 10th-magnitude star that marks the southwest corner of a 10′-wide trapezoid of similarly bright suns (c).

The quick view

At 23×, NGC 1788 is a small (2′) bright glow with a sideways V-shaped concentration; the V's arms point east and southeast. Dark matter causes the variation in the cloud's brightness. A thin and diffuse halo of material surrounds this bright core. A magnification of 72× enhances the central condensation but dims the surrounding diffuse material. Several nebulous knots or dim stars are superimposed on the cloud.

3. NGC 2024 (H V-28)

Type	Con	RA	Dec	Mag	Dim	Rating
Emission nebula	Orion	05^h 41.9^m	−01° 51′	∼7th	30′× 30′	4

10′

General description

NGC 2024 is a large, reasonably bright, yet elusive diffuse nebula in Orion. It literally kisses 2nd-magnitude Zeta (ζ) Orionis, the easternmost star in Orion's Belt. Appropriately, because of its photographic image, which shows the bright nebula sliced in two by a dark lane, it is called, among other things, the Lips Nebula. If you can block Zeta Orionis with a building or some distant structure, NGC 2024 can be seen in 7×50 binoculars from a dark sky. The glow is large, so it is best seen at low powers in a telescope.

Directions

Use Chart 2 to locate 2nd-magnitude Zeta Orionis. NGC 2024 is just 15′ to the northeast. Refer to the photograph to see the nebula's extent.

The quick view

At 23× in the 4-inch, the nebula's core is quite bright, and its dark lane is dim but apparent. Although the nebula shines at about 7th-magnitude, you have to imagine that light spread out over an area the size of the full Moon. The brightest segments of the cloud lie on either side of the north–south oriented dark lane.

4. NGC 1980 (H V-31)

Type	Con	RA	Dec	Mag	Dim	Rating
Emission nebula	Orion	05^h 35.4^m	−05° 54′	–	14′× 14′	3

5′

General description

NGC 1980 is a beautiful emission nebula $\frac{1}{2}$° due south of M42. The nebula is associated with Iota (ι) Orionis, the brightest star in Orion's Sword, and it is connected to the Orion Nebula by a long, faint loop of nebulosity. It can be seen in binoculars under a dark sky and is best appreciated at low power in a rich-field telescope.

Directions

Use Chart 2 to find Iota Orionis, then switch to Chart 2e. The brightest portion of NGC 1980 is immediately northwest of that star.

The quick view

At 23× in the 4-inch, Iota Orionis, otherwise known as Struve 752, is a wonderful triple star. It is also the brightest member of a 15′-wide cluster of some 30 suns (Collinder 72) embedded in the milky haze of NGC 1980. The nebula is most apparent just north and west of Iota Orionis. If you have difficulty seeing it, place Iota at the south end of the eyepiece field, use averted vision, and gently jiggle the tube. Under a dark sky at low power, the nebula is a graceful wisp of angel hair that seems to brush past Iota like a whisper. At 72×, the brightest part of the nebula is a patch of dim light northwest of Iota, dimmer beyond. Large telescopes will reveal the the nebula's gauzelike structure.

Stop. Do not move the telescope. Your next target is nearby!

5.   NGC 1999 (H IV-33)

Type	Con	RA	Dec	Mag	Dim	Rating
Reflection nebula	Orion	05^h 36.5^m	−06° 42′	9.5	2′×2′	3

2′

General description

NGC 1999 is a tiny but surprisingly bright reflection nebula about 1$\frac{1}{2}$° southeast of the Orion Nebula.

Directions

NGC 1999 is only 50′ southeast of Iota (ι) Orionis, the brightest star in Orion's sword. But since it is tiny (2′), you'll have to star-hop to it. From city and suburban locations, start at Iota Orionis. Using Chart 2e as a guide, move your telescope 20′ southeast to Struve 747, a pair of 5th-magnitude stars separated by 36″. Another 20′-hop south of Struve 747 is 7.4-magnitude Star a. NGC 1999 lies only 15′ south and slightly east of Star a.

Observers under dark skies can start by locating 5th-magnitude 49 Orionis, about 1$\frac{1}{2}$° southeast of Iota Orionis. Center that star, then move 40′ north–northwest to 6th-magnitude Star b. A close pair of 8th- and 9th- magnitudes stars (c) lies 15′ to the southwest. NGC 1999 is 20′ west of Pair c. As with a small planetary nebula, once you locate the field, use averted vision and wait for a "star" to swell (in this case by a full magnitude) in brightness.

The quick view

At 23× in the 4-inch, NGC 1999 looks simply like a faint fuzzy 9.5-magnitude star. At 72× to 303×, NGC 1999 is bright enough, and condensed enough, to handle high magnifications well. So do not be afraid to push the power. Under high magnification, the nebula looks like a star surrounded by a circular milky glow with short rays and other remarkable shapes. With time, that roundish glow becomes a broken annulus, one that appears more elliptical than round. The ellipse is tilted northwest to southeast, with the northeastern segment being brightest.

Star charts for fifth night

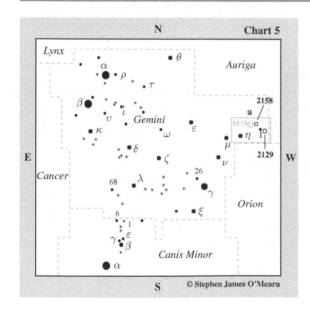

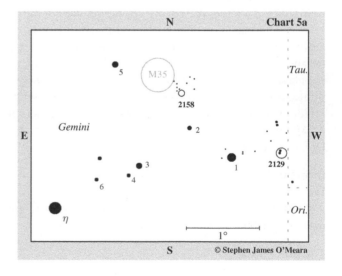

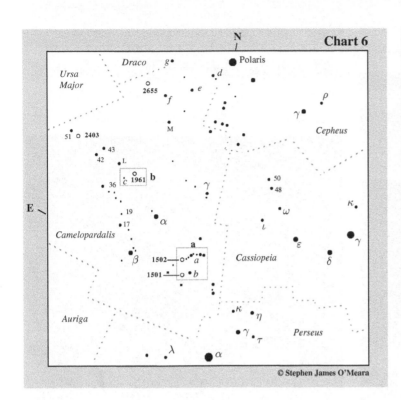

Chart 6

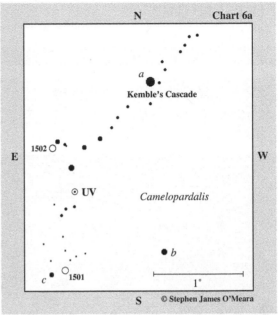

Chart 6a

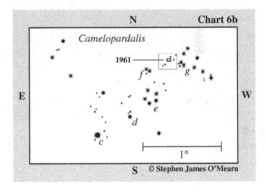

Chart 6b

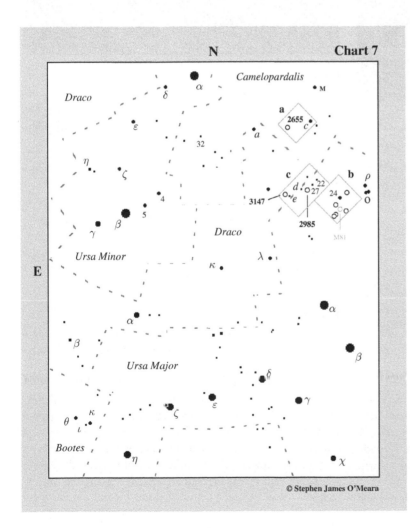

Chart 7

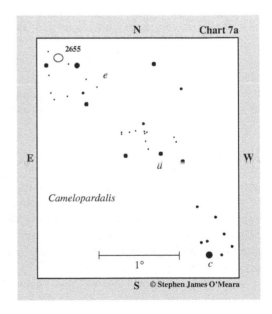

Chart 7a

© Stephen James O'Meara

FIFTH NIGHT

1. NGC 2158 (H VI-17)

Type	Con	RA	Dec	Mag	Diam	Rating
Open cluster	Gemini	06^h 07.4^m	+24° 06′	8.6	5.0′	3

General description
NGC 2158 is a small but rich open star cluster near the magnificent, naked-eye, open star cluster M35 in the far western reaches of Gemini, the Twins.

Directions
NGC 2158 is a cinch to find under dark skies, though it requires a telescope. Use Chart 5 to locate M35 about $2\frac{1}{2}°$ northwest of 3rd-magnitude Eta (η) Geminorum. Center M35 in your telescope, then look about 25′ to the southwest for NGC 2158. If you are under light-polluted skies, use Chart 5a to help you pinpoint the cluster's exact location.

The quick view
At 23× in the 4-inch, NGC 2158 is simply a small (5′) ghostly glow. Light pollution will definitely affect its visibility. At 72×, the cluster is readily apparent as a dim mottled haze. Larger scopes are required to resolve this extremely rich cluster, which packs nearly 1,000 suns in a 5′-wide disk. Alas, the brightest members of this rich star swarm shine at a very dim 15th magnitude!

Stop. Do not move the telescope. Your next target is nearby!

2. NGC 2129 (H VIII-26)

Type	Con	RA	Dec	Mag	Diam	Rating
Open cluster	Gemini	06^h 01.1^m	+23° 19′	6.7	6.0′	4

General description
NGC 2129 is a bright but small open cluster near M35 and NGC 2158 in western Gemini. A few of its brightest stars can be seen in binoculars but you will need a telescope to recognize it as a cluster.

Directions
Using Chart 5a, from NGC 2158, swing your scope 30′ to the south–southwest, where you'll find 7th-magnitude 2 Geminorum. A 40′ sweep southwest will bring you to 4th-magnitude 1 Geminorium. NGC 2129 lies less than 45′ west and slightly north of 1 Geminorium.

The quick view
At 23× in the 4-inch, NGC 2129 appears at first as a tight double star (oriented north–northwest and south–southeast), with a magnitude 7.5 primary and a magnitude 8.0 companion. Averted vision will show these stars superimposed on a small round haze of suns. A magnification of 72× reveals about a dozen suns coarsely scattered around a central triangle of bright stars. A dim glow of unresolved stars seemingly lie in wait "behind" the cluster's brightest members, which shine at 10th magnitude. Larger telescopes will reveal more than 70 stars in an area only 6′ across.

Field note

Another pair of 7th-magnitude stars lies 25′ to the north–northeast, while a slightly dimmer pair, oriented east–west, lies about 10′ to the southeast of the cluster.

3. NGC 1502 (H VII-47)

Type	Con	RA	Dec	Mag	Diam	Rating
Open cluster	Camelo-pardalis	04ʰ 07.8ᵐ	+62° 20′	6.0	20.0′	5

General description

The bright open star cluster NGC 1502 is just visible to the unaided eye (with difficulty) under a dark sky. It is easily seen in binoculars at the southeastern end of the beautiful asterism known as Kemble's Cascade. And though the cluster measures 20′ across, its core is very tight and bright. Four of the cluster's stars are immediately obvious in small binoculars. It is a wonderful sight in telescopes of all sizes.

Directions

To find NGC 1502, you need first to locate Kemble's Cascade. And to find Kemble's Cascade, you need to find 5th-magnitude Star *a* (in Box a) on Chart 6. Note that Star *a* forms the eastern apex of a near equilateral triangle with 4.5-magnitude Iota (*ι*) Cassiopeiae and 4th-magnitude Eta (*η*) Persei. Your fist held at arm's length (10°) will fill the triangle. Even if you cannot see Star *a* with the unaided eye, just sweep your binoculars across its location. There will be no mistaking the correct 5th-magnitude star, because Kemble's Cascade flows right by it – from the northwest to

the southeast. Center Star *a* and Kemble's Cascade in your telescope, then use Chart 6a to navigate that cataract of stars. The cascade forks at the southeastern end. NGC 1502 lies at the end of the fork's eastern branch.

The quick view

At 23× in the 4-inch, the cluster is a tight ball of suns surrounded by a coarse sprinkling of outliers. A brilliant 7th-magnitude double star (Struve 485) shines like finely cut topaz near the cluster's center. More than a dozen suns, many in pairs, surround NGC 1502's tight core. At 72×, the cluster's brightest members form a Lazy X shape centered on Struve 485. NGC 1502 has some 60 members, so those with larger telescopes should be prepared for an even more outstanding display.

Stop. Do not move the telescope. Your next target is nearby!

4. NGC 1501 (H IV-53)

Type	Con	RA	Dec	Mag	Dim	Rating
Planetary nebula	Camelo-pardalis	04ʰ 07.0ᵐ	+60° 55′	10.6	56″ × 48″	4

General description

Popularly known as the Oyster Nebula, NGC 1501 is a bright planetary nebula that's easy to see in small apertures. It's slightly fainter and a little smaller than the more famous planetary nebula M76 in Perseus. Under a very dark sky, NGC 1501 can be seen with 7 × 50 binoculars – if you know *exactly* where to look and brace the binoculars well. In a 2.4-inch telescope, it looks like a conspicuous 11th-magnitude "star."

Directions

Using Chart 6a, return to the fork in Kemble's Cascade. Now follow the southeastern branch, which dips sharply south to the 8th-magnitude semi-regular variable star UV Camelopardalis just $\frac{1}{2}°$ away. A 1° hop to the south–southeast will bring you to 7.5-magnitude Star c. NGC 1501 is just 10' northwest of Star c.

The quick view

At 23× in the 4-inch, NGC 1501 looks like an 11th-magnitude star in a rich starfield. So if you're just sweeping the area at low power, it's doubtful that you'll take notice of it. But if you use Chart 6a to pinpoint its location, even a brief perusal of the area with averted vision will reveal that NGC 1501 is not stellar; it's a tad swollen and fuzzy. At 101×, the planetary nebula immediately reveals itself as a glorious orb of dense light – even with direct vision. Fainter stars can be seen around the nebula. With averted vision, a faint annulus can be seen quite readily; at high power, NGC 1501 looks like the Ring Nebula at low power. In small apertures, NGC 1501 should look best at 227×. In larger telescopes, the nebula is a glorious sight, displaying a sharply defined ring at high power.

5. NGC 1961 (H III-747)

Type	Con	RA	Dec	Mag	Dim	Rating
Mixed spiral galaxy	Camelo-pardalis	05^h 42.1^m	+69° 23'	11.0	4.2' × 4.0'	1

4'

General description

NGC 1961 is a very faint mixed spiral galaxy about 5° northeast of 4th-magnitude Alpha (α) Camelopardalis, and 3° west of 5th-magnitude L Camelopardalis. It is extremely difficult to see in small telescopes, even under dark skies. You need to know exactly where to look and use averted vision. Be prepared to make a careful star-hop to its location, then plan to spend some time working with various magnifications to see it.

Directions

Use Chart 6 to locate the 4th-magnitude stars Alpha and Beta (β) Camelopardalis. Now use your unaided eyes or binoculars to find 5th-magnitude 17 Camelopardalis, which forms the eastern apex of a roughly 4°-wide triangle with Alpha and Beta Camelopardalis; 17 Camelopardalis also lies almost midway between Beta Camelopardalis and another 5th-magnitude star, 36 Camelopardalis, 5° to the northeast. Next look for L Camelopardalis, which lies only about 4° north–northeast of 36 Cam. Finally, use your binoculars to locate 6th-magnitude Star c, about 2° southwest of L Camelopardalis. You want to take the time to center Star c in your telescope at low power, then switch to Chart 6b and confirm its identity. Once you've confirmed Star c, move about 30' northwest to 8th-magnitude Star d, which is flanked to the north by two roughly 11th-magnitude suns. Now hop 25' northwest to a pair of 8th-magnitude stars (e), which forms the southwestern base of a roughly 5'-wide pentagram of stars. Another 25' hop to the north will bring you to a tight triangle of 9th- and 10th-magnitude stars (f). NGC 1961 is 20' northwest of Triangle f, and 10' northeast of a pair of 10th-magnitude suns (g). The box in Chart 6b shows the area covered in the photograph to the left, which should assist you further in identifying this difficult galaxy.

The quick view

I could not see the galaxy in my 4-inch at 23×. The object becomes dimly visible at 101×. It has a tiny core (2') of dim light, surrounded by a 2' halo of hyperfine light. Concentrate on the object's location and use averted vision, which will cause the galaxy's small core to blossom a bit into a very faint glow. Take your time ... and breathe rhythmically!

Stop. Do not move your telescope.

6. NGC 2403 (H V-44)

Type	Con	RA	Dec	Mag	Dim	Rating
Mixed spiral galaxy	Camelo-pardalis	07^h 36.9^m	+65° 36'	8.5	25.5' × 13.0'	4

General description

NGC 2403 is a large and reasonably bright galaxy in a star-poor corridor between 5th-magnitude Omicron (*o*) Ursae Majoris (the Great Bear's nose – see Chart 7) and 6th-magnitude 51 Camelopardalis. Under dark skies, the galaxy is immediately apparent in 7×50 binoculars, looking much like the head of a comet. Telescopically, NGC 2403 is a well-defined elliptical glow trapped between two 11th-magnitude stars.

Directions

Using Chart 6, from NGC 1961, return your gaze to L Camelopardalis. Now, using your unaided eyes or binoculars, look for the 5th-magnitude stars 43 and 42 Camelopardalis 3° to the east and southeast, respectively. Nearly one binocular field (~5°) further is the lone 6th-magnitude star 51 Camelopardalis. Center that star in your telescope. NGC 2403 is only 1° west of 51 Camelopardalis.

The quick view

At 23× in the 4-inch, NGC 2403 is a bright elliptical glow squeezed between two 11th-magnitude stars. The galaxy has nearly the same apparent size as the full Moon. Its core is an amorphous lens with a gradual brightening toward the center. At 72×, much the same detail is revealed, but the galaxy appears larger. Large telescopes should have little trouble resolving some of the galaxy's spiral structure, as well as a multitude of foreground stars sprinkled across the disk and some bright H II regions.

Field note

Be aware that a roughly 13th-magnitude star lies near the galaxy's lens-shaped core. It can be mistaken for a supernova.

7. NGC 2655 (H I-288)

Type	Con	RA	Dec	Mag	Dim	Rating
Mixed spiral galaxy	Camelo-pardalis	08ʰ 55.6ᵐ	+78° 13′	10.2	5.9′ × 5.3′	4

General description

Lying a little more than 10° from Polaris, NGC 2655 is circumpolar from most of the Northern Hemisphere and is visible on any clear night. Another bright wonder, this galaxy is visible in a 60-mm refractor from a dark sky. It has a bright core and a soft but small halo of light. Overall, NGC 2655 looks more like a planetary nebula than a galaxy, so magnification will enhance its visibility.

Directions

Use Chart 7 and plan to spend some time on this hunt. Unless you are under a dark sky, be sure to have binoculars at your side. To start, imagine a line from Alpha (α) Ursae Majoris (Dubhe) to Alpha Ursae Minoris (Polaris). About one-third of the way from Polaris to Dubhe, and a tad west, is 4th-magnitude Star a. It is the brightest star in the area, so there is no mistaking it. Now use your unaided eyes or binoculars to locate 5th-magnitude Star M 4° to the west–southwest. It is the brightest of three equally spaced stars oriented north–northwest to south–southeast. Next look about 4° south–southeast of Star M for 5th-magnitude Star c; it is the brightest star in a little hook of seven suns. (Use your binoculars to confirm that Stars a, M, and c, form a roughly $6\frac{1}{2}°$-wide isosceles triangle.)

Once you confirm Star *c* in your binoculars, place it in your telescope at low power and switch to Chart 7a. The entire hook asterism mentioned above fits nicely in a 1°-wide field of view. If you move 1½° northeast from the northeastern tip of the hook, you should see a 45′-long arc of three 7th- and 8th-magnitude suns (d), oriented east to west. A 1° sweep northeast will bring you to a 1°-wide, isosceles triangle comprising three 7th- and 8th-magnitude suns (e); center the triangle's northeasternmost star, which is also the brightest. NGC 2655 is only 15′ east–northeast of that star; another 7th-magnitude star lies about 10′ to its southeast.

The quick view

At 23× in the 4-inch, NGC 2655 is a soft round glow with a delicate core, all of which is visible with direct vision. The galaxy swells into a slight egg-shaped wonder with averted vision, with the major axis oriented east to west. At 72×, the galaxy's intensity is enhanced. The core transforms into a fuzzy bead of light surrounded by an elliptical inner lens with a dim and diffuse outer halo. The galaxy takes magnification well. In larger scopes, the core becomes ever more intense and its disk displays some superimposed stars that may make the otherwise milky texture appear curdled.

Star charts for sixth night

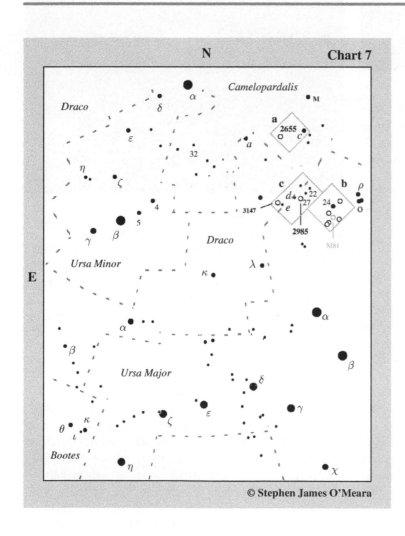

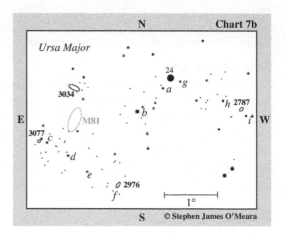

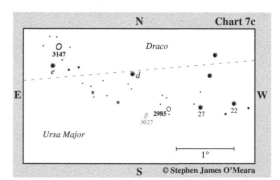

SIXTH NIGHT

1. NGC 3034 (H IV-79) = M82

Type	Con	RA	Dec	Mag	Dim	Rating
Spiral galaxy	Ursa Major	09^h 55.8^m	$+69°$ $41'$	8.4	$11.2' \times$ $4.3'$	5

General description

NGC 3034 (M82) is a bright irregular galaxy about $2°$ east of 4.5-magnitude 24 Ursae Majoris and only about $35'$ north of the stunning spiral galaxy M81. This galaxy pair is one of the most sought after objects in the northern heavens. M81 is the more brilliant object of this dynamic duo, but NGC 3034 can be seen in 7×50 binoculars from a dark-sky site. The telescopic view is nothing short of splendid. It is uncertain as to why William Herschel cataloged this object, since he was careful not to include known Messier objects – unless he had a notable reason.

Directions

Use Chart 7 to locate 2nd-magnitude Alpha (α) Ursae Majoris. Now look $12°$ to the north–northwest for 4.5-magnitude 24 Ursae Majoris. If you live under light-polluted skies, you may need to confirm the star with binoculars; notice that it forms the northeastern apex of a roughly $3°$-wide acute triangle with the similarly bright stars Rho (ρ) and Omicron (o) Ursae Majoris – actually this star marks the position of Omicron^1 and Omicron^2 Ursae Majoris. M81 and NGC 3034 (M82) are only about $2°$ east–southeast of 24 UMa. If necessary, use Chart 7b and low power to star hop to your target. From 24 UMa, move $25'$ southeast to 7.5-magnitude Star a. Next, move about $35'$ further to the southeast to 6th-

magnitude Star b. Bright M81 lies about $1\frac{1}{4}°$ east–southeast of Star b. NGC 3034 is only about $35'$ due north of M81.

The quick view

At 23×, NGC 3034 is a beautiful ellipse of mottled light – one that swells to prominence with averted vision. At 72×, the cigar-shaped galaxy becomes even more prominent and disheveled looking. It's quite the spectacle and deserving of much more time. But it's time to move on to your next target.

Stop. Do not move the telescope. Your next target is nearby!

2. NGC 3077 (H I-286)

Type	Con	RA	Dec	Mag	Dim	Rating
Peculiar galaxy	Ursa Major	10^h 03.3^m	$+68°$ $44'$	9.8	$5.5' \times$ $4.1'$	4

General description

NGC 3077 is a small but bright peculiar galaxy about $45'$ southeast of M81. It is best seen with magnification and looks like a small tailless comet.

Directions

From NGC 3034, return to M81. Now use Chart 7b to locate the pair of 8th-magnitude stars (c) $40'$ to the east–southeast. NGC 3077 is immediately southeast of the northeastern star in Pair c.

The quick view

At 23×, NGC 3077 is a bright but small ($5'$-wide) uniform glow kissing an 8th-magnitude star. At 72×, the galaxy displays a very dense central core that gradually gets brighter toward the middle. The galaxy has a slightly mottled texture when seen with averted vision.

Field note

This overlooked galaxy is just as bright and large as M89, an elliptical galaxy in Virgo, which shows you how powerful an attraction M81 and M82 are to observers.

Stop. Do not move the telescope. Your next target is nearby!

3. NGC 2976 (H I-285)

Type	Con	RA	Dec	Mag	Dim	Rating
Spiral galaxy	Ursa Major	09^h 47.3^m	+67° 55′	10.2	5.0′ × 2.8′	2

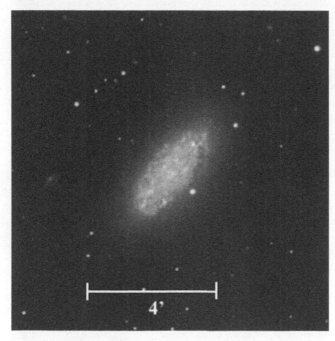

General description

NGC 2976, a member of the M81 group of galaxies, is a dim ellipse of light. It's best viewed under a dark sky at low power. You'll need to take your time when star-hopping to it, because the field stars are rather dim.

Directions

Using Chart 7b, from NGC 3077 and Pair *c*, move 30′ southwest to solitary 8th-magnitude Star *d*. Solitary 9.5-magnitude Star *e* lies equidistant to the south–southwest. Now move 35′ southwest, where you will find a 15′-long line of three 11th-magnitude stars (oriented northeast to southwest); the northeasternmost star is a double. NGC 2976 lies less than 10′ northwest of the middle star in that line (f).

The quick view

At 23×, NGC 2976 is a faint ellipse of diffuse light (3′ × 4′) that almost vanishes with increased magnification. Larger

scopes will reveal a mottled texture to the galaxy, which has no distinct central condensation.

Stop. Do not move your telescope

4. NGC 2787 (H I-216)

Type	Con	RA	Dec	Mag	Dim	Rating
Barred spiral galaxy	Ursa Major	09^h 19.3^m	+69° 12′	10.8	3.5′ × 2.4′	2

General description

NGC 2787 is a small and dim galaxy $1\frac{1}{2}°$ southwest of 24 Ursae Majoris. Seeing it in a small telescope requires a dark sky and magnification. It's important to get the surrounding star field, then look for the galaxy with moderate magnification and averted vision.

Directions

Using Chart 7b, from NGC 2976, return to 4.5-magnitude 24 Ursae Majoris. Using low power, move about 10′ southwest to 8.5-magnitude Star *g*. Next, sweep 55′ further to the southwest to 7.5-magnitude Star *h*. One more hop 30′ southwest will bring you to a pair of 8.5-magnitude stars (i). NGC 2787 is a little more than 5′ northeast of the easternmost star in Pair i.

The quick view

Once you get Pair i I recommend using 72× immediately to search for the tiny galaxy, which should be visible as a 3′-wide diffuse glow with averted vision. There is little central condensation, though the galaxy is slightly brighter toward the middle than at the edges. Light pollution will greatly affect its visibility.

5. NGC 2985 (H I-78)

Type	Con	RA	Dec	Mag	Dim	Rating
Spiral galaxy	Ursa Major	09ʰ 50.4ᵐ	+72° 17′	10.4	3.9′ × 3.0′	3

General description

NGC 2985 is a very small and very condensed glow – much like a dim stellarlike planetary nebula – about 35′ due east of 5th-magnitude 27 Ursae Majoris. In small telescopes, moderate to high power is required to see it as a definite fuzzy extragalactic wonder.

Directions

Use Chart 7 to locate 4.5-magnitude 24 Ursae Majoris. Now use your binoculars to locate a 1°-wide triangle of 5th- and 7th-magnitude stars 2½° to the north–northeast; it is composed of 5th-magnitude 27 Ursae Majoris, 6th-magnitude 22 Ursae Majoris, and a 7th-magnitude star that marks the triangle's northern apex; see Chart 7c. An extra 7.5-magnitude star between the northern apex and 27 Ursae Majoris makes the asterism Y shaped. You want to center 27 Ursae Majoris in your telescope. NGC 2985 is only 35′ due east of 27 Ursae Majoris. Again, you can confirm its position among the stars by referring to Chart 7c.

The quick view

At 23×, NGC 2985 is barely visible as a "galaxy," looking very condensed and nearly stellar. At 72×, the galaxy is a 1′-wide ball of faint nebulous matter surrounding a starlike nucleus. The galaxy's fainter extensions cannot be seen, though the galaxy does seem to swell a bit with averted vision.

Stop. Do not move the telescope. Your next target is nearby!

6. NGC 3147 (H I-79)

Type	Con	RA	Dec	Mag	Dim	Rating
Spiral galaxy	Draco	10ʰ 16.9ᵐ	+73° 24′	10.6	4.3′ × 3.7′	2

General description

NGC 3147 is a small, dim, low-surface-brightness (13.5) galaxy with a bright starlike nucleus (the saving grace for small-telescope users). It resides in Draco roughly midway, and a bit west, of a line joining Alpha (α) Ursae Minoris (Polaris) and Alpha Ursae Majoris (Dubhe). It is only about 2° northeast of NGC 2985.

Directions

Using Chart 7c, from NGC 2985, move 50′ northeast to 6.5-magnitude Star d at the border of Draco. Now make a slow and careful 1½° sweep to 7th-magnitude Star e, which is the bright star at the eastern end of a 30′-wide arc of stars, oriented east to west. NGC 3147 lies just 20′ north–northwest of Star e.

The quick view

At 23×, under very dark skies in the 4-inch, the galaxy is just visible with averted vision as a faint, hazy "star." At 72×, the galaxy is extremely small and dim, perhaps only about 2′ in extent with a starlike core that is most prominent with averted vision. In larger telescopes the galaxy becomes very condensed and obvious, though not much more detail is suspected.

Star charts for seventh night

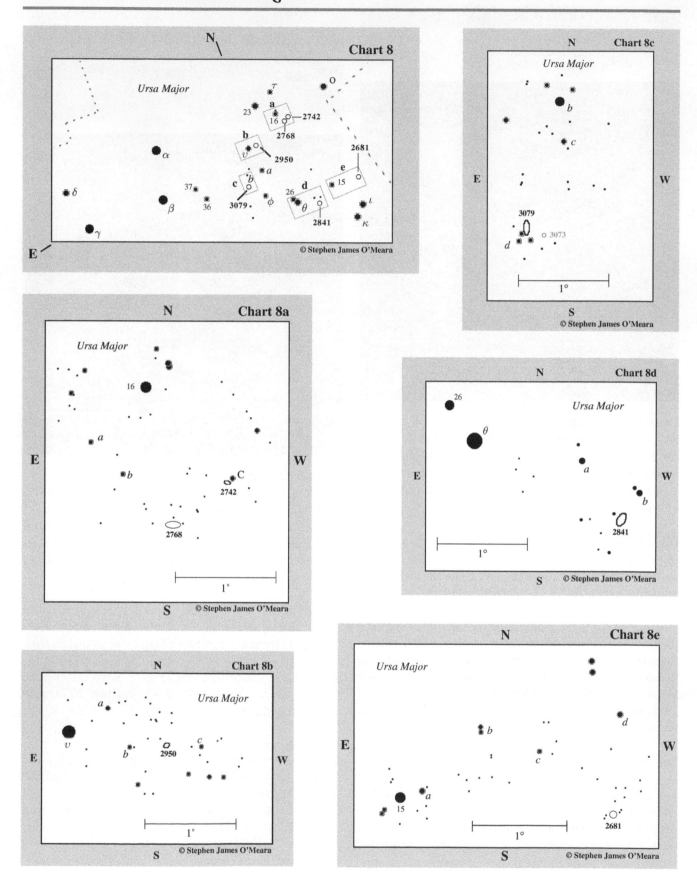

© Stephen James O'Meara

SEVENTH NIGHT

1. NGC 2768 (H I-250)

Type	Con	RA	Dec	Mag	Dim	Rating
Elliptical galaxy	Ursa Major	09^h 11.6^m	+60° 02′	9.9	6.6′ × 3.2′	4

4′

General description

NGC 2768 is a very nice, elliptical galaxy, elongated and bright, about $1\frac{1}{2}°$ south–southwest of 5th-magnitude 16 Ursae Majoris.

Directions

Use Chart 8 to locate 2nd-magnitude Alpha (α) Ursae Majoris. Now look 10° (a fist held at arm's length) to the west for 3.5-magnitude 23 Ursae Majoris. Now use your unaided eyes or binoculars to find 5th- magnitude 16 Ursae Majoris $2\frac{1}{2}°$ to the southwest. Center 16 Ursae Majoris in your telescope at low power, then switch to Chart 8a. From 16 Ursae Majoris, move 45′ southeast to solitary 8.5-magnitude Star a. Now look for similarly bright Star b less then 30′ to the southwest. NGC 2768 lies about 45′ further to the southwest.

The quick view

At 23× in the 4-inch under a dark sky, NGC 2768 is very visible as a bright 3′-long elongated glow, oriented east to west. At 72×, the galaxy's core appears elliptical and very condensed, getting much brighter toward the middle. A fine sight.

Stop. Do not move the telescope. Your next target is nearby!

2. NGC 2742 (H I-249)

Type	Con	RA	Dec	Mag	Dim	Rating
Spiral galaxy	Ursa Major	09^h 07.6^m	+60° 29′	11.4	3.0′ × 1.6′	2

4′

General description

NGC 2742 is a very small and dim spiral galaxy less than 40′ northwest of NGC 2768. Small-telescope users will need to be under a dark sky and use moderate magnification to see it. Averted vision, magnification, and patience is also required.

Directions

Chart 8a. From NGC 2768, move a little more than 40′ northwest to 8th-magnitude Star c. NGC 2742 is less than 5′ southeast of Star c.

The quick view

At 72×, NGC 2742 is a very dim object about 1′ in length. You'll need to use averted vision to separate the galaxy from its neighboring star. If you spend some time with the galaxy, it should appear as a dim circular glow with a slightly brighter center.

Stop. Do not move your telescope.

3. NGC 2950 (H IV-68)

Type	Con	RA	Dec	Mag	Dim	Rating
Barred spiral galaxy	Ursa Major	09^h 42.6^m	+58° 51′	10.9	3.3′ × 2.4′	3

4'

General description

NGC 2950 is a very small but very condensed barred spiral galaxy a little more than 1° west–southwest of 4th-magnitude Upsilon (υ) Ursae Majoris.

Directions

Chart 8. From NGC 2742, return your gaze to 3.5-magnitude 23 Ursae Majoris. Now look $4\frac{1}{2}°$ southeast for Upsilon Ursae Majoris; note too that Upsilon Ursae Majoris marks the western apex of a 9°-long isosceles triangle with Alpha (α) and Beta (β) Ursae Majoris. Center Upsilon Ursae Majoris in your telescope at low power and switch to Chart 8b. From Upsilon Ursae Majoris, move about 30′ northwest to 8.5-magnitude Star a. Now hop 30′ south–southwest to 9.5-magnitude Star b, which has a 10.5-magnitude companion immediately to its west. NGC 2950 is 25′ due west of Star b, about midway between Star b and similarly bright Star c.

The quick view

At 23× in the 4-inch, NGC 2950 appears as very small star-like object that swells slightly with averted vision. At 72×, the galaxy is a 1′-wide oval of condensed light (oriented west–northwest to east–southeast) with a very bright nucleus. Larger instruments make the galaxy appear brighter but do little else to show detail.

Stop. Do not move the telescope. Your next target is nearby!

General description

NGC 3079 lies about $3\frac{1}{2}°$ southeast of Upsilon (υ) Ursae Majoris. It is a somewhat difficult galaxy to see, given that it is a long and narrow streak of fairly faint light. Light pollution will certainly affect the visibility of this galaxy. If so, try using moderate magnifications.

Directions

Chart 8. From NGC 2950, return your gaze to Upsilon Ursae Majoris. Now look 2° to the southwest where you'll find 5th-magnitude Star a. Star a marks the western apex of a nearly 2° wide triangle with two slightly fainter suns to the east. Use binoculars if you must to identify this triangle. You want to center the southeasternmost of those two stars (Star b in Box c) in your telescope at low power, then switch to Chart 8c. From Star b, move 20′ south–southwest to 7.5-magnitude Star c. NGC 3079 is a little more than 50′ southeast of Star c, just 6′ north of a triangle of roughly 8.5- to 9.5-magnitude stars (d).

The quick view

At 23× in the 4-inch, this barred spiral galaxy is visible as a weak, 5′-long, streak of light, oriented north to south. Try centering Triangle d, then using averted vision to see the galaxy pop into view. At 72×, the galaxy displays sharp edges and a mottled nuclear region. A very pretty sight.

4. NGC 3079 (H V-47)

Type	Con	RA	Dec	Mag	Dim	Rating
Barred spiral galaxy	Ursa Major	10ʰ 02.0ᵐ	+55° 41′	10.9	8.0′ × 1.5′	3

5. NGC 2841 (H I-205)

Type	Con	RA	Dec	Mag	Dim	Rating
Spiral galaxy	Ursa Major	09ʰ 22.0ᵐ	+50° 59′	9.0	6.6′ × 3.4′	5

4'

General description

NGC 2841 is a small but deceivingly bright spiral galaxy about $1\frac{2}{3}°$ west–southwest of 3rd-magnitude Theta (θ) Ursae Majoris in the Great Bear's right foreleg. From a dark-sky site, it can be seen in 7×50 binoculars, looking like a dim companion to a 9th-magnitude star about 5' to its northeast.

Directions

Use Chart 8 to locate Theta (θ) Ursae Majoris, which lies about 6° southwest of NGC 3079. Center Theta Ursae Majoris in your telescope at low power, then switch to Chart 8d. From Theta UMa, sweep $1\frac{1}{4}°$ west to 6th-magnitude Star a. Next move 40' southwest to 6th-magnitude Star b, which has an 8th-magnitude companion about 4' to the northeast. NGC 2841 is only 20' south–southeast of that pair.

The quick view

At 23×, NGC 2841's central lens is so bright it can be seen in astronomical twilight. Otherwise, under a dark sky, it is an elongated glow with a sharp stellar nucleus, which looks like a moonlit stone centered in a pool of vapors. At 72×, the galaxy's starlike nucleus lies within a slightly elliptical inner lens; it looks like a comet with a tail seen nearly head-on. With concentration, the galaxy's northwestern side appears mottled. The rest of the disk is milky smooth.

Field note

Be aware that a prominent 11th-magnitude star can be seen northwest of NGC 2841's nucleus; that star can cause you to think that a supernova has erupted.

Stop. Do not move your telescope.

6. NGC 2681 (H I-242)

Type	Con	RA	Dec	Mag	Dim	Rating
Mixed spiral galaxy	Ursa Major	08^h 53.5^m	+51° 19'	10.3	3.5'× 3.5'	3

4'

General description

NGC 2681 is another small but reasonably bright (because it is condensed) mixed spiral galaxy $2\frac{1}{2}°$ west–southwest of 4.5-magnitude 15 Ursae Majoris.

Directions

Chart 8. From 2841, return your gaze to 3rd-magnitude Theta (θ) Ursae Majoris. Now look $3\frac{1}{2}°$ west for 15 Ursae Majoris. You want to center this star in your telescope at low power, then switch to Chart 8e. From 15 Ursae Majoris, move 15' west–northwest to 7th-magnitude Star a. Next, move 55' northwest to a pair of stars (b), oriented north to south, whose components shine at 8.5 and 9th magnitude. Now move 40' southwest to 9th-magnitude Star c. NGC 2681 lies 1° further to the southwest. Note that it marks the southern apex of an equilateral triangle with Star c and 7th-magnitude Star d.

The quick view

At 23× in the 4-inch under dark skies, the galaxy is a round, 2'-wide glow with a condensed core. The galaxy gradually brightens inward. A magnification of 72× does little to enhance the view.

Field note

NGC 2841 is nestled between some dim suns that may be confused as knots or H II regions.

2 · February

Star charts for first night

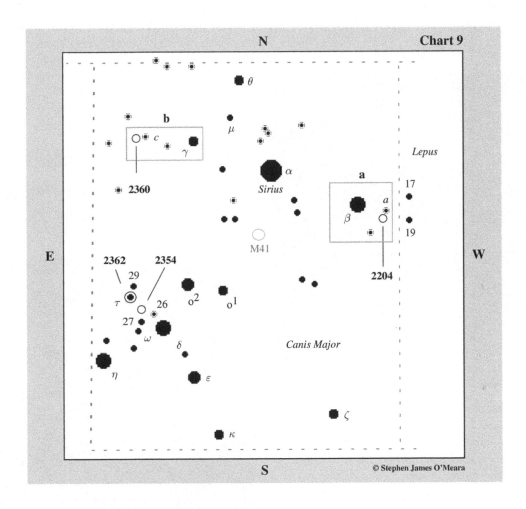

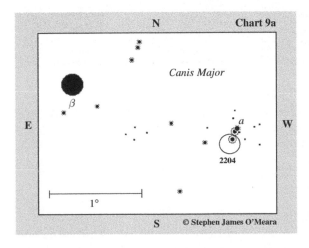

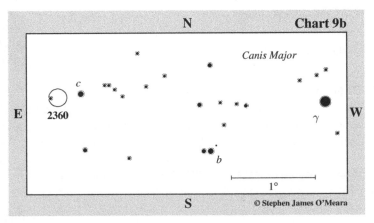

FIRST NIGHT

1. NGC 2204 (H VII-13)

Type	Con	RA	Dec	Mag	Diam	Rating
Open cluster	Canis Major	06^h 15.5^m	−18° 40′	8.6	10.0′	2

General description

NGC 2204 is a faint but rich open cluster in Canis Major discovered by William, not Caroline, Herschel. Many amateur astronomers find NGC 2204 a challenge to see as a well-defined open cluster, even in a 10-inch scope. In a small telescope it looks more like a globular cluster than an open star cluster. Add light pollution, and this southern cluster becomes a veritable demon to detect. The key to seeing this object is knowing exactly where to look and using magnification.

Directions

Use Chart 9 to locate 2nd-magnitude Beta (β) Canis Majoris about 5$\frac{1}{2}$° west–southwest of brilliant Alpha (α) Canis Majoris (Sirius). Only about 2° further to the west–southwest – about halfway between Beta Canis Majoris and 19 Leporis – is the 6th-magnitude Star a. NGC 2204 is just 10′ south–southeast of Star a. Use Chart 9a to identify the exact location of this dim cluster.

The quick view

When I tried to see NGC 2204 in my 4-inch at 23× from Hawaii, I first mistook it for a close gathering of 9th-magnitude and fainter suns surrounding a 6th-magnitude star little more than 10′ to the north–northwest of my target. I ultimately did see NGC 2204 with averted vision, shimmering inside a triangle of 9th- and 10th-magnitude suns like a caged comet. The brightest of these three stars shines at magnitude 8.8 and is superimposed on the northern fringe of the cluster, which, at low power, resembles an unresolved globular cluster with a glowing central orb surrounded by a halo of diffuse light. I call it the Angel Starfish cluster. At 72×, NGC 2204 looks like a globular cluster just starting to be resolved. The cluster packs 383 stars of 13th magnitude and fainter in a disk no larger than 10′. It is a difficult object to resolve even in 10- to 12-inch telescopes.

2. NGC 2354 (H VII-16)

Type	Con	RA	Dec	Mag	Diam	Rating
Open cluster	Canis Major	07^h 14.3^m	−25° 41′	6.5	18.0′	3

General description

NGC 2354 is a large, bright, but low-surface-brightness open star cluster in the hind quarters of the Great Dog, just 1$\frac{1}{2}$° northeast of 2nd-magnitude Delta (δ) Canis Majoris. The cluster is rich, but its stars are loose and scattered. It is best appreciated under a dark sky.

Directions

Use Chart 9 to find Delta Canis Majoris. Now locate 6th-magnitude 26 Canis Majoris just 1° to the northeast. Center 26 Canis Majoris in your telescope. NGC 2354 is only 30′ to the east–northeast. At low power it should be quite large (18′) and obvious. Note that the cluster also marks the northeastern arm of a north–south-oriented Y comprising NGC 2354, 26 CMa, 27 CMa, and Omega (ω) CMa.

The quick view

At 23× in the 4-inch, NGC 2354 is a large and diffuse glow bespeckled with a coarse scattering of about a half-dozen stars. A magnification of 41× seems to be the best for this cluster, which shows more than two dozen suns spread over 18'. There is no central condensation, and its brightest stars are scattered in groups (most to the west and south) across a "diffuse" background of unresolved suns. While the cluster contains some 300 stars of 9th-magnitude and fainter, the vast majority of them are tangled in a rich Milky Way background.

Stop. Do not move the telescope. Your next target is nearby!

3. NGC 2362 (H VII-17)

Type	Con	RA	Dec	Mag	Diam	Rating
Open cluster	Canis Major	07^h 18.7^m	−24° 57'	3.8	6.0'	4

General description

NGC 2362, the famous Tau (τ) Canis Majoris cluster, is an inconspicuous but visually stunning open cluster in the lower backbone of the Great Dog. Although the cluster's brightest member, 3.8-magnitude Tau Canis Majoris, is visible to the naked eye, the rest of the cluster is so small (6') that it is a challenge to see in binoculars owing to the overwhelming brightness of Tau. It is a beautiful object in telescopes of all sizes.

Directions

Use Chart 9 to locate Tau Canis Majoris, which is nearly 3° northeast of Delta Canis Majoris, or a little more than 1°

northeast of open cluster NGC 2354. Once you pinpoint Tau Canis Majoris, center it in your telescope; the cluster immediately surrounds it.

The quick view

At 23× in the 4-inch, the cluster is a pyramid of starlight that encompasses icy blue Tau Canis Majoris. I call it the Liquid Pyramid, because when seen with direct vision, the brightest stars (which form a cube around Tau) appear to "melt" into a pool of dimmer starlight when seen with averted vision. At 72×–101×, the cluster is a magnificent array of dim suns encircling Tau. Averted vision also shows sinuous arms of starlight spiraling outward toward the main cardinal points. The cluster contains 60 stars of 8th magnitude and fainter packed into a disk 6' across. It is is a dazzling sight in both small and large telescopes.

4. NGC 2360 (H VII-12)

Type	Con	RA	Dec	Mag	Diam	Rating
Open cluster	Canis Major	07^h 17.7^m	−15° 38'	7.2	14.0'	4

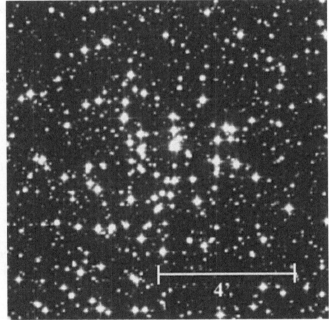

General description

NGC 2360 is a fairly bright and rich open star cluster nearly $3\frac{1}{2}°$ east of 4th-magnitude Gamma (γ) Canis Majoris, and about 20' east of a 6th-magnitude star. It is easily spied in binoculars as a small "puff" of light a little less than 25' east of 5.5-magnitude Star c. It is a grand sight in telescopes of all sizes.

Directions

Use Chart 9 to locate Alpha (α) Canis Majoris, then Gamma Canis Majoris a little more than 4° to the east–northeast.

NGC 2360 is $3\frac{1}{2}°$ due east of Gamma Canis Majoris and a little less than 25′ east of 5.5-magnitude Star c, which, under a dark sky, can be seen with the unaided eye. If you have difficulty, use Chart 9b to star hop to it from Gamma Canis Majoris. Using low power, make a slow and careful $1\frac{1}{2}°$ sweep southeast to Star b, which has a 6.6-magnitude companion immediately to its east. Now make another slow and careful sweep a little more than $1\frac{1}{2}°$ northeast of Star b to Star c. Again, NGC 2360 is a little less than 25′ east of Star c.

The quick view

At 23× in the 4-inch, NGC 2360 is a rich and dynamic cluster with a dense, elliptical core surrounded by long limber arms of stars. At 72×, the cluster's core is composed of several parallel rays of stars (oriented north to south). The surrounding starscape is one of a vast array of geometrical patterns. More than 90 stars can be seen in this 14′-wide cluster. The brightest members shine at 10th magnitude.

Star charts for second night

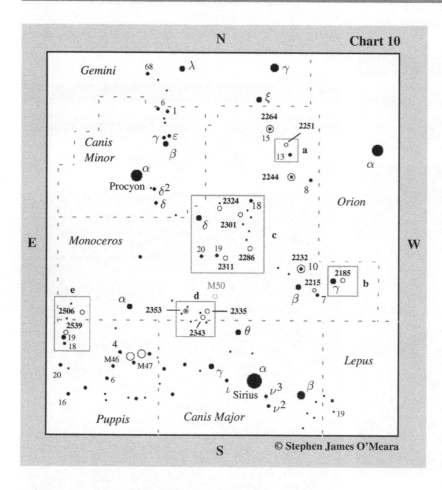

Chart 10

© Stephen James O'Meara

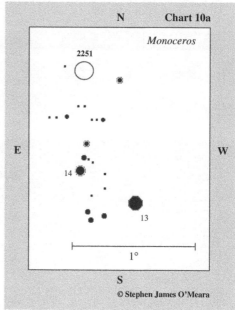

Chart 10a

© Stephen James O'Meara

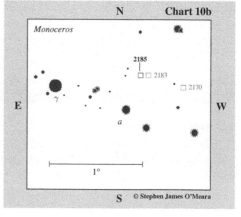

Chart 10b

© Stephen James O'Meara

SECOND NIGHT

1. NGC 2264 (H VIII-5)

Type	Con	RA	Dec	Mag	Diam	Rating
Open cluster	Mono-ceros	06^h 41.0^m	+09° 54'	4.4	40.0'	5

General description

NGC 2264 is a bright and obvious open star cluster in Monoceros – one visible to the naked eye and well resolved in binoculars. It is a popular winter target, especially in late December, because its shape resembles that of a well trimmed Christmas tree. Under a dark sky, 15 stars are immediately apparent in 7×50 binoculars. It is a marvel to see in telescopes of all sizes.

Directions

Use Chart 10 to locate 2nd-magnitude Gamma (γ) Geminorum and 3rd-magnitude Xi (ξ) Geminorum in the feet of the Twins. Xi Geminorum also marks the northeastern end of a 10°-long line of four stars that flows to the southwest; the other three stars are 5th-magnitude 15 Monocerotis, 4.5-magnitude 13 Monocerotis, and 4th-magnitude 8 Monocerotis. You want to raise your binoculars to 15 Monocerotis. The 4th-magnitude cluster will be immediately apparent.

The quick view

At 23× in the 4-inch, NGC 2264 is a brilliant triangular-shaped cluster (the Christmas Tree) with 5th-magnitude 15 Monocerotis marking the tree's base. The 40'-wide cluster has 40 stars of 5th-magnitude and fainter. It is associated with swaths of bright nebulosity. One dim region of nebulosity surrounds the 7th-magnitude star at the tip of the Christmas Tree asterism. The dark Cone Nebula resides in it, but that feature is a target best suited for larger telescopes.

Stop. Do not move the telescope. Your next target is nearby!

2. NGC 2244 (H VII-2)

Type	Con	RA	Dec	Mag	Diam	Rating
Open cluster	Mono-ceros	06^h 32.3^m	+04° 51'	4.8	30.0'	5

10'

General description

NGC 2244 is a bright and beautiful open star cluster just southeast of the midpoint between the unicorn's nose and eye (8 and 13 Monocerotis, respectively). Under dark skies, the cluster is a reasonable target for the unaided eye and a fine sight in binoculars. NGC 2244 is a spectacle in rich-field telescopes, especially at low power, which may also reveal the enormous Rosette Nebula surrounding the cluster.

Directions

Using Chart 10 as your guide, continue down the line of naked-eye stars that flows southwest of 15 Monocerotis to 8 Monocerotis. NGC 2244 is 2° east–northeast of 8 Monocerotis; it should be obvious in binoculars. If you place NGC 2264 at the 11-o'clock-position angle in a pair of 7×50 binoculars when the object is high in the south, NGC 2244 will be at the 5-o'clock-position angle in the same field of view.

The quick view

At 23×, under dark skies, NGC 2244 is a 30′-wide rectangle of six binocular suns, which is surrounded by the extensive, though dim, Rosette Nebula. Aside from the bright rectangle of suns, about a dozen fainter stars populate the core of the gaseous wreath. Another dozen or so fainter stars form loose sinuous arms that flow away from the cluster. The brightest star in the rectangle is 6th-magnitude 12 Monocerotis, and the two middle stars in the rectangle are nice doubles.

Stop. Do not move the telescope. Your next target is nearby!

3. NGC 2251 (H VIII-3)

Type	Con	RA	Dec	Mag	Diam	Rating
Open cluster	Mono-ceros	06^h 34.6^m	+08° 22′	7.3	10.0′	3

General description

NGC 2251 is a small and indistinct open cluster a little more than 1° north–northeast of 13 Monocerotis. The cluster is visible in 7 × 50 binoculars from a dark-sky site, looking like a piece of mottled lint. Through a telescope it is a coarse scattering of suns, elongated, with no central condensation. It may be difficult to see it well defined from a city or light-polluted suburb.

Directions

Use Chart 10 to locate 4.5-magnitude 13 Monocerotis. Center that star in your telescope at low power then switch to Chart 10a. From 13 Monocerotis, sweep 30′ northeast to 6th-magnitude 14 Monocerotis, which marks the southern end of a 25′-long Y-shaped asterism of 8th- and 9th-magnitude suns oriented north to south. NGC 2251 is 45′ due north of 14 Monocerotis, or about 20′ from the northern end of the Y.

The quick view

At 23× in the 4-inch with direct vision, there's a long row of resolved stars, oriented northwest to southeast. With averted vision, the center swells, so it looks like the snake that swallowed the elephant. At 72×, the cluster is an irregularly long gathering of some two dozen suns of mixed magnitudes. A "tail" of four stars extends to the north. These stars are centered on a hazy, elongated, and partially resolved core. With imagination, the cluster looks like a Mexican Freetail bat in flight. The cluster's brightest sun (a 9.1-magnitude star) has a roughly 10.5-magnitude companion to its west. The other bright stars in the cluster form irregular patterns around it, especially to the southeast, north, and southwest. In larger telescopes, NGC 2251 appears only a bit more dynamic; the cluster contains about 90 stars of 9th magnitude and fainter.

4. NGC 2232 (H VIII-25)

Type	Con	RA	Dec	Mag	Diam	Rating
Open cluster	Mono-ceros	06^h 27.2^m	−04° 45′	4.2	53.0′	5

General description

NGC 2232 is a bright but coarse cluster of 20 stars associated with 5th-magnitude 10 Monocerotis. In 7 × 50 binoculars, its six brightest members are easily visible from a city. It is a nice sight in a small telescope at low power.

Directions

Use Chart 10 to locate 4.6-magnitude Beta (β) Monocerotis, which is about 10° (a fist held at arm's length) northwest of Alpha (α) Canis Majoris (Sirius). Use binoculars to confirm the star, which is also about $3\frac{1}{2}°$ east–southeast of similarly bright Gamma (γ) Monocerotis. The star 10 Monocerotis is just $2\frac{1}{2}°$ north, and slightly west of Beta Monocerotis. The cluster is right there, with its brightest members streaming to the south of 10 Monocerotis. If you have an equatorially mounted telescope, you can simply center Beta Monocerotis at low power and use setting circles to move $2\frac{1}{2}°$ to the north.

The quick view

At 23× in the 4-inch, NGC 2232 is a nice assortment of about a dozen suns hugging 10 Monocerotis; the brightest stars (7th- to 11th-magnitude) form a jagged stream to the south like blood dripping from an open wound. At 72×, the cluster is still well defined and rich. This magnification is especially good for revealing some 13 suns immediately around 10 Monocerotis.

Stop. Do not move the telescope. Your next target is nearby!

5. NGC 2215 (H VII-20)

Type	Con	RA	Dec	Mag	Diam	Rating
Open cluster	Mono-ceros	06^h 20.8^m	−07° 17′	8.4	8.0′	3

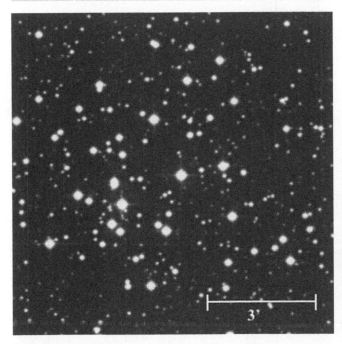

General description

NGC 2215 is a moderately small and weakly condensed cluster near 5th-magnitude 7 Monocerotis. From a dark-sky site it is visible in 7×50 binoculars as a dim concentration of light – especially from a bar of stars at the cluster's center. Telescopically, it is a rather nice cluster for one so dim. It may be a challenge for light-polluted areas.

Directions

Using Chart 10, from 10 Monocerotis, return your gaze to 4.5-magnitude Beta Monocerotis. Now look for 5th-magnitude 7 Monocerotis, which is south of the midpoint between Beta and 4.5-magnitude Gamma (γ) Monocerotis. All three stars should be in the same binocular field of view. You want to center 7 Monocerotis in your telescope. There should be no mistaking it, since it has a 7th-magnitude companion about 6′ to the southeast. NGC 2215 is 35′ northeast of 7 Monocerotis; therefore, the cluster should be in the same low-power field of view with 7 Monocerotis, just 15′ east of an 8th-magnitude star.

The quick view

At 23× in the 4-inch, four moderately bright stars (all ~11th magnitude) populate the cluster's core, forming a distinct line oriented east to west. This row, or bar, of stars appears surrounded by an 8′-wide halo of fainter, glittering gems. At 72×, the cluster is an attractive sight, with the straight row of stars surrounded by two dozen suns of various brightnesses in haphazard arrangements. Larger scopes should reveal twice as many fainter suns in this tight but relatively sparse cluster.

Stop. Do not move the telescope. Your next target is nearby!

6. NGC 2185 (H IV-20)

Type	Con	RA	Dec	Mag	Dim	Rating
Reflection nebula	Mono-ceros	06^h 11.1^m	−06° 13′	–	1.5′ × 1.5′	2

General description

NGC 2185 is a an extremely small reflection nebula nearly 1° west of Gamma (γ) Monocerotis. It is associated with several other patches of nebulosity that extend about 1° to the west–southwest of NGC 2185. In small telescopes NGC 2185 is visible under a dark sky, if you know where to look. Think small, and be prepared to look for a fuzzy star. It will be very difficult to see under light-polluted skies.

Directions

Use Chart 10 (and your binoculars, if necessary) to locate 4th-magnitude Gamma Monocerotis. Center Gamma Monocerotis in your telescope at low power then switch to Chart 10b. From Gamma Monocerotis, move 50′ west–southwest to 5th-magnitude Star a. NGC 2185 lies 22′ north–northwest of Star a.

The quick view

At 23× in the 4-inch under very dark skies, NGC 2185 is just visible if you know exactly where to look, with averted vision. Once you get the field, I suggest switching immediately to a moderate or high magnification. At 72×, the nebula looks like an 11th-magnitude comet just beginning to shine with a double nucleus. That double nucleus is actually two nebulae: the "brighter" glow to the east is NGC 2185; the weaker haze to the west is NGC 2183.

Star charts for third night

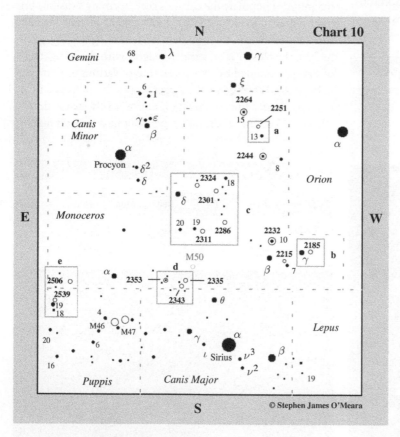

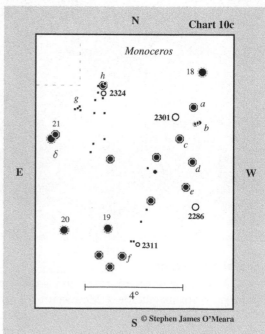

THIRD NIGHT

1. NGC 2301 (H VI-27)

Type	Con	RA	Dec	Mag	Diam	Rating
Open cluster	Mono-ceros	06^h 51.8^m	+00° 28′	6.0	15.0′	4

General description

NGC 2301 is a surprisingly bright and obvious object. Shining at 6th-magnitude, it is a stunning gem, perhaps the finest open cluster in Monoceros for small-telescope users. In 7×50 binoculars, it looks like a wavy line (oriented north to south) centered on a stunning double star. Telescopically, the cluster lies almost exactly on the galaxy's mid plane, and, as a consequence, is seen against a rich stellar backdrop.

Directions

Use Chart 10 to locate 2nd-magnitude Gamma (γ) Geminorum and 3rd-magnitude Xi (ξ) Gem in the feet of the Twins. Next, follow the 10°-long line of four stars comprising Xi Geminorum, 5th-magnitude 15 Monocerotis, 4.5-magnitude 13 Monocerotis, and 4th-magnitude 8 Monocerotis. Just 6° southeast of the midpoint between 8 and 13 Mon is 4.5-magnitude 18 Monocerotis, which forms the southeast apex of a triangle with these other two stars; note too that the Rosette Cluster (NGC 2244) lies in the northwestern side of that triangle. Center 18 Monocerotis in your telescope with low power, then switch to Chart 10c. From 18 Monocerotis, make a generous sweep 1½° south–southeast to 6th-magnitude Star a. NGC 2301 is a little more than 50′ southeast of Star a.

The quick view

At 23× in the 4-inch, NGC 2301 is immediately striking in form and bountiful in stars. A beautiful string of stars can be seen oriented north to south, the southern half of which is composed of four roughly 9th-magnitude stars that form a brilliant arc. With imagination, I see a mythical dragon in flight, so I have dubbed it Hagrid's Dragon. At 72×, the cluster's core is a tight bundle of five suns with a bright double star at the eastern end. The northern extension is sharply linear with a double star at the northern tip. The more southerly one has a warm hue.

Stop. Do not move the telescope. Your next target is nearby!

2. NGC 2286 (H VIII-31)

Type	Con	RA	Dec	Mag	Diam	Rating
Open cluster	Mono-ceros	06^h 47.7^m	−03° 09′	7.5	15.0′	3

General description

NGC 2286 is a moderately big and somewhat dim open cluster in an obscure Milky Way field about halfway between Beta (β) and Delta (δ) Monocerotis. From a dark sky, it is visible in 7×50 binoculars. It will be a challenge to find, and see, from light-polluted skies. Through a telescope, its stars seem to blend with the surrounding Milky Way.

Directions

To find NGC 2286, all you need to do is take your time and make five 1° star-hops, using Chart 10c as your guide. Start from NGC 2301 and move 1° west–southwest to the obvious 5′-wide string of 7th- and 8th-magnitude suns (b). Now sweep 1° to the southeast to 6th-magnitude Star *c*. Another hop 1° to the southwest brings you to 6th-magnitude Star *d*. And one more 1° sweep south–southeast brings you to 6th-magnitude Star *e*. NGC 2286 lies 1° south–southwest of Star *e*.

The quick view

At 23× in the 4-inch, NGC 2286 appears diffuse and scattered. The cluster is the size of a quarter Moon and the brightest dozen or so stars form a starfish pattern against a rich Milky Way background. Fainter unresolved suns form a background haze that would disappear under light-polluted skies. At 72×, the cluster's starfish pattern is more clearly defined, with pairs of stars marking all the tips, except the one to the northeast. Other pairs and groupings of stars are scattered about. The cluster contains 80 stars of 9th magnitude and fainter, and larger telescopes do a fine job in resolving the fainter members. Large scopes also show more stars in the cluster's loose core.

3. NGC 2311 (H VIII-60)

Type	Con	RA	Dec	Mag	Diam	Rating
Open cluster	Mono-ceros	06^h 57.8^m	−04° 37′	9.6	7.0′	3

General description

NGC 2311 is a small and dim open cluster near 5th-magnitude 19 Monocerotis. It is best seen in moderate-sized telescopes under a dark sky.

Directions

Use Chart 10 to locate 4th-magnitude Delta (δ) Monocerotis, which is almost 10° (a fist) southwest of 0-magnitude Alpha

(α) Canis Minoris (Procyon). Confirm Delta Monocerotis in your binoculars, which is easy since 5th-magnitude 21 Monocerotis (not shown) lies 15′ to the north–northwest. Now look 4° south–southwest for the 5th-magnitude stars 20 and 19 Monocerotis. Center 19 Monocerotis in your telescope at low power then switch to Chart 10c. From 19 Monocerotis, make a slow and careful sweep 1¼° south to a 1°-wide triangle of three 5th- and 6th-magnitude stars; the brightest of these stars is almost due south of 19 Monocerotis. You want to center the westernmost star in the triangle (6th-magnitude Star f). NGC 2311 is 1° northwest of Star f, just 10′ west–southwest of a roughly 7′-long arc of three 8th- to 9th-magnitude suns oriented mostly east to west.

The quick view

At 23× in the 4-inch under a dark sky, the cluster is a small puff of faint light that scintillates with tiny bits of starlight when averted vision is used. At 72×, the cluster is loose and sparse with no central condensation. Its five brightest members (about 12th-magnitude) form a broad Z-shaped asterism, which is elongated northwest to southeast. The fainter members look like ornamental filigree. The cluster contains about 50 members, but only about two dozen of them are visible in the 4-inch.

Stop. Do not move your telescope.

4. NGC 2324 (H VII-38)

Type	Con	RA	Dec	Mag	Diam	Rating
Open cluster	Mono-ceros	07^h 04.1^m	+01° 03′	8.4	8.0′	2

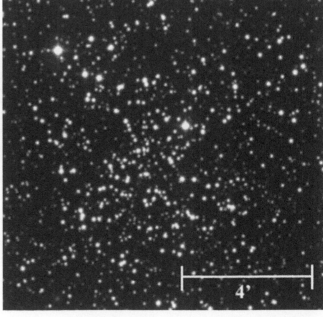

General description

NGC 2324 is a small, low-surface-brightness open cluster near 4th-magnitude Delta (δ) Monocerotis and the south-western border of Canis Minoris. It is best seen in moderate-sized telescopes under dark skies with magnification.

Directions

Using Chart 10c, from NGC 2311, return your gaze to to 4th-magnitude Delta Monocerotis. Using binoculars, now locate 6.6-magnitude Star h about $2\frac{1}{2}°$ to the northwest. Center that star in your telescope at low power; note that it is a double, with a 7.6-magnitude companion to the north-west. NGC 2324 is almost $\frac{1}{2}°$ due south of Star h. You can also star-hop to Star h. From 21 Monocerotis, move 1° northwest to a roughly 6′-wide, Y-shaped asterism (g) lying on its side. Now make a slow and careful $1\frac{1}{4}°$ sweep northwest to Star h.

The quick view

At 23× in the 4-inch under a dark sky, NGC 2324 is visible with averted vision as a small diffuse haze. It lies about 10′ southwest of a tight 5′-wide gathering of about a dozen suns (many in pairs) shining between 8th- and 12th-magnitude. At 72×–101×, the cluster is largely a uniform glow of dim starlight. Look for a nice arc of suns at the north-eastern edge of the cluster and another bright stud near the western edge. The remaining stars seem to dangle between these stars to the south, making the cluster appear like a dew-spangled cobweb. Many of these dimmer suns are in wavy rows, oriented northeast to southwest. The cluster has more than 130 stars, of 12th magnitude and fainter, in a disk 8′ across. Large telescopes will have a finer view of them.

Star charts for fourth night

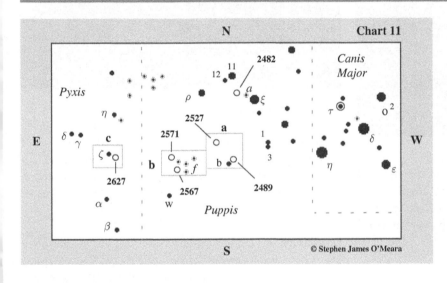

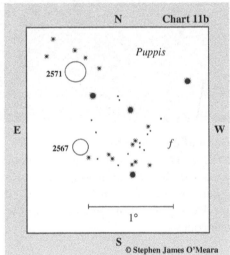

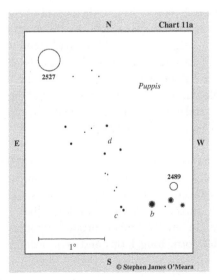

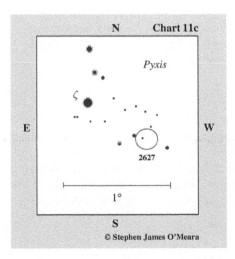

FOURTH NIGHT

1. NGC 2482 (H VII-10)

Type	Con	RA	Dec	Mag	Diam	Rating
Open cluster	Puppis	07ʰ 55.2ᵐ	−24° 15′	7.3	10.0′	3

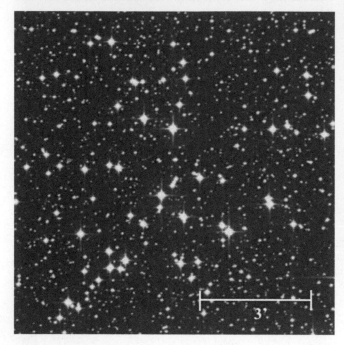

General description

NGC 2482 is a moderately rich though relatively faint open star cluster near 3rd-magnitude Xi (ξ) Puppis. From a dark-sky site, it is visible in 7×50 binoculars as a dim speckled glow. It is best seen at low power in small telescopes.

Directions

Use Chart 11 to locate 3rd-magnitude Xi Puppis, which is 7° northeast of 2nd-magnitude Eta (η) Canis Majoris in the Great Dog's tail. Center Xi Puppis in your telescope. Now sweep 30′ northeast of Xi Puppis to 6.4-magnitude Star *a*. NGC 2482 is about 55′ east–northeast of Star *a*.

The quick view

At 23× in the 4-inch, NGC 2482 is a loose and coarse cluster. A pretty peppering of fainter stars surrounds a bright Y-shaped core. The brightest stars shine at 10th magnitude, and the cluster contains some 40 members. Larger scopes show its structure to resemble that of yet another starfish. This appearance is shown in the accompanying photograph.

2. NGC 2489 (H VII-23)

Type	Con	RA	Dec	Mag	Diam	Rating
Open cluster	Puppis	07ʰ 56.2ᵐ	−30° 04′	7.9	5.0′	4

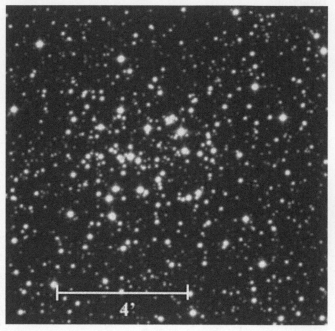

General description

NGC 2489 is a small but beautiful open cluster about 3° southeast of 4th-magnitude 3 Puppis, just 20′ northwest of solitary 5th-magnitude Star *b* (see Box a). From a dark-sky site, NGC 2489 is a very faint glow with averted vision in 7×50 binoculars, but it is a tiny wonder in telescopes of all sizes.

Directions

Use Chart 11 to locate 2nd-magnitude Eta (η) Canis Majoris. Just 4° east–northeast is a close paring of two 4th-magnitude stars – 1 and 3 Puppis. A 3° hop to the southeast will bring you to Star *b* – the brightest star in the region. Use binoculars to confirm the Star *b*, noting that it should be the brightest star in a 30′-wide arc of three suns, oriented east–west. NGC 2489 is just 25′ northwest of Star *b*, or about 12′ north–northeast of the middle star in the arc. All three stars and the cluster should fit easily and comfortably in a 1° field of view. You can also refer to Chart 11a, if needed.

The quick view

At 23× in the 4-inch, NGC 2489 is a small (5′) but beautiful round object, very concentrated and bright. The stars' brightnesses are well mixed, making them look like a stellar gum-ball machine. At 72× many bright members (of ~11th magnitude) form long loops and arcs. Fainter members are peppered about, but there's no clear central condensation. Quite a lovely sight.

Stop. Do not move the telescope. Your next target is nearby!

3. NGC 2527 (H VIII-30)

Type	Con	RA	Dec	Mag	Diam	Rating
Open cluster	Puppis	08ʰ 04.9ᵐ	−28° 08′	6.5	10′ (20′?)	4

General description
NGC 2527 is another fairly bright but loose open cluster in Puppis. It forms the southern apex of a near-equilateral triangle with the 3rd-magnitude stars Rho (ρ) and Xi (ξ) Puppis. It can be seen in 7×50 binoculars under a dark sky as a large and diffuse glow. Although some modern sources list the cluster's size as 10′, the cluster's visual extent appears twice that size, and this larger size is still listed in some reference books.

Directions
Using Chart 11a, from NGC 2489, return to Star b. Now use low power and move 30′ east and slightly south to a close pair of 7.5-magnitude stars (c). Now move 1° north and slightly east to a 30′-wide isosceles triangle composed of three roughly 7.5-magnitude stars (d). NGC 2527 lies 1¼° northeast of the northernmost star in Triangle d.

The quick view
At 23× in the 4-inch, NGC 2527 is a somewhat confusing sight. The cluster displays a small core of loosely knit suns at the center of a larger ellipse of dim stars. At 72×, the cluster's

core is a loose, detached assortment of two dozen similarly bright suns shining at 8.6 magnitude and fainter. Overall, the core is round in form and it is surrounded by weaker patches of irregular starlight. The cluster contains some 45 members.

Stop. When you have completed your observation, return to Star b.

4. NGC 2571 (H VI-39)

Type	Con	RA	Dec	Mag	Diam	Rating
Open cluster	Puppis	08ʰ 18.9ᵐ	−29° 45′	7.0	7.0′	3

General description
NGC 2571 is a large and scattered open cluster about 6° southeast of 3rd-magnitude Rho (ρ) Puppis. It is a telescopic object and may be difficult to extract from the surrounding field stars because it is highly elongated and sparse.

Directions
Use Chart 11 and your binoculars to locate a squat, 1°-wide Y-shaped asterism of four 6th- and 7th-magnitude suns (f) about 4° due east of 5th-magnitude Star b. Center this asterism in your telescope, then switch to Chart 11b. NGC 2571 is only 10′ northeast of the 6th-magnitude star that marks the northern arm of the Y.

The quick view
At 23× in the 4-inch, NGC 2571 is a sprinkling of 10th- to 11th-magnitude suns that seem to orbit around a bright pair of 9th-magnitude stars at the cluster's "center." With

averted vision, the 7′-wide cluster appears elongated north-west–southeast. At 72×, the cluster is a long, boxy array of two dozen suns with linear extensions at each corner of the box. Overall the stars are irregularly scattered and loose. Larger telescopes will show some 40 members here.

Stop. Do not move the telescope. Your next target is nearby!

5. NGC 2567 (H VII-64)

Type	Con	RA	Dec	Mag	Diam	Rating
Open cluster	Puppis	08ʰ 18.5ᵐ	−30° 39′	7.4	11.0′	3

General description
A near twin to NGC 2571, NGC 2567 is another large and scattered open cluster about 7° southeast of 3rd-magnitude Rho (ρ) Puppis. It is a telescopic object and may be difficult to extract from the surrounding field stars because it too is highly elongated and sparse.

Directions
Use Chart 11b; NGC 2567 is a 1° sweep almost due south of NGC 2571.

The quick view
At 23× in the 4-inch, NGC 2567 is sparse aggregation of suns (of 11th magnitude and fainter) centered on a beautiful north–south-trending string of about a half-dozen similarly bright suns. The cluster itself is oriented northeast to south-west; the northeast end is tapered. At 72×, the central line of stars looks like a string of pearls; it is the cluster's most

glorious aspect. Like NGC 2571, only about two dozen suns are visible in the 4-inch, though larger telescopes will show its 40 or so members.

6. NGC 2627 (H VII-63)

Type	Con	RA	Dec	Mag	Diam	Rating
Open cluster	Pyxis	08ʰ 37.2ᵐ	−29° 57′	8.4	9.0′	3

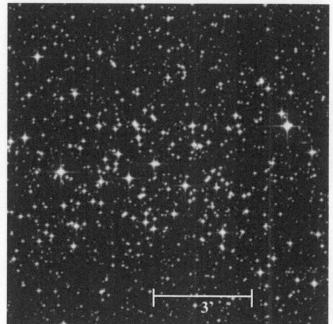

General description
NGC 2627 is a broad and scattered open cluster 40′ south-west of 5th-magnitude Zeta (ζ) Pyxidis. Under a dark sky it is visible in 7×50 binoculars as a soft glow tucked inside in a box of 8th- and 9th-magnitude stars. Telescopically it is somewhat bright, extended, and highly mottled. It may be difficult to see under light-polluted skies.

Directions
Use Chart 11 to locate 4th-magnitude Gamma (γ) Pyxidis, which is 10° (a fist) east–southeast of 3rd-magnitude Rho (ρ) Puppis. Fifth-magnitude Delta (δ) Pyxidis lies just 1° to the east. Gamma Pyxidis also marks the eastern apex of a 3°-wide triangle with the 5th-magnitude stars Eta (η) and Zeta (ζ) Pyxidis. Use binoculars to confirm these stars, then center Zeta Pyxidis in your telescope at low power. NGC 2627 lies only 40′ southwest of Zeta Pyxidis. Use Chart 11c if you need to pinpoint the cluster's exact location.

The quick view
At 23× in the 4-inch under a dark sky, NGC 2627 is a loose and scattered grouping of 17 stars of 11th-magnitude and

fainter spread across nearly 10′ of sky. The cluster is very mottled along an east–west line. Multiple pairs of stars can be seen, the brightest of which lies to the west. At 72×, the cluster remains elongated along the east–west line, but a fainter extension of stars branches to the southeast from the central region – which is not concentrated. Overall the cluster looks like a sideways Y composed of braided strings of stars.

Star charts for fifth night

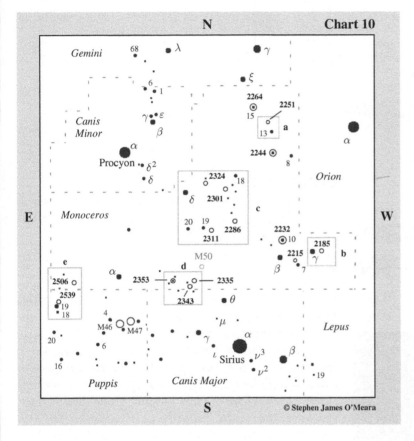

Chart 10

© Stephen James O'Meara

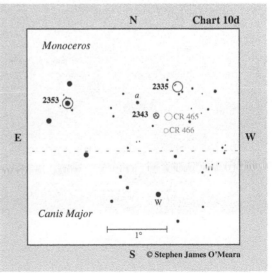

Chart 10d

© Stephen James O'Meara

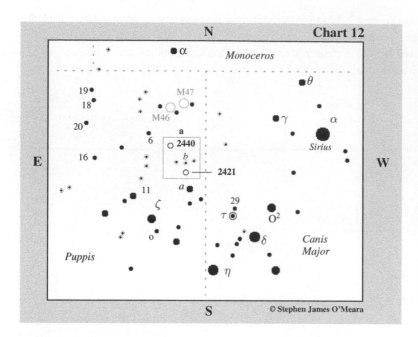

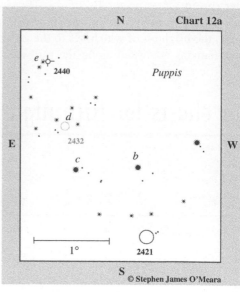

© Stephen James O'Meara

FIFTH NIGHT

1.   NGC 2421 (H VII-67)

Type	Con	RA	Dec	Mag	Diam	Rating
Open cluster	Puppis	07^h 36.2^m	−20° 37′	8.3	8.0′	3

General description

NGC 2421 is a moderately small and dim open cluster in Puppis, about 6° northeast of 4th-magnitude Tau (τ) Canis Majoris. It is a telescopic object.

Directions

Use Chart 12 to locate 2nd-magnitude Delta (δ) Canis Majoris, in the Great Dog's hind quarters. Tau Canis Majoris (and its associated cluster) is only 2⅔° to the northeast. From Tau Canis Majoris, use your unaided eyes or binoculars to look 4½° northeast for 4th-magnitude Star *a*. Star *a* marks the northeast apex of a 1°-wide triangle with two slightly fainter suns. Now look 2½° north–northeast for a nearly 2°-wide arc of three 6th-magnitude stars oriented east–west. You want to center the arc's middle star, Star *b*, in your telescope, then switch to Chart 12a. NGC 2421 is a little less than 1° south and slightly west of Star *b*.

The quick view

At 23×, NGC 2421 is a small puff of irregular light with a strong central condensation and faint spiral extensions. At 72×, the cluster displays a mix of suns, of 11th magnitude and fainter, scattered irregularly around a slightly elliptical core. The core itself has about seven prominent suns in a roughly triangular pattern about 1.5′ across. These brighter stars are seen against a circular backdrop of fainter suns. The cluster contains some 70 stars seen against the Milky Way.

   Stop. Do not move the telescope. Your next target is nearby!

2. NGC 2440 (H IV-64)

Type	Con	RA	Dec	Mag	Dim	Rating
Planetary nebula	Puppis	07^h 41.9^s	−18° 12′ 31″	9.1	74″× 42″	3.5

General description

NGC 2440 is a bright, stellarlike planetary nebula about 3° northeast of NGC 2421 in Puppis. It is also 4.4° almost exactly due south of another planetary nebula − NGC 2438 in the bright open star cluster M46, also in Puppis. If you have an equatorial mount, you could try centering M46 in your telescope at low power, then, after switching to a moderate magnification, swing the telescope 4.4° to the south.

Directions

Using Chart 12a, from NGC 2421 move 1° north and slightly east to Star b. Now move 50′ east to 6th-magnitude Star c. Next, swing about 40′ north and slightly east to a 20′-wide triangle of 8th-magnitude stars (d); larger telescopes may also see the dim (10th-magnitude) open star cluster NGC 2432 immediately to the southeast. Now hop another 50′ northeast, where you will encounter a 10′-wide arc of three 8th- and 9th-magnitude stars (e), oriented northwest to southeast. The faint "star" at the northwest end of the arc is NGC 2440.

The quick view

At 23×, the 1.2′-wide planetary nebula is virtually stellar. At 72×, NGC 2440 displays a bright core surrounded by a dim halo of light, especially with averted vision. With averted vision, see if you cannot discern a slight bluish

cast to the nebula. In the 4-inch, I find the best view is at 301×, which shows the nebula's inner "ring" or annulus very well.

3. NGC 2353 (H VIII-34)

Type	Con	RA	Dec	Mag	Diam	Rating
Open cluster	Monoceros	07^h 14.5^m	−10° 16′	7.1	18.0′	5

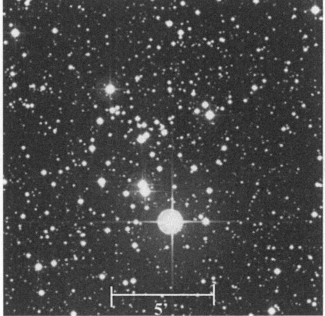

General description

NGC 2353 is a bright and condensed open cluster in Monoceros 10° (a fist) northeast of Alpha (α) Canis Majoris (Sirius). From a dark-sky site, it is visible in 7×50 binoculars as a 6th-magnitude star caught in a web of nebulosity − in this case, the "nebulosity" comprises unresolved starlight. If you accept that the 6th-magnitude star superimposed on the cluster marks the position of NGC 2353, then the cluster, arguably, can be seen with the unaided eye.

Directions

Use Chart 10 to locate NGC 2353. First locate Alpha (α) Canis Majoris (Sirius) and the two 4th-magnitude stars marking the top of the Great Dog's head: Gamma (γ) and Theta (θ) Canis Majoris. Fifth-magnitude Mu (μ) Canis Majoris lies midway between, and a little southwest of, those two stars. Using your binoculars, draw an imaginary line from Sirius though Mu Canis Majoris, then extend that line about 6° to the northeast; you should encounter three 6th-magnitude stars in an arc only $\frac{3}{4}$° long and oriented north–northwest to south–southeast. The middle of these three suns is NGC 2353. You can confirm the view by referring to Chart 10d.

The quick view

At 23×, open cluster NGC 2353 is bright, coarse, and scattered. The 6th-magnitude star, which is superimposed on the southern outskirts of the cluster, has a slight orange hue. The dozen or so surrounding suns seem to flow radially away from that orange sun. The core is weak and two long arms extend to the northwest and southeast; three weak arms fan out to the north west. At 72×, the orange, 6th-magnitude star is centered on a pretty arc of four 9th- to 10th-magnitude suns, whose two northeastern members are the beautiful silvery double star Struve 1052; the primary shines at a magnitude of 8.8, and its 9.2-magnitude companion lies 20″ to the north–northeast. This arc is itself part of a larger gathering of about two dozen suns in an area spanning only about 5′. The entire region out to 20′ is awash with 11th-magnitude and fainter suns. I counted about 80 stars at 72×.

Stop. Do not move the telescope. Your next target is nearby!

of four clusters in a 1° area of sky. NGC 2343 is visible in 7 × 50 binoculars from a dark-sky site. The cluster is a very pretty sight in small telescopes and a rich delight in larger ones.

Directions

Using Chart 10d, from NGC 2353, move your telescope slowly and carefully $1\frac{1}{4}°$ due west to 6.5-magnitude Star a. NGC 2343 is just 35′ southwest of that star.

The quick view

At 23×, NGC 2343 is a small but bright wedge of starlight that consists of about a half dozen moderately bright suns, with some outliers. At 72×, the cluster is a bright triangle of roughly 8th-magnitude stars, with a meandering, north–south-oriented stream of fainter starlight flowing through it. The cluster contains about 55 stars in an area only 6′ across. Larger scopes will have a better view of the dimmer companions.

Stop. Do not move the telescope. Your next target is nearby!

4. NGC 2343 (H VIII-33)

Type	Con	RA	Dec	Mag	Diam	Rating
Open cluster	Monoceros	07^h 08.1^m	−10° 37′	6.7	6.0′	4

General description

NGC 2343 is a fairly bright and rich cluster about $1\frac{1}{2}°$ west–southwest of NGC 2353. It is the most conspicuous

5. NGC 2335 (H VIII-32)

Type	Con	RA	Dec	Mag	Diam	Rating
Open cluster	Monoceros	07^h 06.8^m	−10° 02′	7.2	7.0′	4

General description

NGC 2335 is a small and somewhat bright open cluster northwest of NGC 2343. The view is rather simple in small telescopes. Light-polluted skies may make seeing this cluster a bit difficult. Look first for the cluster's tiny concentrated nucleus of stars.

Directions
Using Chart 10d, NGC 2335 is 40′ northwest of NGC 2343.

The quick view
At 23×, NGC 2335 is a little (∼5′) Y-shaped grouping of stars about 7′ southwest of a 7th-magnitude star. At 72×, about 14 stars of 10th-magnitude and fainter are readily visible in an area less than 5′. The stars have the shape of a rocking horse. Larger scopes will show nearly 60 stars in an area 7′ across.

Star charts for sixth night

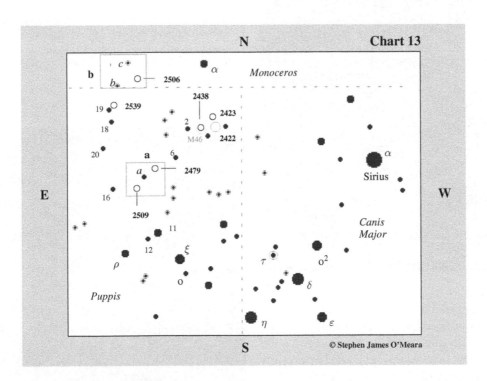

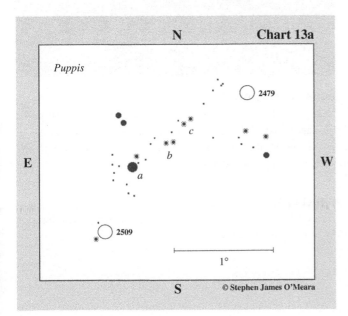

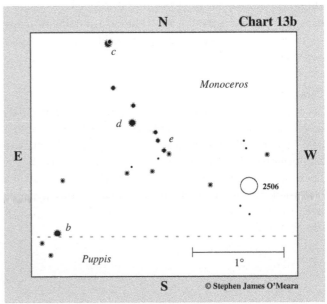

SIXTH NIGHT

1. NGC 2422 (H VIII-38) = M47

Type	Con	RA	Dec	Mag	Diam	Rating
Open cluster	Puppis	07^h 36.6^m	−14° 29′	4.4	25.0′	5

General description

When William Herschel discovered NGC 2422, he was unaware that Charles Messier (or Giovanni Batista Hodierna before him) had already discovered it. The reason for the confusion is that Messier, in his catalog, published an incorrect position for the object. But NGC 2422 is, in fact, M47. The cluster is brilliant, being visible to the unaided eye under a dark sky. It is is a marvel in 7 × 50 binoculars and a stunning sight in telescopes of all sizes, preferably those that give generous wide-field views.

Directions

To find NGC 2422 (see Chart 13), use your binoculars to sweep either 15° east of Alpha (α) Canis Majoris (Sirius) or 5° south and slightly west of 4th-magnitude Alpha (α) Monocerotis. The cluster is paired with another bright open star cluster, M46, which is less than 1$\frac{1}{2}$° to the east–southeast. There will be no mistaking this dynamic duo; together they are one of the binocular highlights of the heavens.

The quick view

At 23×, NGC 2422 is a rich assortment of three dozen stellar jewels splashed across 25′ of sky – nearly the apparent diameter of the full Moon. The cluster contains about 120 members of 5th-magnitude and fainter arranged in a most haphazard manner. The brightest suns form a 10′-wide pyramid of light. At 72×, the cluster offers several fine double stars, the most dynamic of which, Struve 1121, lies near the cluster's center; it consists of two opalescent 8th-magnitude suns separated by 7.4″.

Stop. Do not move the telescope. Your next target is nearby!

2. NGC 2423 (H VII-28)

Type	Con	RA	Dec	Mag	Diam	Rating
Open cluster	Puppis	07^h 37.1^m	−13° 52′	6.7	12.0′	4

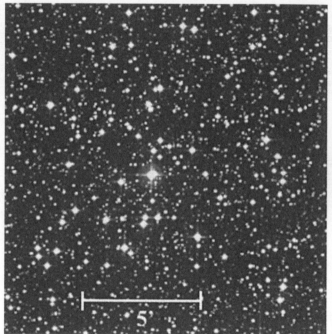

General description

NGC 2423 is a fairly bright and rich cluster 35′ north and a bit east of NGC 2422 (M47). In a dark sky, it is visible in 7 × 50 binoculars, although its stars seem to blend with the surrounding Milky Way.

Directions

Using Chart 13, from NGC 2422, just move your telescope 35′ to the north–northeast.

The quick view

At 23×, NGC 2423 is round and rather rich. A 9th-magnitude star at the cluster's core dominates the scene. The remaining stars look like an enhancement of the surrounding Milky Way. At 72×, the cluster displays about two dozen members centered on the 9th-magnitude star. They cover an area about 10′ across, though it's hard to tell which stars in the region are members; the Milky Way background is so rich, and the stars are not of uniform brightness. The cluster contains 86 stars in an area 12′ across. Larger telescopes will bring out the dimmer members.

Stop. Do not move the telescope. Your next target is nearby!

3. NGC 2438 (H IV-39)

Type	Con	RA	Dec	Mag	Diam	Rating
Planetary nebula	Puppis	07^h 41.8^m	−14° 44′	11.0	>66″	4

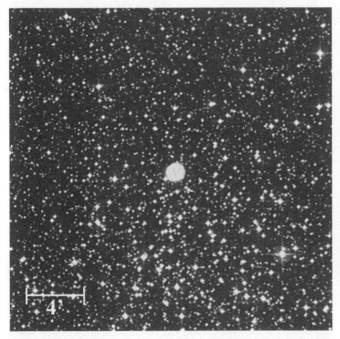

General description

NGC 2438 is a relatively small but bright planetary nebula superimposed on the northern flank of the round and rich open cluster M46.

Directions

Using Chart 13 as a guide, move your telescope 1¼° east–southeast to 6th-magnitude M46. Your target, planetary

nebula NGC 2438, is about 5′ north of the cluster's center. Think, "small and diffuse."

The quick view

At 23× in the 4-inch, NGC 2438 is only suspected. At 72×, the nebula is clearly seen as a tiny, 1′-wide ghostly orb. At 101×, the annulus is clearly seen with averted vision. Those with larger apertures should use higher magnifications and look for a 13th-magnitude star *near* the nebula's center; this is not the planetary nebula's central star, which shines at 18th magnitude!

4. NGC 2509 (H VIII-1)

Type	Con	RA	Dec	Mag	Diam	Rating
Open cluster	Puppis	08^h 00.8^m	−19° 03′	9.3	12.0′	3

General description

NGC 2509 is a rather dim and patchy open cluster about 2° west–northwest of 5th-magnitude 16 Puppis. It can be glimpsed with averted vision as a dim, elongated glow in 7 × 50 binoculars under a dark sky. It is best seen in telescopes.

Directions

Use Chart 13 to locate 3rd-magnitude Xi (ξ) Puppis. About 2½° to the northeast is 4th-magnitude 11 Puppis. Now imagine a line from Xi Puppis to 11 Puppis, which continues on for a little more than $4\frac{1}{2}°$ to 4th-magnitude 16 Puppis. Use binoculars to confirm these stars. Note also that 16 Puppis marks the southeast end of a 5°-long arc of three roughly 4th-magnitude stars – the other stars being 6 Puppis at the northwest end and 4.6-magnitude Star *a*. Center Star *a* in your telescope, then switch to Chart 13a. NGC 2509 is 45′ south–southeast of Star *a*.

The quick view

At 23×, NGC 2509 is an elongated mist of ten dim suns with a tiny central condensation just 6′ northwest of an 8th-magnitude star. A fainter ribbon of stars threads through the cluster from the northeast to the southwest. At 72× the cluster is very patchy. A prominent right-angle of stars extends out from the cluster's core like the hands of a clock – the clock's face being the faint stars around the tiny central core. NGC 2509 contains some 70 stars in an area 12′ wide, though the 4-inch shows only about half of them.

Stop. Do not move the telescope. Your next target is nearby!

General description

NGC 2479 is another dim and patchy Puppis open cluster about $1\frac{1}{2}°$ northwest of NGC 2509. In a dark sky, NGC 2479 is a very difficult binocular object. It is best seen in a telescope.

Directions

Using Chart 13a, from NGC 2509 return to Star *a*. NGC 2479 is almost $1\frac{1}{2}°$ northwest of Star *a*. To star-hop to it, first move 25′ northwest to a pair of 9th-magnitude stars (*b*). Now hop 15′ north–northwest to another pair of 9th-magnitude stars (*c*). NGC 2479 is less than 40′ to the northwest of Pair *c*.

The quick view

At 23×, NGC 2479 is a diffuse ellipse of dim stars with no central condensation. At 72×, the cluster is not only loose and scattered but very patchy and stringy. With averted vision, it looks like broken glass blown by the wind. The cluster contains 45 members, the brightest of which shine at about 12th magnitude.

5. NGC 2479 (H VII-58)

Type	Con	RA	Dec	Mag	Diam	Rating
Open cluster	Puppis	07^h 55.1^m	−17° 42.5′	9.6	11.0′	3

6. NGC 2539 (H VII-11)

Type	Con	RA	Dec	Mag	Diam	Rating
Open cluster	Puppis	08^h 10.6^m	−12° 49′	6.5	15.0′	4

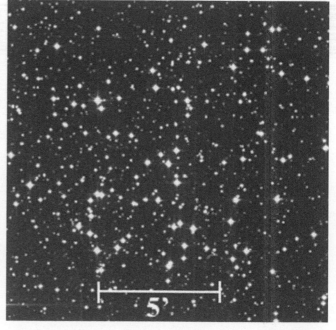

General description

NGC 2539 is a moderately condensed open cluster tucked away in the obscure northwestern quadrant of Puppis, just south of the point where the southeastern corner of Monoceros meets Hydra. The cluster is immediately northeast of 5th-magnitude 19 Puppis. Under a very dark sky, the cluster teeters on the verge of naked-eye visibility. In 7×50 binoculars, it looks like a veil of morning mist rising before a burning yellow Sun (19 Puppis); with just a little concentration and averted vision, the 6th-magnitude glow appears mottled.

Directions

Use Chart 13 to locate 19 Puppis, which marks the eastern end of a right triangle with 4th-magnitude Alpha (α) Monoceros and the naked-eye double star 2 and 4 Puppis (not shown). The southeastern lip of NGC 2539 kisses 19 Pup.

The quick view

At 23×, with direct vision, the cluster looks like a tapered candle flame. With averted vision, its oval shape and prickly body remind me of NGC 3532 (Caldwell 91), the Pincushion Cluster. The cluster's brightest stars shine at 9th magnitude, and a beautiful 11th-magnitude double star lies near its center, which is about 10′ northwest of 19 Puppis – which, by the way, has a pretty powder-blue companion to the west–southwest. With averted vision, the entire cluster is broken into patches of starlight. At 72×, NGC 2539 has about 60 stars in an area 15′ wide, which is about what most visual observers see. The cluster contains many pairs of stars. The others are arranged in all manner of geometrical patterns, which appear strewn helter-skelter across the field like jacks. It's a virtual playground for the eye, so let your imagination fly.

Stop. Do not move the telescope. Your next target is nearby!

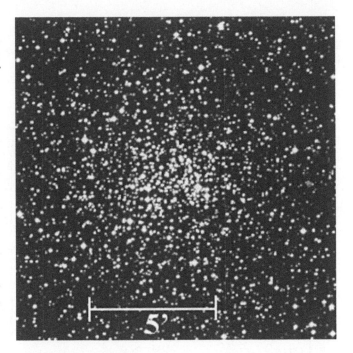

General description

NGC 2506 is a fine open cluster in Monoceros, about $3\frac{1}{2}°$ northwest of 19 Puppis and NGC 2539. It is visible from a dark-sky site in 7×50 binoculars as a faint glow with a bright core, which becomes apparent only with averted vision.

Directions

Using Chart 13, from NGC 2539, return to 19 Puppis. Now use your unaided eyes or binoculars to locate the 6th-magnitude stars b and c. Center Star c in your telescope at low power, then switch to Chart 13b. Note that Star c is a fine telescopic double. From Star c, move 55′ south–southwest to 7th-magnitude Star d. About 25′ further to the southwest is a little chain of 7th-magnitude stars (e), oriented northeast to southwest. NGC 2506 is about 1° further to the southwest.

The quick view

At 23×, NGC 2506 is a very rich assortment of suns of 11th-magnitude and fainter. It looks like a knot in the Milky Way, appearing more like some dim and distant globular cluster than an open star cluster. At 72×, the cluster's brightest stars form a loose arrangement of arms that "spiral" out from the cluster's bright triangular core. The stars here are plenty. Indeed, the cluster contains some 800 members in an area only 12′ across. Most form meandering star chains and clumps, all of which are seen against a rich and mottled background haze.

7. NGC 2506 (H VI-37)

Type	Con	RA	Dec	Mag	Diam	Rating
Open cluster	Monoceros	08ʰ 00.0ᵐ	−10° 46′	7.6	12.0′	4

Star charts for seventh night

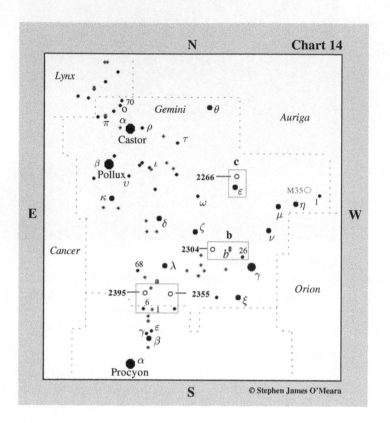

Chart 14

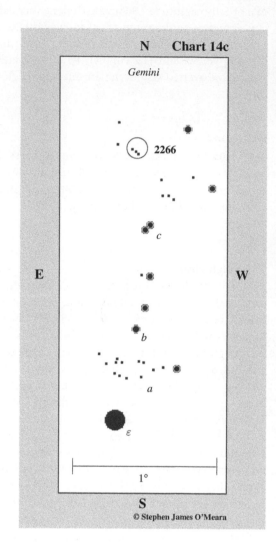

Chart 14c

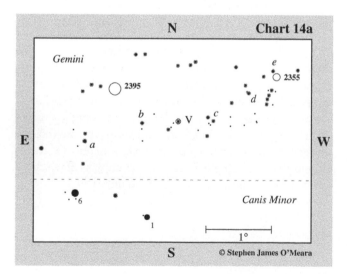

Chart 14a

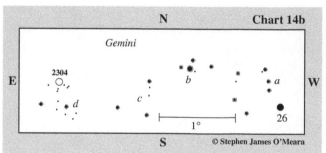

Chart 14b

SEVENTH NIGHT

1. NGC 2395 (H VIII-11)

Type	Con	RA	Dec	Mag	Diam	Rating
Open cluster	Gemini	07ʰ 27.2ᵐ	+13° 36′	8.0	15.0′	3

General description

NGC 2395 is a dim, low-surface-brightness open cluster in Gemini, $1\frac{3}{4}°$ north–northwest of 4.5-magnitude 6 Canis Minoris. Light pollution will affect its visibility, especially in a small telescope. It is best seen at low power with averted vision.

Directions

Use Chart 14 to locate 6 Canis Minoris, which is about 4° north and slightly east of 3rd-magnitude Beta (β) Canis Minoris. Center 6 Canis Minoris in your telescope at low power, then switch to Chart 14a. From 6 Canis Minoris, move your telescope about 45′ north–northwest to 8th-magnitude Star *a*. NGC 2395 is 1° northwest of Star *a*.

The quick view

At 23× in the 4-inch under a dark sky, NGC 2395 is a pale elliptical glow as seen with averted vision. It's a very loose and scattered congregation of stars of 10th magnitude and fainter. At 72×, the cluster displays about 30 suns in an elliptical form, which is elongated northwest to southeast. The stars are of mixed brightnesses. A dim trapezoid-shaped clustering of faint suns forms what could be considered NGC 2395's core. The cluster contains more than 50 stars in an area 15′ across.

Stop. Do not move the telescope. Your next target is nearby!

2. NGC 2355 (H VI-6)

Type	Con	RA	Dec	Mag	Diam	Rating
Open cluster	Gemini	07ʰ 17.0ᵐ	+13° 45′	9.7	8.0′	3

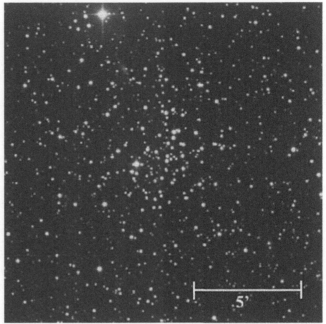

General description

NGC 2355 is another dim, low-surface-brightness open cluster in Gemini, roughly $2\frac{1}{2}°$ west, and slightly north of NGC 2395. Although NGC 2355 is fainter than NGC 2395, small-telescope users may have an easier time in seeing it; its roughly 40 stars are compressed in an area only 8′ across.

Directions

Use Chart 14a; if your telescope has setting circles, try sweeping $2\frac{1}{2}°$ west of NGC 2395 to NGC 2355. Otherwise, from NGC 2395, move 35′ southwest to 7.5-magnitude Star *b*. A little more than 1° west is a pair of 7.5-magnitude suns (*c*). A roughly 55′ leap to the northwest brings you to the 8th-magnitude double star *d*. NGC 2355 is 30′ further to the northwest, just 7′ southwest of 8th-magnitude Star *e*.

The quick view

At 23×, NGC 2355 is a small but seemingly rich assortment of very dim suns with some central condensation. It's essentially a low-surface-brightness patch of nervous starlight. At 72×, the cluster is a flurry of dim suns with similar brightnesses (13th magnitude and fainter). While the cluster's core is concentrated, its outer flanks are ragged and dim. The core also has a sinuous string of stars running through it north to south.

3. NGC 2304 (H VI-2)

Type	Con	RA	Dec	Mag	Diam	Rating
Open cluster	Gemini	06^h 55.2^m	+17° 59′	10.0	3.0′	2

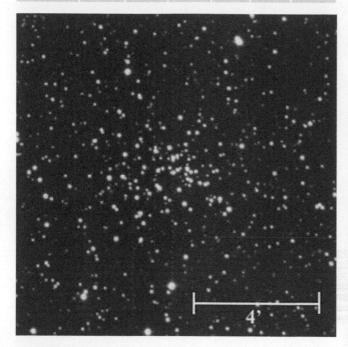

General description

NGC 2304 is a very small and very dim open cluster about 5° east–northeast of 2nd-magnitude Gamma (γ) Geminorum. Light pollution will greatly affect its visibility in a small telescope. The cluster is part of a more obvious chain of stars oriented northeast to southwest. Think, "very small and faint," as you hunt.

Directions

Use Chart 14 to locate Gamma Geminorum. Now use your unaided eyes or binoculars to locate 5th-magnitude 26 Geminorum 1$\frac{3}{4}$° to the northeast. Center this star in your telescope at low power, then switch to Chart 14b. From 26 Geminorum move 20′ north–northeast to a north–south oriented line of three 8th-magnitude suns (a). Center the northernmost of these three stars, then make a careful 1° sweep due east to 6th-magnitude Star b, which is the brightest star in an asterism that looks like a flattened diamond. From Star b, swing 50′ southeast to a $\frac{1}{2}$°-wide isosceles triangle of three 8th-magnitude suns (c). Center the southeastern star in Triangle c, then move 40′ east to 8th-magnitude Star d. NGC 2304 is a little more than 20′ north–northeast of Star d.

The quick view

At 23×, NGC 2304 is but a glint of ill-defined haze stuck to a 10′-long chain of 10th- to 12th-magnitude suns, which is oriented northeast to southwest. The cluster is extremely small (3′) and dim. While its brightest star shines at a magnitude of 11.5, the other members are mere suggestions in my 4-inch. At 72× the cluster is a breath of starlight, with perhaps only a half dozen or so stars sparkling in and out of view.

4. NGC 2266 (H VI-21)

Type	Con	RA	Dec	Mag	Diam	Rating
Open cluster	Gemini	06^h 43.3^m	+26° 58′	9.5	5.0′	3

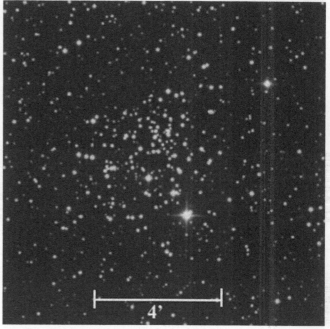

General description

NGC 2266 is a small and relatively faint open cluster nearly 2° north of 3rd-magnitude Epsilon (ε) Geminorum.

Directions

Use Chart 14 to locate Epsilon Geminorum, then switch to Chart 14c. From Epsilon Geminorum, use low power and move your telescope 20′ north and slightly west to a charming ellipse of 10th-magnitude suns (a). Just 15′ further to the north–northwest is 8.5-magnitude Star b. A 40′ hop to the north–northwest brings you to a pair of 9th- and 10th-magnitude suns (c). NGC 2266 is only 30′ north of Pair c.

The quick view

At 23× in the 4-inch, NGC 2266 is a small but beautiful fan of dim suns whose northwestern edge tapers to a point. The southeastern edge of the fan is a bright arc of about a half-dozen suns. At 72×, the cluster's brightest stars form an obvious Y-shaped asterism around which swarms a very fuzzy halo of unresolved starlight. NGC 2266 contains some 50 stars of 11th magnitude and fainter, though most of these are better seen in larger apertures. Still the cluster is a very nice sight given its isolated position and relatively dim brightness.

3 · March

Star charts for first night

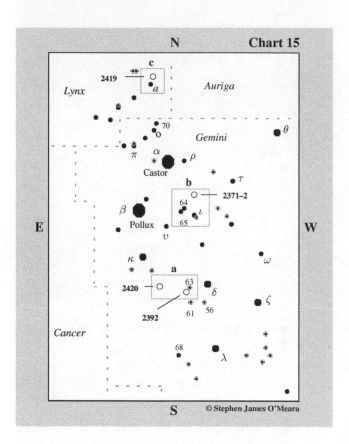

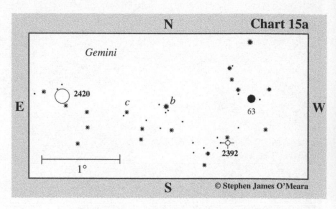

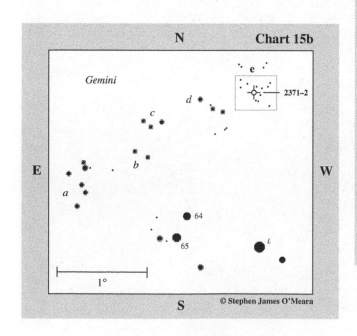

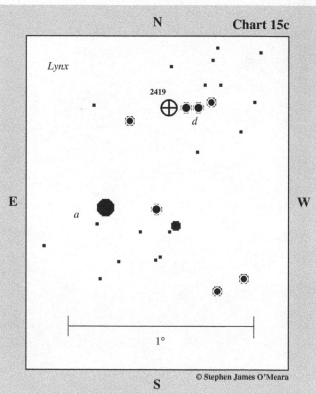

FIRST NIGHT

Type	Con	RA	Dec	Mag	Dim	Rating
Planetary nebula	Gemini	07ʰ 29.2ᵐ	+20° 55′	9.2	47″× 43″	5

General description

NGC 2392 is the famous Eskimo planetary nebula. It is almost $2\frac{1}{2}°$ southeast of 3.5-magnitude Delta (δ) Geminorum. The smallest of telescopes will show it, and it is a gem in telescopes of all sizes, especially at high magnifications. Note, however, that at low power in a small telescope, the nebula will look nearly stellar, so be sure to move your telescope slowly and carefully to its location. Any moderate magnification will reveal the planetary's tiny disk.

Directions

Use Chart 15 to locate Delta Geminorum. Note that Delta Geminorum is the bright star at the northwest end of a $2\frac{1}{2}°$-wide kite-shaped asterism with the 5th-magnitude stars 56, 61, and 63 Geminorum. Use binoculars to verify this appearance. Now center 63 Geminorum at low power in your telescope, then switch to Chart 15a. NGC 2392 lies just 40′ southeast of 63 Geminorum and 1.6′ due south of an 8th-magnitude star.

The quick view

At 23×, NGC 2392 looks like a slightly bloated star when viewed with averted vision. Otherwise, one could easily sweep over it, thinking it is a star. At 72×, the planetary

nebula is an obvious disk of greenish light nearly 1′ in diameter. At 130×, the planetary's central star can be seen inside a sharp inner annulus surrounded by a fainter halo of light. The higher the magnification, the easier it is to see these features.

Stop. Do not move the telescope. Your next target is nearby!

Type	Con	RA	Dec	Mag	Diam	Rating
Open cluster	Gemini	07ʰ 38.4ᵐ	+21° 34′	8.3	6.0′	2

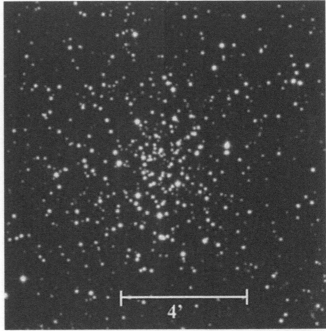

General description

NGC 2420 is a beautiful yet subtle open cluster a little more than 2° northeast of NGC 2392. In a small telescope at low power, its round fuzzy form looks like a comet just beginning to shine. It's a very rich cluster and a delight to see with larger apertures.

Directions

Using Chart 15a, from NGC 2392, make a slow and careful sweep about 55′ northeast to 7th-magnitude Star b. Now look about 35′ east and slightly south of Star b for 9th-magnitude Star c with a roughly 10th-magnitude companion. NGC 2420 is about 50′ east–northeast of Star c.

The quick view

At 23×, NGC 2420 is a ghostly round glow of largely uniform light, like a tailless comet that gets gradually brighter in the middle. With averted vision, some two dozen suns of between 11th- and 12th-magnitude can be seen forming a net around a dazzling array of dimmer suns at the core; it's as if these fainter central jewels have been dredged up from the depths of space. At 72×, the cluster looks more fractured; countless dimmer suns also waft in and out of view. Indeed, this cluster is very rich in stars, having more than 300 members populating a mere 6′ of sky. The cluster is quite stunning in larger telescopes.

3 AND 4   NGC 2371–2 (H II-316/17)

Type	Con	RA	Dec	Mag	Diam	Rating
Planetary nebula	Gemini	07ʰ 25.6ᵐ	+29° 29′	11.3	>55″	1

5′

General description

NGC 2371–2 is a dim, low-surface-brightness planetary nebula 1½° north of 4th-magnitude Iota (ι) Geminorum. Its shell appears broken into two parts (thus the double NGC number). That is also why NGC 2371–2 counts as two Herschel 400 objects. To claim this double prize, you must resolve the nebula into two bright patches. Be warned, seeing the twin parts will be a challenge (but not impossible) in a small telescope. The key is to be under a dark sky, know *exactly* where to look, and use *high* power with averted vision. If you still have trouble, try gently tapping the tube. Plan to spend some time with this object. If you're patient, chances are you'll succeed.

Directions

Use Chart 15 to locate Iota Geminorum, which forms the southwestern apex of an isosceles triangle with the 1st-magnitude stars Alpha (α) and Beta (β) Geminorum. Less than 1° northeast of Iota Geminorum are the 5th-magnitude

stars 64 and 65 Geminorum. Center 64 Geminorum in your telescope at low power then switch to Chart 15b. From 64 Geminorum, make a generous 1° sweep east–northeast to a crooked-Y-shaped asterism of five 6th- and 7th-magnitude stars (a). Now move 50′ northwest to a pair of roughly 8.5-magnitude stars (b). A 25′ hop to the north–northwest will bring you to an arc of three 7.5- to 8.5-magnitude stars (c). Now move 30′ northwest to 7.5-magnitude Star d, which lies at the northeast end of a 15′-long crooked line of slightly fainter suns. NGC 2371–2 lies only 35′ west and slightly north of Star d. Box **e** shows roughly the same area of sky as that in the photograph. Use the stars plotted in Box **e** and the accompanying photograph to identify the field of faint stars that lie around the little planetary nebula.

The quick view

I could see NGC 2371–2 in my 4-inch at 23× but only *after* I had located it at *high* power. At 23×, it looks like a 1′-wide "lint ghost." I suggest you use low power to locate the object's field, then immediately switch to high power. At 72× and 101×, the planetary nebula is visible as a dim 1′-wide binary glow oriented northeast to southwest. With averted vision, a clean separation between the two halves is apparent. With time, the dual object becomes more and more obvious. For those with large telescopes, the planetary nebula's central star shines at a magnitude of 14.8.

5.   NGC 2419 (H I-218)

Type	Con	RA	Dec	Mag	Diam	Rating
Globular cluster	Lynx	07ʰ 38.1ᵐ	+38° 52′	10.3	4.6′	2

4′

General description

NGC 2419, the famous Intergalactic Wanderer or Tramp, is a dim and distant (275,000 light-years) globular cluster in Lynx, about 7° north–northeast of 1.6-magnitude Alpha (α) Geminorum (Castor). It is a small, diffuse, and difficult object for small telescopes, even under dark skies. It is best to find the field first at low power, then use moderate magnification and averted vision to see it.

Directions

Use Chart 15 to locate 4th-magnitude Alpha Geminorum. About 3° north–northeast of Alpha Geminorum you'll find the roughly 5th-magnitude stars 70 and Omicron (o)

Geminorum. Now use your unaided eyes or binoculars to look $3\frac{1}{2}°$ north and slightly east of 70 Geminorum for 6th-magnitude Star a. Center Star a in your telescope at low power then switch to Chart 15c. From Star a, move 45′ northwest to a pair of 8th-magnitude stars (d). NGC 2419 lies about 3′ due east of the easternmost star in Pair d.

The quick view

At 23× NGC 2419 is a dim and diffuse, 5′-wide glow at the verge of visibility. There is no central condensation, just a moist pool of faint light. At 72× the globular cluster is a weak circular glow with a hint of central condensation and a ragged outer shell.

Star charts for second night

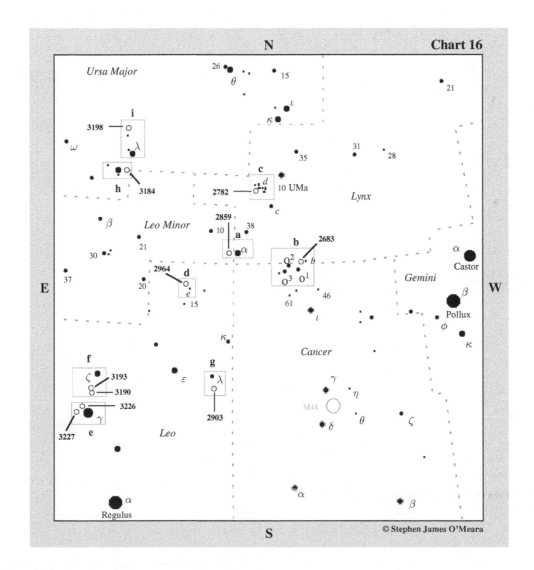

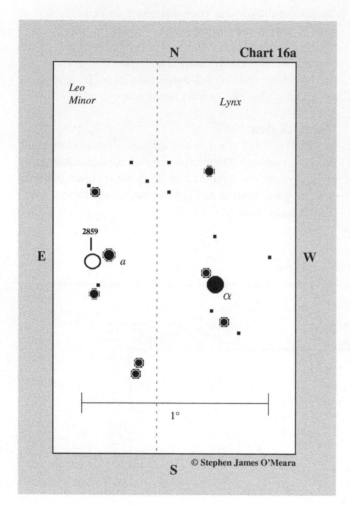

Chart 16a

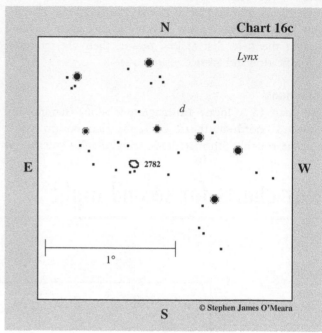

Chart 16c

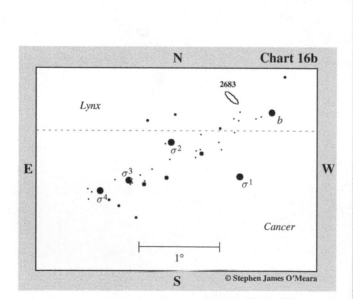

Chart 16b

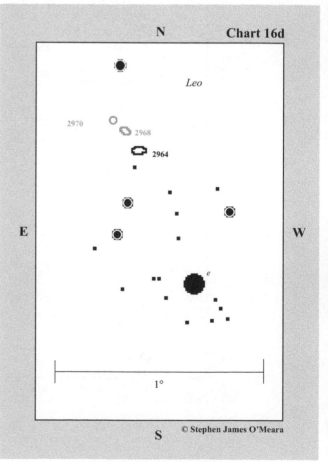

Chart 16d

SECOND NIGHT

1. NGC 2859 (H I-137)

Type	Con	RA	Dec	Mag	Dim	Rating
Barred spiral galaxy	Leo Minor	09^h 24.3^m	+34° 31′	10.9	4.6′ × 4.1	2

4′

General description
NGC 2859 is a small and dim barred spiral galaxy in Leo Minor that lies 40′ east and slightly north of Alpha (α) Lyncis. It is a difficult object for small telescopes and requires a dark sky and averted vision.

Directions
Use Chart 16 to locate 3rd-magnitude Alpha (α) Lyncis. Center this star in your telescope at low power, then switch to Chart 16a. NGC 2859 is only 40′ east and slightly north of Alpha Lyncis, which places it in Leo Minor. The galaxy is only 5′ southeast of 7.5-magnitude Star a.

The quick view
At 23×, NGC 2859 is not readily visible. I could just barely detect it. At 72×, the galaxy remains difficult. With averted vision, it appears as a 2′ × 1′ glow. With some attention, a very subtle core becomes apparent.

2. NGC 2683 (H I-200)

Type	Con	RA	Dec	Mag	Dim	Rating
Spiral galaxy	Lynx	08^h 52.7^m	+33° 25′	9.7	9.1′ × 2.7′	5

4′

General description
NGC 2683, the UFO Galaxy, is the brightest of some two dozen galaxies populating Lynx. It is just visible in 7×50 binoculars under a dark sky; it should be more readily visible in 10×50 binoculars. Even the smallest of apertures under a dark sky will show NGC 2683 as an elliptical haze, just 10′ north of a little triangle of 10th- and 11th-magnitude stars and about 3′ west of a 12th-magnitude sun.

Field note
Don't be fooled by Sigma^2 (σ^2) Cancri; it is the brightest star in a tight triangle of suns, which, like M40 in Ursa Major, can look fuzzy.

Directions
Using Chart 16 to find NGC 2683, draw an imaginary line from 1st-magnitude Beta (β) Geminorum (Pollux) to 3rd-magnitude Epsilon (ε) Leonis in the Lion's Sickle. Now look about half way along that line and a little north, where you will find 4th-magnitude Iota (ι) Cancri. Almost 10° (a fist) north of Iota Cancri is 3rd-magnitude star Alpha (α) Lyncis, which marks the northern apex of an isosceles triangle with Iota Cancri and Epsilon Leonis. Now set your gaze halfway between Alpha Lyncis and Iota Cancri, and a tad northwest. There you'll find a roughly 1°-long chain of three 5th- and 6th-magnitude suns: Sigma^2 (σ^2), Sigma^3 (σ^3), and Sigma^4 (υ^4 = not shown) Cancri; from a dark sky site, these stars are visible to the unaided eye as a mottled stellar haze. Center this grouping in your telescope at low power, then switch to Chart 16b. Sigma^2 Cancri, marks the eastern end of a 1$\frac{1}{2}$°-wide triangle with two other 6th-magnitude stars – Sigma^1 (σ^1) Cancri, which lies to the southwest, and Star b, to the west–northwest, just across the border in Lynx. NGC 2683 is about 30′ east–northeast of Star b.

The quick view

At 23×, NGC 2683 is a a beautiful silver needle of light, nearly 10′ long and elongated northeast to southwest. It has a distinct sheen, like moonlight glinting off a sword. The galaxy is easier to see than the edge-on systems NGC 891 (Caldwell 23) in Andromeda, and NGC 4565 (Caldwell 38) in Coma Berenices; although NGC 4565 and NGC 2683 are equals in magnitude and surface brightness, NGC 2683 is slightly more open, so we see more light from its disk. Details start to appear at 72×, which, in small apertures, is about the maximum power you'll want for comfortable study. The nucleus gleams like a brilliant diamond flanked on either side by cultured pearls. Spiral structure in the inner lens may be inferred in larger telescopes, but in the 4-inch, all I see can see are irregular patches of light and dark.

3. NGC 2782 (H I-167)

Type	Con	RA	Dec	Mag	Dim	Rating
Mixed spiral galaxy	Lynx	09ʰ 14.1ᵐ	+40° 07′	11.6	3.8′× 2.9′	2

4'

General description

NGC 2782 is a small and dim spiral galaxy in a stellar void 3° southeast of 4th-magnitude 10 Ursae Majoris, which is now in Lynx. Be prepared for a bit of sweeping. Small-telescope users will need to use keen averted vision and moderate magnifications to see it well – even under a dark sky.

Directions

Use Chart 16 to locate 3.6-magnitude Kappa (κ) Ursae Majoris in the right forepaw of the Great Bear. You'll find 10 Ursae Majoris (10 UMa) about 5½° south of Kappa Ursae Majoris. Now use your unaided eye or binoculars to find 4.5-magnitude Star c 3½° to the south–southeast. If you look 2° east of the midpoint between 10 Ursae Majoris and Star c, you should see a wonderful 1½°-wide W-shaped asterism (d) of 8th-magnitude suns; it looks like a miniature version of the famous W in the constellation Cassiopeia. Center Asterism d in your telescope at low power, then switch to Chart 16c. NGC 2782 is about 35′ southeast of the midpoint of Asterism d.

The quick view

At 23×, NGC 2782 is a very dim and very small (1′) circular glow just north of two 13th-magnitude stars. There is no central condensation and averted vision is needed to see it in small telescopes. At 72×, the galaxy is still a low-surface-brightness object and averted vision is required to see it. But with a little time, a dim stellar component can be seen at or near the galaxy's core; it is surrounded by faint mottlings. The galaxy appears brighter and more condensed in larger telescopes.

4. NGC 2964 (H I-114)

Type	Con	RA	Dec	Mag	Dim	Rating
Mixed spiral galaxy	Leo	09ʰ 42.9ᵐ	+31° 51′	11.3	3.2′× 1.8′	2

4'

2968

2964

General description

NGC 2964 is a very small and dim spiral galaxy 2° north of 5.5-magnitude 15 Leonis. Small-telescope users under a dark sky will need to star-hop to its location, then use high power and averted vision to see it.

Directions

From Chart 16, using the unaided eye or binoculars, locate 5.6-magnitude 15 Leonis, which is about 6° due north of 3rd-magnitude Epsilon (ε) Leonis at the western tip of the Lion's mane. 15 Leonis is easy to identify in binoculars because it has a 7th-magnitude companion about 10′ to the northwest. 15 Leonis is also the southernmost star in a $1\frac{1}{2}°$-wide isosceles triangle with two equally bright stars. You want to center the northwesternmost star in that triangle (Star e) in your telescope at low power, then switch to Chart 16d; although NGC 2964 lies less than 40′ northeast of Star e, small-telescope users will probably need to carefully seek it out it.

The quick view

At 23× in the 4-inch, NGC 2964 is a very faint, very small (1′), anemic glow; it is barely visible, even under a dark sky. It's best to get the field, then increase your magnification. At 72×, the galaxy displays a round core in a little elliptical disk about 2′ in length. The galaxy is easier to see at this magnification, especially with averted vision.

Field note

NGC 2964 is the brightest of three NGC galaxies in the area. The others are 12th-magnitude NGC 2968 and 13th-magnitude NGC 2970. I did not see these other two galaxies at a quick glance. Fainter anonymous galaxies are also nearby.

Star charts for third night

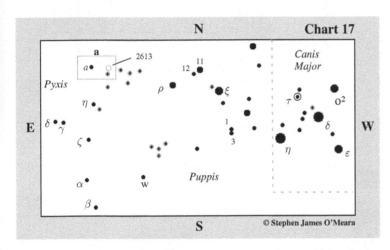

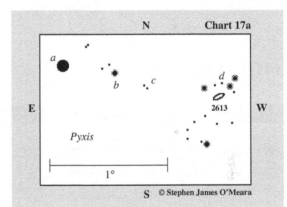

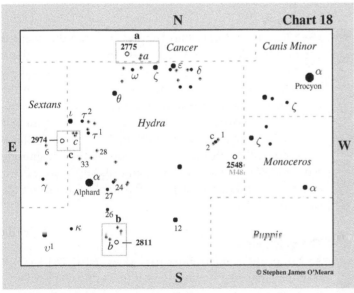

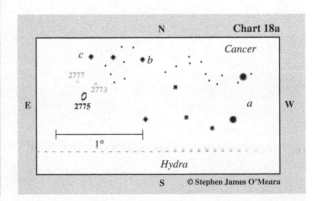

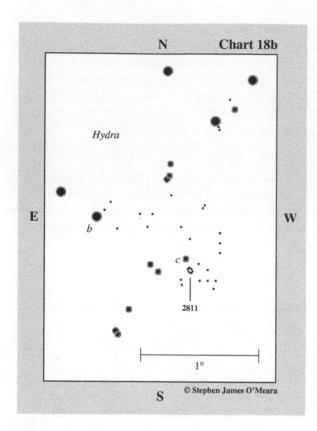

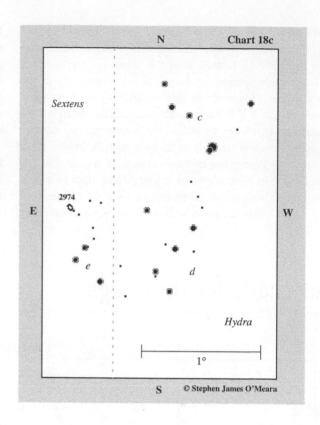

THIRD NIGHT

1. NGC 2613 (H II-266)

Type	Con	RA	Dec	Mag	Dim	Rating
Spiral galaxy	Pyxis	08^h 33.4^m	−22° 58′	10.5	7.6′ × 1.9′	3

General description

NGC 2613 is a nearly edge-on galaxy in Pyxis about 6° east–northeast of 3rd-magnitude Rho (ρ) Puppis. It is moderately faint for a 4-inch telescope under a dark sky, mainly because it is a long and thin spindle of light. Seeing it in a small telescope requires a dark sky and moderate magnification.

Directions

Use Chart 17 to locate Rho Puppis. Then use the unaided eye or binoculars to look for 5th-magnitude Star a, about $7\frac{1}{2}°$ further to the east–northeast. There is no mistaking Star a because it is the only star of that magnitude in the immediate vicinity. You want to center Star a in your telescope at low power, then switch to Chart 17a. From Star a, move 25′ west and slightly south to 8.5-magnitude Star b. Next, hop 18′ southwest to a close pair of 10.5-magnitude stars (c). NGC 2613 is 40′ west and slightly south of Pair c and 10′ southeast of a pair of roughly 9th-magnitude stars (d).

The quick view

At 23× in the 4-inch under dark skies, NGC 2613 is not readily visible. At 72×, the galaxy is just a thin ellipse of dim light, oriented northwest to southeast; look for the galaxy wedged inside a triangle of roughly 12th- to 13th-magnitude suns. Larger telescopes will reveal NGC 2613's tiny, pinpoint nucleus.

2. NGC 2548 (H VI-22 = M48)

Type	Con	RA	Dec	Mag	Diam	Rating
Open cluster	Hydra	08ʰ 13.7ᵐ	−05° 45′	5.8	30.0′	5

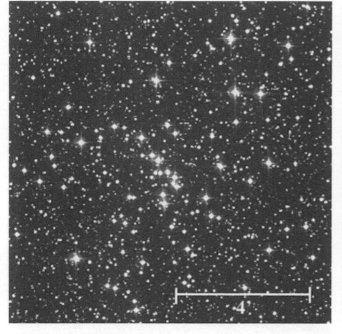

3. NGC 2775 (H I-2)

Type	Con	RA	Dec	Mag	Dim	Rating
Spiral galaxy	Cancer	09 h 10.3 m	+07° 02′	10.1	4.6′ × 3.7′	4

General description

When Caroline Herschel discovered NGC 2548 on March 8, 1783, she and her brother William were unaware that Charles Messier had already discovered it in February 1771. The reason is that Messier, in his catalog, cited an incorrect position for the object. But NGC 2548 is in fact open cluster M48. The cluster can be seen in 7×50 binoculars as a partially resolved cloud of starlight with averted vision. It is a dynamic sight in any telescope.

Directions

Use Chart 18. NGC 2548 (M48) marks the southern tip of a near equilateral triangle with 4th-magnitude Zeta (ζ) Monocerotis and the equally bright grouping of 1, 2, and c Hydrae. Just raise your binoculars to this position and you will see the cluster. Otherwise, just imagine this point on the sky and point your telescope to it and gently sweep the sky until you see its form, which has the same apparent diameter as the full Moon.

The quick view

At 23×, NGC 2548 is a loose collection of about 80 stars with magnitudes between 8 and 13. The brighter stars form a distinct arrowhead with a tight, elliptical, off-axis core.

General description

NGC 2775 is a small and moderately dim spiral galaxy about $3\frac{3}{4}°$ east–northeast of 3rd-magnitude Zeta (ζ) Hydrae, which places it just over the Hydra border and into southern Cancer. The galaxy is visible in 7×50 binoculars from a dark-sky site (with effort). Look for a 5′-wide oval glow in small telescopes.

Directions

Use Chart 18 to locate 3rd-magnitude Zeta Hydrae in the back of the Hydra's head. It also lies about $1\frac{3}{4}°$ east of a pair of 6th- and 7th-magnitude stars (a); use binoculars to confirm these stars, then center them in your telescope at low power. Now using Chart 18a as a guide, center the northernmost star in Pair a, and move $1\frac{1}{4}°$ east and slightly north to 7.5-magnitude Star b, which is the westernmost of three similarly bright stars in a 35′- long line. NGC 2775 is 30′ south and slightly east of the easternmost star in that line (Star c).

The quick view

At 23×, NGC 2775 is an egg-shaped glow with a bright central condensation. The disk is 5′-wide and is oriented northwest to southeast. At 72×, the galaxy has essentially the same appearance as it does at low power, but its oval form is more obvious to see.

4. NGC 2811 (H II-505)

Type	Con	RA	Dec	Mag	Dim	Rating
Barred spiral galaxy	Hydra	09^h 16.2^m	−16° 19′	11.3	1.9′× 0.6′	2

General description

NGC 2811 is a very small and faint barred spiral galaxy about 8° south–southeast of 2nd-magnitude Alpha (α) Hydrae (Alphard). It resides in a remote section of the constellation, so you will need to take the time to star-hop to its position using binoculars and your telescope. Small-telescope users need to be under a dark sky and will have to rely on using averted vision; you'll mainly be seeing the galaxy's circular core and nucleus.

Directions

Use Chart 18 to locate Alpha Hydrae, then with your unaided eyes or binoculars, look 2° southwest for 5th-magnitude 27 Hydrae. Next look 3° due south for the equally bright star 26 Hydrae. Now use your binoculars to make a slow and careful search 3° due south of 26 Hydrae for a 1$\frac{1}{2}$°-wide rectangle of four 6th-magnitude stars. Once you confirm the rectangle, center its southeasternmost star (b) in your telescope at low power, then switch to Chart 18b. From Star b, make a slow and careful sweep 40′ southwest to a 25′-wide acute triangle of 9.5- to 10th-magnitude suns. NGC 2811 is only 5′ south and slightly west of the westernmost star in the triangle (Star c).

The quick view

At 23×, NGC 2811 is a dim circular glow that becomes ever-so slightly elongated northeast to southwest with averted

vision. Higher powers do not help its appearance much in small telescopes; it's best to keep its light concentrated.

5. NGC 2974 (H I-61)

Type	Con	RA	Dec	Mag	Dim	Rating
Spiral galaxy	Sextans	09^h 42.6^m	−03° 42′	10.9	3.0′× 1.7′	2

General description

NGC 2974 is a small and faint spiral galaxy in Sextans, a little more than 6° northeast of 2nd-magnitude Alpha (α) Hydrae (Alphard) and about 2$\frac{1}{2}$° south–southeast of 4th-magnitude Iota (ι) Hydrae. Small-telescope users may not see this galaxy at low power, especially from light-polluted areas. It's best to follow the directions below to locate the galaxy's precise position, which is extremely close to a 10th-magnitude star, and look for it with moderate magnifications.

Directions

Use Chart 18 to locate Iota Hydrae, which is 8° north–northeast of Alpha Hydrae. Now use your unaided eyes or binoculars to look 3° southwest for 4.5-magnitude Tau^1 (τ^1) Hydrae Now use your binoculars to look 2° due east of Tau^1 Hydrae for a 40′-wide triangle of three roughly 7th-magnitude suns (c). Center Triangle c in your telescope at low power then switch to Chart 18c. Note that the southernmost star in Triangle c is a fine telescopic double. Center this double, then move about 45′ south–southeast to a 30′-long L-shaped asterism of 7.5- and 8th-magnitude stars (d). Center the southeasternmost star in the L, which is also a double star, then move 35′ east to an acute triangle of three

8th-magnitude suns (*e*). NGC 2874 is 20′ north–northeast of the northernmost star (another double) in Triangle *e*. Follow the arc of 10th-magnitude and fainter suns leading to the galaxy from that star as depicted on the chart. Again, you want to use moderate magnification once you find the field to identify the galaxy, which is only about 1′ northeast of a 10th-magnitude star.

The quick view

At 23×, NGC 2974 is very tiny and faint with averted vision. The 10th-magnitude star nearby is a significant distraction. At 72×, the galaxy is more defined, displaying a starlike core with a tight and rather obvious circular core. I could not detect any elongation, though larger telescopes should show the 3′-long galaxy elongated in a northeast–southwest direction.

Star charts for fourth night

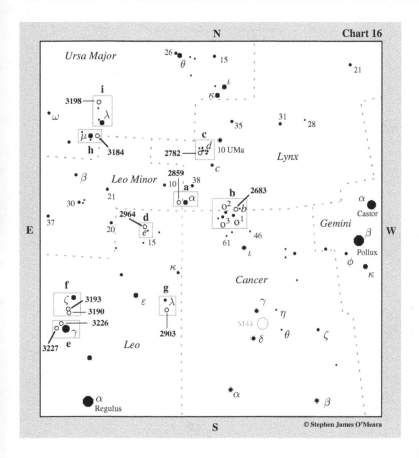

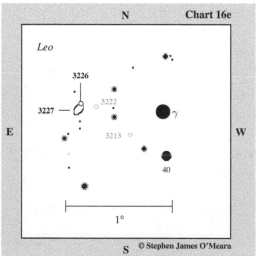

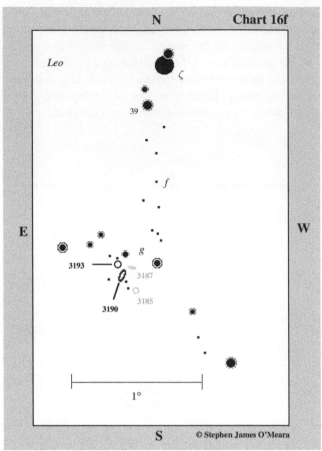

Chart 16f

N

S

E

W

Leo

ζ

39

f

g

3193

3187

3190

3185

1°

© Stephen James O'Meara

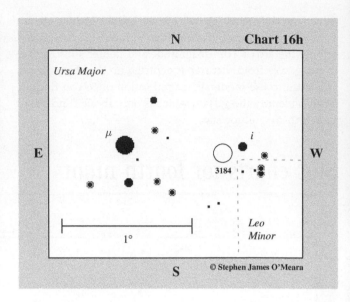

Chart 16h

N

S

E

W

Ursa Major

μ

i

3184

Leo
Minor

1°

© Stephen James O'Meara

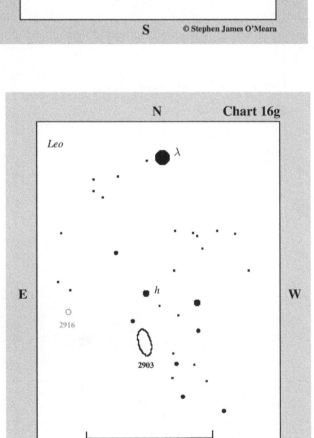

Chart 16g

N

S

E

W

Leo

λ

h

2916

2903

1°

© Stephen James O'Meara

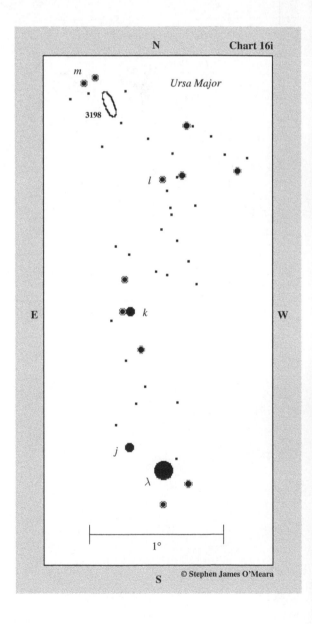

Chart 16i

N

S

E

W

Ursa Major

m

3198

l

k

j

λ

1°

© Stephen James O'Meara

FOURTH NIGHT

1. NGC 3227 (H II-29)

Type	Con	RA	Dec	Mag	Dim	Rating
Spiral galaxy	Leo	10^h 23.5^m	+19° 52′	10.3	6.9′ × 5.4′	4

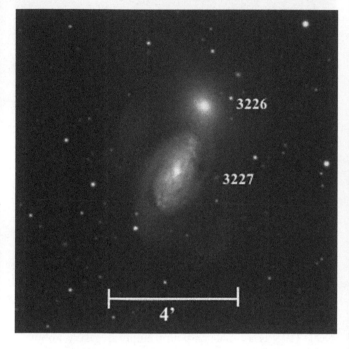

General description

NGC 3227 is a moderately bright and elongated spiral galaxy 50′ due east of 2.6-magnitude Gamma (γ) Leonis. It is tightly paired (and interacting) with our next Herschel object, NGC 3226, which lies 2′ to the north–northwest of NGC 3227's center. A Seyfert galaxy, NGC 3227's core is easy to see at moderate magnifications.

Directions

Use Chart 16 to locate Gamma Leonis. Center that star in your telescope at low power, then switch to Chart 16e. NGC 3227 lies only 50′ east of Gamma Leonis.

The quick view

At 23×, NGC 3227 is simply an elongated haze, 2′-long with no discernible details. At 72×, NGC 3227 and NGC 3226 become visible as two nearly equal galaxies kissing. NGC 3227 is the more southerly of the two and appears as an amorphous elliptical glow with a starlike nucleus in a circular lens of light.

Stop. NGC 3226 is described below.

2. NGC 3226 (H II-28)

Type	Con	RA	Dec	Mag	Dim	Rating
Elliptical galaxy	Leo	10^h 23.4^m	+19° 54′	11.4	2.5′ × 2.2′	3

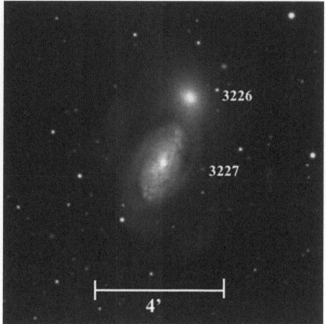

General description

NGC 3226 is a small and dim elliptical galaxy 2′ north–northwest of NGC 3227, with which it is interacting.

Directions

Using Chart 16e, NGC 3226 is 2′ north–northwest of NGC 3227.

The quick view

At 72×, the galaxy is a 1′-wide circular glow with a reasonably bright core in a faint halo of light.

Stop. Do not move your telescope.

3. NGC 3190 (H II-44)

Type	Con	RA	Dec	Mag	Dim	Rating
Spiral galaxy	Leo	10^h 18.1^m	+21° 50′	11.2	4.1′ × 1.6′	2

General description

NGC 3190 is a small and dim spiral galaxy midway between the 3rd-magnitude stars Gamma (γ) and Zeta (ζ) Leonis. It is paired with our next Herschel object, NGC 3193, and is the brightest of four NGC objects in the immediate area. NGC 3190 was not readily visible at 23× in my 4-inch, but it was distinct at 72×. So be prepared to find the area, then use moderate magnification to see it.

Directions

Use Chart 16 to locate Zeta Leonis. Place that star in your telescope at low power, then switch to Chart 16f. From Zeta Leonis, move 20′ south–southeast to 6th-magnitude 39 Leonis. There are two ways to make the next step: (1) make a slow and careful sweep $1\frac{1}{4}°$ to the south to a pair of 7.5-magnitude stars (g), or (2) follow a line of 9th- and 10th-magnitude stars (f) that meanders to Pair g. Once you find Pair g, center the pair's easternmost star and switch to a moderate magnification. NGC 3190 is less than 10′ south of that easternmost star.

The quick view

At 72×, NGC 3190 is a 2′-long ellipse, oriented northwest to southeast, with a slightly brighter center. With averted vision, the galaxy's disk is not uniformly lit but appears disturbed. Larger telescopes will show that the disturbance is caused by a warped dust lane that mars one side of the galaxy's disk.

Stop. Do not move the telescope. Your next target is nearby!

4. NGC 3193 (H II-45)

Type	Con	RA	Dec	Mag	Dim	Rating
Elliptical galaxy	Leo	10^h 18.4^m	+21° 54′	10.9	2.5′× 2.5′	2

General description

NGC 3193 is a small and dim elliptical galaxy about 5′ northeast of NGC 3190.

Field note

Be careful how you move the telescope. Another, albeit slightly dimmer, galaxy, NGC 3185, lies 10′ southwest of NGC 3193.

Directions

Using Chart 16f, NGC 3193 is 5′ northeast of NGC 3190.

The quick view

At 72×, NGC 3193 is a very tiny (~1′-wide) circular glow with no defining characteristics. Note that it is about 1.5′ south of an 8.5-magnitude star.

5. NGC 2903 (H I-56)

Type	Con	RA	Dec	Mag	Dim	Rating
Mixed spiral galaxy	Leo	09^h 32.2^m	+21° 30′	9.0	11.6′× 5.7′	4

General description

NGC 2903 is a fairly bright and extensive spiral galaxy $1\frac{1}{2}°$ south of Lambda (λ) Leonis. From a dark-sky site, the galaxy is visible in 7×50 binoculars. Under the same dark-sky conditions, the smallest of telescopes will show it as a slightly fuzzy star that swells with averted vision. Light pollution will greatly affect the galaxy's visibility.

Directions

Use Chart 16 to find 3rd-magnitude Epsilon (ε), then 4th-magnitude Lambda (λ) Leonis, at the northwest tip of Leo's Sickle. Center Lambda Leonis in your telescope at low power, then switch to Chart 16g. From Lambda, make a slow and careful sweep $1\frac{1}{4}°$ south to 7.4-magnitude Star h. NGC 2903 is only 20′ due south of Star h.

The quick view

At 23×, NGC 2903 displays a sharp nucleus inside a bright central lens, which is surrounded by a diffuse, elliptical glow. At 72× and 101×, the core remains quite bright, like a sunlit jewel lying in an elliptical bed of cotton. This soft inner lens is quite defined, having a uniform texture until it reaches a sharp outer boundary. Larger telescopes will show several of its 70 H II regions and the galaxy's more extensive outer arms.

General description

NGC 3184, the Little Pinwheel Galaxy, is a reasonably bright, nearly face-on galaxy tickling the left hind toes of Ursa Major, the Great Bear. From a dark-sky site, it is just visible in 7×50 binoculars as a threshold object, appearing as a very dim haze bordered by two stars. It is best seen in small telescopes at low power.

Directions

Use Chart 16 to locate 3rd-magnitude Mu (μ) Ursae Majoris. Center that star in your telescope at low power, then switch to Chart 16h. NGC 3184 is 45′ west of Mu Ursae Majoris, just 10′ east–southeast of 6th-magnitude Star i.

The quick view

At 23× in the 4-inch, NGC 3184 is a round, diffuse glow with a bright core. Its size swells with averted vision. It does appear very similar to a comet just beginning to shine and would be easily swept up in a comet hunt, as long as the sweep rate is slow. At 72×, the galaxy appears to have a very bright and starlike nucleus. But with a little bit of study, it becomes apparent that this bright star is not at the galaxy's core; it is, in fact, an 11th-magnitude field star immediately north of the galaxy's true nucleus, which is visible as a tiny, and much fainter, pip of light.

6. NGC 3184 (H I-168)

Type	Con	RA	Dec	Mag	Dim	Rating
Mixed spiral galaxy	Ursa Major	10ʰ 18.3ᵐ	+41° 25′	9.4	7.5′× 7.0′	4

7. NGC 3198 (H I-199)

Type	Con	RA	Dec	Mag	Dim	Rating
Barred spiral galaxy	Ursa Major	10ʰ 19.9ᵐ	+45° 33′	10.3	9.2′× 3.5′	3

4'

General description

NGC 3198 is a somewhat dim and extended barred spiral galaxy about $2\frac{1}{2}°$ north–northeast of Lambda (λ) Ursae Majoris in the Great Bear's left hind foot. Users of small telescopes will want to first star-hop to the field, then use moderate magnifications to sight it. Light pollution will certainly destroy this object.

Directions

Use Chart 16 to locate 3rd-magnitude Lambda Ursae Majoris. Center Lambda Ursae Majoris in your telescope at low power, then switch to Chart 16i. From Lambda Ursae Majoris, move 18′ northeast to 6.5-magnitude Star j. Exactly 1° due north of Star j, is 6.5-magnitude Star k, which has a 9.5-magnitude companion immediately to its east. Now make a slow and careful sweep 1° north–northeast to a pair of 7.5 and 8.0-magnitude stars (l). Another 1° sweep, this time to the northeast, will bring you to a tighter pair of 9th-magnitude suns (m). NGC 3198 is about 15′ southeast of Pair m. Center these stars in your telescope, then use moderate magnification to find the galaxy.

The quick view

At 72× under a dark sky, NGC 3198 is a beautiful phantom of elongated light nearly 10′ in length. The ellipse is very diffuse and ill defined. Averted vision shows a slight brightening at the center, but this is in no way dynamic. Larger telescopes, however, will reveal the galaxy's sharp core and mottlings throughout the disk.

Star charts for fifth night

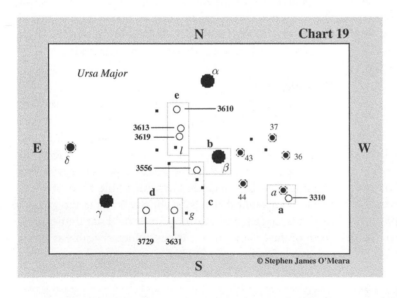

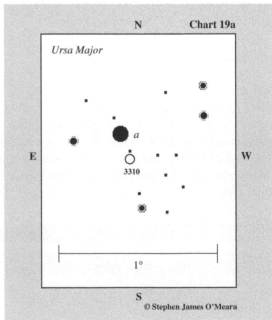

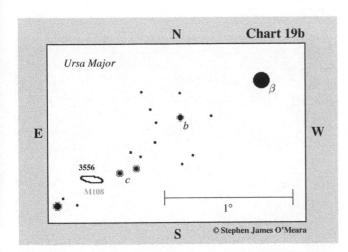

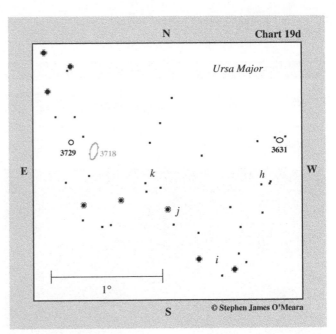

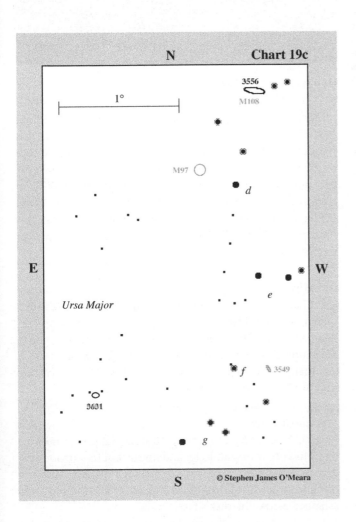

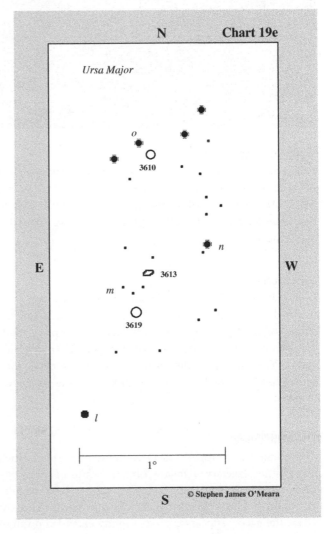

Fifth night

1. NGC 3310 (H IV-60)

Type	Con	RA	Dec	Mag	Dim	Rating
Mixed spiral galaxy	Ursa Major	10^h 38.7^m	+53° 30'	10.8	3.5' × 3.2'	4

General description
NGC 3310 is a small but very condensed spiral galaxy almost $4\frac{1}{2}°$ southwest of 2nd-magnitude Beta (β) Ursae Majoris and 10' south–southwest of a 5th-magnitude star. It is easy to detect under a dark sky and is probably a good object for suburban skies.

Directions
Use Chart 19 to locate Beta Ursae Majoris. Now use your unaided eyes or binoculars to look 3° west for a 3°-wide circlet of 5th-and 6th-magnitude stars – comprising 43, 37, 36, 43, and 44 Ursulae Majoris and 5.5-magnitude Star a. You want to center Star a in your telescope, then switch to Chart 19a. NGC 3310 is only 10' south–southwest of Star a.

The quick view
At 23×, the galaxy appears as a slightly swollen star – much like a little planetary nebula – directly south of a roughly 12.5-magnitude star. At 72×, the galaxy is a circular glow, about 1' in diameter, with a bright, starlike core. At 101×, the galaxy appears ever-so-slightly extended north to south.

2. NGC 3556 (H V-46 = M108)

Type	Con	RA	Dec	Mag	Dim	Rating
Barred spiral galaxy	Ursa Major	11^h 11.5^m	+55° 40'	10.0	8.1' × 2.1'	4

General description
NGC 3556 is the 108th object included in modern version's of Messier's catalog. In 1960 Owen Gingerich identified NGC 3556 with an object mentioned by Messier in that position, and it has ever since been known as M108. Messier's original published catalog ended with M103; the first published position of NGC 3556 appeared in William Herschel's first catalog of 1000 new nebulae in 1784. Under a dark sky it is visible as a faint gray streak $1\frac{1}{2}°$ southeast of Beta (β) Ursae Majoris. It is a marvel in telescopes of all sizes.

Directions
Use Chart 19 to locate 2nd-magnitude Beta Ursae Majoris, which you should center in your telescope at low power, then switch to Chart 19b. From Beta Ursae Majoris, move 45' southeast to 7.5-magnitude Star b. NGC 3556 (M108) is less than 50' further to the southeast, just 15' east of two roughly 8.5-magnitude stars (c).

The quick view
At 23×, NGC 3556 is a nearly 10'-long ellipse of mottled light. Its core is bright, long, and linear, but its surrounding disk is a phantom haze. At 72× and 101×, the galaxy is a patchwork quilt of bright and dark patches, irregularly arranged across the disk's face.

Stop. Do not move the telescope. Your next target is nearby!

3. NGC 3631 (H I-226)

Type	Con	RA	Dec	Mag	Dim	Rating
Spiral galaxy	Ursa Major	11^h 21.0^m	+53° 10′	10.4	5.5′ × 4.6′	2

4'

General description

NGC 3631 is a very faint, low-surface-brightness galaxy about 2$\frac{3}{4}$° southeast of NGC 3556 (M108). Seeing it will be a challenge for small-telescope users unless they are under a dark sky. Be prepared to make a careful star-hop to the galaxy's field. Averted vision and magnification may be required to see it.

Directions

Use Chart 19c and low power. From NGC 3556 (M108), move 45′ to the south–southeast to 7th-magnitude Star d; M97, the Owl planetary nebula, is less than 20′ northeast of Star d. Now slowly move 50′ south–southwest to a pair of 7th-magnitude stars (e); the westernmost star in Pair e has an 8th-magnitude companion about 5′ to the northwest. Another 50′ sweep southeast will bring you to 8.5-magnitude Star f, which has an 11th-magnitude companion immediately to the northeast. Note too that the 12th-magnitude galaxy NGC 3549 lies 20′ to Star f's west. A shorter 35′ hop to the southeast will bring you to a roughly 20′-wide acute triangle (g) comprising three 7th-magnitude suns. NGC 3631 lies 50′ northeast of the easternmost (and brightest) star in Triangle g.

The quick view

At 23× under a dark sky, NGC 3631 is a very faint, 2′-wide, circular glow with a sharp, bright core. At 72×, the galaxy is a bit more defined, and its core is a tad sharper, but little else can be seen.

Stop. Do not move the telescope. Your next target is nearby!

4. NGC 3729 (H I-222)

Type	Con	RA	Dec	Mag	Dim	Rating
Barred spiral galaxy	Ursa Major	11^h 33.8^m	+53° 08′	11.4	3.1′ × 2.2′	2

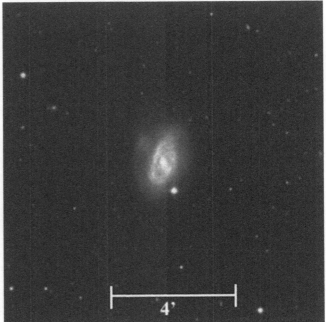

4'

General description

NGC 3729 is a small and dim barred spiral galaxy almost 2° due east of NGC 3631. It also lies about 12′ east–northeast of the 10.8-magnitude galaxy NGC 3718, with which it is interacting. Note that even though NGC 3718 is plotted as a larger galaxy and that its total magnitude is greater than that of NGC 3729, the light of NGC 3718 is also spread across a much larger area. In a small telescope, then, smaller and more condensed NGC 3729 appears brighter and more obvious than NGC 3718 – the reverse is true in larger instruments.

Directions

Use Chart 19d. If you have an equatorially mounted telescope, move your scope just 12.8 m east in right ascension, because NGC 3729 and NGC 3621 are at essentially the same declination. Otherwise, you can star-hop to it. From NGC 3631, move about 25′ southeast to 3′-wide Triangle h, which is composed of three 10th- and 11th-magnitude stars. Now look 50′ southeast for two 8th-magnitude stars (i), oriented east–northeast to west–southwest, and separated by 20′. Almost 30′ northeast of the easternmost star in Pair i is solitary 9.5-magnitude Star j. Next, hop about 12′ northeast to 6′-wide Triangle k, which is composed of three 10th- and 11th-magnitude suns. NGC 3718 lies 40′ to the northeast of Triangle k, and NGC 3729 lies about 12′ to the east–northeast of NGC 3718, which is a large (10.8′×4′) amorphous glow with a gradual brightening toward the center.

The quick view

At 23× under a dark sky, NGC 3729 is a small round glow. Averted vision is needed to see it well because its light seems to blend with that of a nearby pairing of stars, which can also look fuzzy. At 72×, the galaxy is a 2'-wide circular glow whose light is moderately compact. With averted vision, the inner core gradually brightens to a starlike center.

5. NGC 3619 (H I-244)

Type	Con	RA	Dec	Mag	Dim	Rating
Spiral galaxy	Ursa Major	11^h 19.4^m	+57° 46'	11.5	3.7'× 2.8'	2

General description

NGC 3619 is a small and very faint spiral galaxy $2\frac{3}{4}°$ north-east of 2nd-magnitude Beta (β) Ursae Majoris.

Directions

Use Chart 19 to locate Beta Ursae Majoris, then use your binoculars to sweep 3° east–northeast to solitary 6.4-magnitude Star l. Center Star l in your telescope at low power, then switch to Chart 19e. NGC 3619 is 45' north-west of Star l and 10' south of a 10'-wide isosceles triangle (m) of 11th- and 12th-magnitude stars.

The quick view

At 23×, NGC 3619 is a small 1'-wide circular glow that can only be seen in a small telescope under a dark sky with averted vision ... if you know exactly where to look. At 72×, it is still very faint and requires averted vision to see. Yet, with some time and patience, a dim starlike nucleus materializes out of that dim mist of light.

Stop. Do not move the telescope. Your next target is nearby!

6. NGC 3613 (H I-271)

Type	Con	RA	Dec	Mag	Dim	Rating
Elliptical galaxy	Ursa Major	11^h 18.6^m	+58° 00'	10.9	3.4'× 1.9'	3

General description

NGC 3613 is a tiny and dim elliptical galaxy 15' north–northwest of NGC 3619; it is easier to see than NGC 3619.

Directions

Use Chart 19e. NGC 3613 is 15' north–northwest of NGC 3619, nestled between two 10th-magnitude stars, which are oriented northwest to southeast and separated by about 10'.

The quick view

At 23× under a dark sky, NGC 3613 is a very small (~1'), very starlike pip of light; it swells ever so slightly with averted vision. At 72×, the galaxy is a small (1.5') ellipse of fuzzy light at the center of which is a brighter, though tiny, elliptical core, which appears mottled with averted vision.

Stop. Do not move the telescope. Your next target is nearby!

7. NGC 3610 (H I-270)

Type	Con	RA	Dec	Mag	Dim	Rating
Elliptical galaxy	Ursa Major	11^h 18.4^m	+58° 47'	10.8	3.2'× 3.2'	4

General description
NGC 3610 is a relatively bright elliptical galaxy nearly 50′ due north of NGC 3613.

Directions
Use Chart 19e. If you have an equatorially mounted telescope, move 47′ north from NGC 3613 to NGC 3610. Otherwise, from NGC 3613, move a little less than 30′ northwest to 8th-magnitude Star n. Now make a slow sweep 50′ northwest to 8th-magnitude Star o, which has a similarly bright companion about 12′ to the southeast. NGC 3610 is less than 10′ southwest of Star o.

The quick view
At 23× under a dark sky, NGC 3610 is a small (2′) but bright circular disk. The disk gets gradually, then suddenly, very much brighter in the middle to a sharp starlike nucleus. At 72×, the galaxy's starlike nucleus is greatly pronounced.

Star charts for sixth night

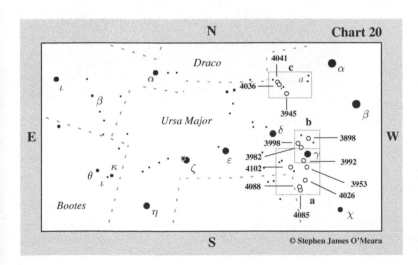

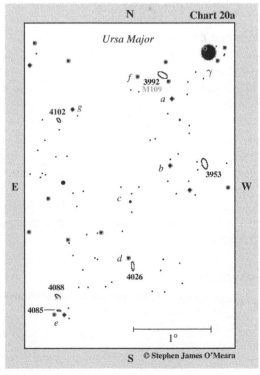

SIXTH NIGHT

1. NGC 3992 (H IV-61) = M109

Type	Con	RA	Dec	Mag	Dim	Rating
Barred spiral galaxy	Ursa Major	11^h 57.6^m	+53° 23′	9.8	7.6′ × 4.3′	4

General description

In 1960, Harvard historian Owen Gingerich identified NGC 3992 as one of two objects that Charles Messier mentioned in his note to M97. William Herschel was not aware of Messier's observation when he independently discovered the "nebula" during his great sky survey. Herschel, then, was the first to bring the world's attention to the existence of NGC 3992. The galaxy is a somewhat bright and compact glow only 40′ southeast of 2nd-magnitude Gamma (γ) Ursae Majoris in the Big Dipper's Bowl. It can be seen under a dark sky (with effort) with 7×50 binoculars and has a dim, comet-like quality when seen through a small telescope.

Directions

Use Chart 20 to locate Gamma Ursae Majoris. NGC 3992 is only 40′ southeast of that star. If necessary, use Chart 20a to identify the field.

The quick view

At 23×, NGC 3992's moderately bright core is nestled inside a dim yet mottled disk that measures about 5′×3′. At 72×, the galaxy is more apparent. With some effort, I can see a central bar and faint, looping spiral arms in my 4-inch.

Stop. Do not move the telescope. Your next target is nearby!

2. NGC 3953 (H V-45)

Type	Con	RA	Dec	Mag	Dim	Rating
Barred spiral galaxy	Ursa Major	11^h 53.8^m	+52° 20′	10.1	6.0′ × 3.2′	4

General description

NGC 3953 is a relatively bright barred spiral galaxy, similar in brightness and appearance to NGC 3992, which lies about $1\frac{1}{4}°$ to the north–northeast.

Directions

Using Chart 20a, from NGC 3992, you can either try making a slow and careful sweep $1\frac{1}{4}°$ south–southwest to NGC 3953 (which will appear similar to NGC 3992 in brightness and size), or you can make smaller star-hops. From NGC 3992, move about 18′ south–southwest to 9th-magnitude Star *a*. Next, make a slow 50′ sweep due south to similarly bright Star *b*. NGC 3953 is about 30′ east and slightly north of Star *b*.

The quick view
At 23× under a dark sky, NGC 3953 is a pretty obvious circular glow that, with time, transforms into an oblong disk 5′×2′ in extent. The core is bright and concentrated. At 72×, the galaxy is a mottled, elongated disk with a bright core. Thin, anemic arms may waver in and out of view.

Stop. Do not move the telescope. Your next target is nearby!

The quick view
At 23× in the 4-inch, NGC 4026's tack-sharp starlike core is an easy target to see once you get Star d. With averted vision, the sharp core expands into a thin 4′-long string of dim and fuzzy light, oriented north to south. At 72×, the galaxy is a dim, needle-like disk with a bright core that intensifies to a starlike nucleus.

Stop. Do not move the telescope. Your next target is nearby!

3. NGC 4026 (H I-223)

Type	Con	RA	Dec	Mag	Dim	Rating
Lenticular galaxy	Ursa Major	11ʰ 59.4ᵐ	+50° 58′	10.8	4.6′× 1.2′	2

4. NGC 4088 (H I-206)

Type	Con	RA	Dec	Mag	Dim	Rating
Mixed spiral galaxy	Ursa Major	12ʰ 05.6ᵐ	+50° 33′	10.6	5.4′× 2.1′	2

General description
NGC 4026 is a relatively faint but very concentrated lenticular galaxy a little more than 1½° southeast of NGC 3953. In small telescopes, it will appear starlike at low power, so it's best to star-hop to the field, then use averted vision to see the galaxy's elongated form at medium powers.

Directions
Using Chart 20a, from NGC 3953, return to 9th-magnitude Star b. Now make a careful slide 40′ southwest to a 30′-long Y-shaped asterism (c), comprising five 9.5- to 10.5-magnitude stars. NGC 4026 is 40′ south of the southernmost star in Asterism c – just 7′ south–southwest of 9.5-magnitude Star d.

General description
NGC 4088 is a dim diffuse glow a little more than 1° southeast of NGC 4026. It is in a more remote region of sky, so finding it will require a bit of patient star-hopping.

Directions
Using Chart 20a, from NGC 4026, make a slow and careful sweep 1° southeast, where you will encounter a pair of roughly 9th-magnitude stars (e). These stars are the brightest in the immediate area. NGC 4088 marks the northern apex of a roughly 15′-wide isosceles triangle with Pair e.

The quick view

At 23×, NGC 4088 is a diffuse and difficult glow, about 5' in diameter. Averted vision is required. Any light pollution will wipe it out. The galaxy is better seen at 72× in the 4-inch. But even under very dark skies, it remains a dim and diffuse glow with a very weak central condensation. Large telescopes will enhance the galaxy's brightness and show some mottled texture.

Stop. Do not move the telescope. Your next target is nearby!

5. NGC 4085 (H I-224)

Type	Con	RA	Dec	Mag	Dim	Rating
Mixed spiral galaxy	Ursa Major	12^h 05.4^m	+50° 21'	12.4	2.5' × 0.8'	2

4'

General description

NGC 4085 is a tiny and very dim galaxy near the brighter and more obvious NGC 4088.

Directions

Use Chart 20a. NGC 4085 lies only about 10' due south of NGC 4088 and 5' north of the midpoint between the stars in Pair e.

The quick view

At 23× in the 4-inch under very dark skies, the galaxy is a 12th-magnitude speck of light seen only with averted vision. If you see a "star" in the galaxy's location, you are seeing the galaxy's core. At 72×, the galaxy is slightly more apparent. Still, it's just a dot of dim light with averted vision. Larger apertures will show a thin halo (2.5' wide) extended east to west.

6. NGC 4102 (H I-225)

Type	Con	RA	Dec	Mag	Dim	Rating
Mixed spiral galaxy	Ursa Major	12^h 06.4^m	+52° 43'	11.2	2.9' × 1.8'	3

4'

General description

NGC 4102 is small but reasonably obvious mixed spiral galaxy 1½° southeast of NGC 3992 (M109). In small telescopes it is best seen at moderate magnifications.

Directions

Using Chart 20a, from NGC 4085, return to Gamma Ursae Majoris, then to NGC 3992. Now move 20' east to 9.5-magnitude Star f. Next, make a careful 1° sweep to 8th-magnitude Star g, the brightest star in the area. NGC 4102 is a little more than 10' southeast of Star g.

The quick view

At 23× in the 4-inch, NGC 4102 is simply a very small (1') soft glow with a bright core, which is not difficult to see. At 72×, NGC 4102 reveals a very bright nucleus surrounded by an 3'-wide elliptical halo, which is oriented northeast to southwest. A 11.5-magnitude sun on the halo's western flank could be mistaken for a supernova.

Star charts for seventh night

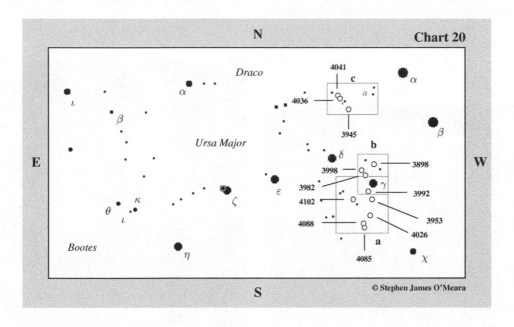

Chart 20

N

Draco

4041

c

a

4036

α

3945

β

Ursa Major

b

3898

δ

3998

γ

3982

3992

4102

3953

4088

4026

4085

a

E

W

ι

β

α

ζ

ε

θ

κ

ι

Bootes

η

χ

S

© Stephen James O'Meara

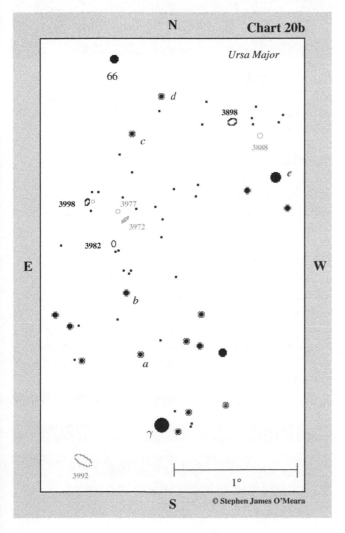

Chart 20b

N

Ursa Major

66

d

3898

c

3888

e

3998   3977

3972

3982

b

a

γ

3992

E

W

1°

S

© Stephen James O'Meara

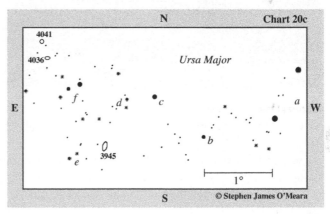

Chart 20c

N

Ursa Major

4041

4036

f

d

c

a

E   W

b

e

3945

1°

S

© Stephen James O'Meara

SEVENTH NIGHT

1.   NGC 3982 (H IV-62)

Type	Con	RA	Dec	Mag	Dim	Rating
Mixed spiral galaxy	Ursa Major	11^h 56.5^m	+55° 08′	11.0	2.2′ × 2.0′	2

General description
NGC 3982 is a very small and very dim spiral galaxy about $1\frac{1}{2}°$ north–northeast of Gamma (γ) Ursae Majoris. Small-telescope users will need to star-hop to its location, then use moderate magnification and averted vision to see it. The galaxy will be extremely difficult to see with any interference from light pollution.

Directions
Use Chart 20 to locate Gamma Ursae Majoris. Center Gamma Ursae Majoris in your telescope at low power, then switch to Chart 20b. From Gamma Ursae Majoris, move 35′ north–northeast to 9th-magnitude Star *a*. Another roughly 35′ hop to the north–northeast will bring you to 8.5-magnitude Star *b*. NGC 3982 is less than 25′ further to the northeast, just 3′ north of two 12th-magnitude stars.

The quick view
At 23×, NGC 3982 is a very small (1.5′ × 1′), and very amorphous, glow that requires averted vision to see even under a dark sky. At 72×, the galaxy remains a featureless glow, but its intensity is somewhat enhanced. Although the galaxy appears more obvious in larger telescopes, it is still largely featureless.
   Stop. Do not move the telescope. Your next target is nearby!

2.   NGC 3998 (H I-229)

Type	Con	RA	Dec	Mag	Dim	Rating
Spiral galaxy	Ursa Major	11^h 57.9^m	+55° 27′	10.7	3.0′ × 2.6′	3

General description
NGC 3998 is a small and very condensed spiral galaxy about 25′ northeast of NGC 3982. It is much easier to see than NGC 3982 and is a fine object for small telescopes.

Directions
Using Chart 20b, from NGC 3982, carefully move your telescope 25′ to the northeast. NGC 3998 should be readily visible immediately east of a 10′-long acute triangle of three 10th- to 10.5-magnitude stars.

The quick view
At 23×, NGC 3998 is a small 2′-wide circular glow with an intense central core. This core is easy to see from a dark sky and should be visible from a suburban location. At 72×, the galaxy still appears as a circular glow that gets gradually brighter to a starlike nucleus. Larger telescopes may be able to see its 12.6-magnitude companion (NGC 3990) 3′ to the west.
   Stop. Do not move the telescope. Your next target is nearby!

3.   NGC 3898 (H I-228)

Type	Con	RA	Dec	Mag	Dim	Rating
Spiral galaxy	Ursa Major	11^h 49.2^m	+56° 05′	10.7	3.3′ × 1.9′	2

General description

NGC 3898 is a moderately small and dim spiral galaxy less than $1\frac{1}{2}°$ northwest of NGC 3998. In small telescopes averted vision is required to see it at low power.

Directions

Using Chart 20b, from NGC 3998, move your telescope slowly 40′ to the northwest, where you will encounter solitary 8.5-magnitude Star c. Now move little more than 20′ northwest to similarly bright Star d. NGC 3898 is less than 40′ southwest of Star d. It is also a little more than 30′ northeast of 6th-magnitude Star e.

The quick view

At 23×, NGC 3898 is a dim, 2′-wide glow that is virtually invisible with direct vision. It does swell remarkably with averted vision, though, appearing as a concentrated ellipse of light. With averted vision and 72×, the galaxy has a stellar nucleus inside a bright and condensed inner lens surrounded by a large circular glow. The tight inner lens is also surrounded by a dim halo of light that makes the galaxy appear slightly out of round.

General description

In a dark sky, NGC 3945 is a moderately conspicuous galaxy about 6° east–southeast of 2nd-magnitude Alpha (α) Ursae Majoris.

Directions

Use Chart 20 to locate Alpha Ursae Majoris. Then use your unaided eyes or binoculars to find Pair a – two 6th-magnitude stars a little more than 3° east–southeast of Alpha Ursae Majoris. Center the southernmost star in Pair a in your telescope at low power, then switch to Chart 20c. Now make a slow 1° sweep southeast to 7th-magnitude Star b. Next, swing 55′ northeast to 6.5-magnitude Star c. Just 25′ east–southeast is a pair of 8th- and 9th-magnitude stars (d). NGC 3945 is 45′ southeast of Pair d, and 30′ northwest of another pair of 8th- and 9th-magnitude stars (e).

The quick view

At 23× and averted vision, NGC 3945 is simply a small but moderately bright oval of light. At 72×, the galaxy appears more circular – like a comet that gets gradually brighter toward the middle. Note that 12th- and 13th-magnitude stars hug the galaxy's southwest and northwest flanks. Larger scopes may reveal the galaxy's dim, stellarlike core. They may also show the galaxy's extensive, though dim, outer halo, as shown in in the photograph above.

Stop. Do not move the telescope. Your next target is nearby!

4. NGC 3945 (III-251)

Type	Con	RA	Dec	Mag	Dim	Rating
Barred spiral galaxy	Ursa Major	11^h 53.2^m	+60° 41′	10.8	5.9′ × 3.7′	3

5. NGC 4036 (H I-253)

Type	Con	RA	Dec	Mag	Dim	Rating
Lenticular galaxy	Ursa Major	12^h 01.4^m	+61° 54′	10.7	3.8′ × 1.9′	3

6. NGC 4041 (H I-252)

Type	Con	RA	Dec	Mag	Dim	Rating
Spiral galaxy	Ursa Major	12^h 02.2^m	+62° 08′	11.3	2.6′ × 2.6′	2

General description

NGC 4036 is a very condensed and obvious lenticular galaxy a little north of the midpoint between the 2nd-magnitude stars Alpha (α) Ursae Majoris and Delta (δ) Ursae Majoris in the Big Dipper's Bowl, and about $1\frac{1}{2}°$ northeast of NGC 3945. NGC 4041 (see below) is in the same field of view. Note that NGC 4036 is only about 1 magnitude fainter than the famous lenticular galaxy NGC 5866 in Draco.

Directions

Using Chart 20c, from NGC 3945 make a slow and careful 1° sweep northeast to a 20′-wide arc of three 6.5- to 7.5-magnitude suns (f); the stars in Arc f get progressively fainter to the southeast. NGC 4036 is about 35′ northeast of the center star in Arc f.

The quick view

At 23× in the 4-inch, NGC 4036 is a dim, yet concentrated elliptical glow (oriented east to west) about 10′ east of two roughly 11th-magnitude stars. At 72×, the galaxy is much more obvious, displaying a bright lens surrounded by a thin elliptical halo.

Stop. Do not move the telescope. Your next target is nearby!

General description

NGC 4041 is a small and dim, face-on spiral galaxy near NGC 4036.

Directions

Using Chart 20c, from NGC 4036 move 15′ north–northeast to NGC 4041.

The quick view

At 23×, NGC 4041 is not readily visible at this magnification, though with concentration, you should be able to see its small, starlike nucleus with averted vision. At 72× in the 4-inch under a dark sky, the galaxy swells to a dim cometary glow. With averted vision, it appears as a roughly 3′-wide circular glow. The low-surface-brightness glow has a uniform luster, though, with concentration and averted vision, a small but faint nuclear region can be perceived The galaxy requires that you locate the field, and sit and stare with averted vision until it looms gradually into view. Through larger telescopes, the galaxy is an irregularly bright glow that becomes gradually brighter to a dim, starlike nucleus.

Spring

4 · April

Star charts for first night

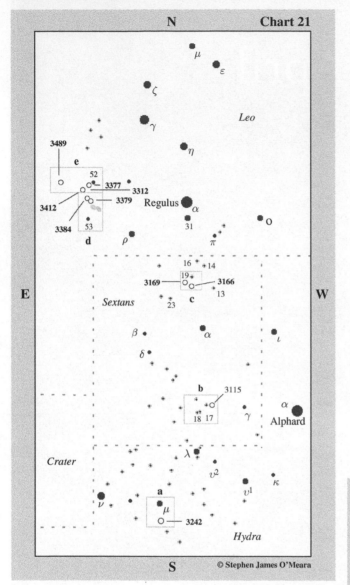

Chart 21

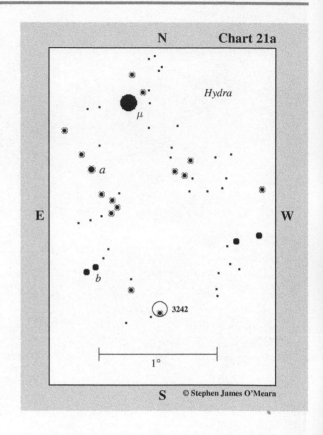

Chart 21a

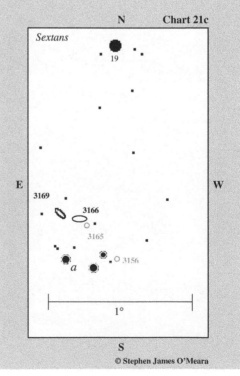

Chart 21c

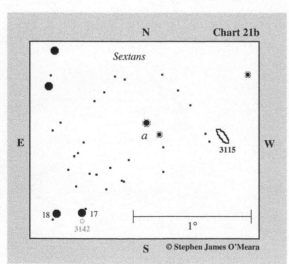

Chart 21b

FIRST NIGHT

1. NGC 3242 (H IV-27)

Type	Con	RA	Dec	Mag	Dim	Rating
Planetary nebula	Hydra	10^h 24.8^m	$-18°$ $38'$	7.3	$45'' \times$ $36''$	5

General description

NGC 3242, the famous Ghost of Jupiter, is a bright planetary nebula 1.8° south–southwest of Mu (μ) Hydrae. It is visible as a 7th-magnitude starlike object in 7×50 binoculars and is a wonder in telescopes of all sizes.

Directions

Use Chart 21 to locate 3.8-magnitude Mu Hydrae. Center this star in your telescope, then switch to Chart 21a. From Mu Hydrae, move about 40' southeast to 7.5-magnitude Star a. Next, move about 50' south to a pair of roughly 7.5-magnitude stars (b). NGC 3242 is 40' southwest of Pair b.

The quick view

At 23×, NGC 3242 is a 7th-magnitude starlike object. At 72×, the nebula shows forth as a bright ring with hints of a halo and a roughly 11.5-magnitude central star. Higher powers magnify the glory of this well-structured planetary nebula, which looks like a translucent aquamarine eye.

2. NGC 3115 (H I-163)

Type	Con	RA	Dec	Mag	Dim	Rating
Lenticular galaxy	Sextans	10^h 05.2^m	$-07°$ $43'$	8.9	$6.9' \times$ $3.4'$	5

General description

NGC 3115, the famous Spindle Galaxy, is a small but intensely bright lenticular system about $4\frac{3}{4}°$ north–northwest of 3.6-magnitude Lambda (λ) Hydrae, or $1\frac{1}{2}°$ northwest of 6th-magnitude 17 Sextantis. The galaxy is easily visible in 7×50 binoculars under a dark sky. Telescopically it is a condensed spindle of light whose intensity increases with aperture. It is a fine object for all telescopes.

Directions

Use Chart 21 to locate Lambda Hydrae, which is about 25° south of 1st-magnitude Alpha (α) Leonis (Regulus) and 11° southeast of 1st-magnitude Alpha (α) Hydrae (Alphard). From Lambda Hydrae, use your unaided eyes or binoculars to look about 4° due north for a close pair of 6th-magnitude stars (17 and 18 Sextantis). Center 17 Sextantis in your telescope, then switch to Chart 21b. From 17 Sextantis, make a slow 1° sweep to the northwest, where you'll find a pair of 6.6- and 7.5-magnitude stars (a) oriented northeast to southwest. NGC 3115 is about $\frac{1}{2}°$ due west of Pair a.

The quick view

At 23× in the 4-inch, NGC 3115 is a 7'-long spindle, oriented northeast to southwest, with a bright central lens and a gently tapered halo. At 72×, the galaxy's core is stunningly bright. It is surrounded by an oval haze that dims gradually outward. The halo tapers to form sharp linear appendages that emanate from each side of the lens. A wonderful object!

3. NGC 3166 (H I-3)

Type	Con	RA	Dec	Mag	Dim	Rating
Mixed spiral galaxy	Sextans	10^h 13.8^m	+03° 26'	10.4	4.6'× 2.6'	4

General description

NGC 3166 is a small but pretty bright spiral galaxy less than $8\frac{1}{2}°$ south–southeast of 1st-magnitude Alpha (α) Leonis (Regulus) and $1\frac{1}{4}°$ south of 6th-magnitude 19 Sextantis. It is paired (and interacting) with the slightly dimmer galaxy, NGC 3169 (see below).

Directions

Use Chart 21 to locate 1st-magnitude Alpha Leonis. Just 25' south of Alpha Leonis is 4th-magnitude 31 Leonis, and about 35' southwest of 31 Leonis is 5th-magnitude Pi (π) Leonis. Now use your unaided eyes or binoculars to locate 14, 16, and 19 Leonis, which form a 25'-wide isosceles triangle about 45' southeast of Pi Leonis. Center 19 Leonis, the southernmost star in that triangle, in your telescope at low power, then switch to Chart 21c. From 19 Leonis, make a generous, but slow and careful $1\frac{1}{2}°$ sweep south and slightly east to an 18'-wide triangle of 7.5- to 8.0-magnitude stars (a). NGC 3166 is less than 20' northwest of the easternmost star in Triangle a.

The Quick View

At 23×, NGC 3166 is a bright circular glow about 2' in diameter that gets increasingly brighter in the middle to a sharp nucleus. The galaxy is brighter at 72× but no other details can be seen. Large-telescope users should be able to detect the galaxy's dimmer and larger halo, which makes the galaxy extend in an east–west direction. Using moderate magnifications, they might also glimpse the galaxy's tiny and dim companion NGC 3165 immediately to the southwest.

Stop. Do not move the telescope. Your next target is nearby!

4. NGC 3169 (H I-4)

Type	Con	RA	Dec	Mag	Dim	Rating
Spiral galaxy	Sextans	10^h 14.2^m	+03° 28'	10.2	5.0'× 2.8'	3

General description

NGC 3169 is slightly smaller and slightly dimmer than NGC 3166. It is also immediately seen as elongated – differentiating it from its more circular-appearing neighbor in small telescopes.

Directions

Use Chart 21c. NGC 3169 is less than 8' east–northeast of NGC 3166.

The quick view

At 23× in the 4-inch, NGC 3169 is a very small (1'), amorphous ellipse of light, oriented northeast to southwest. At 72×, the galaxy's nuclear region is less defined than that of NGC 3166. The lens seems slightly misshapen, uneven, or mottled.

Star charts for second night

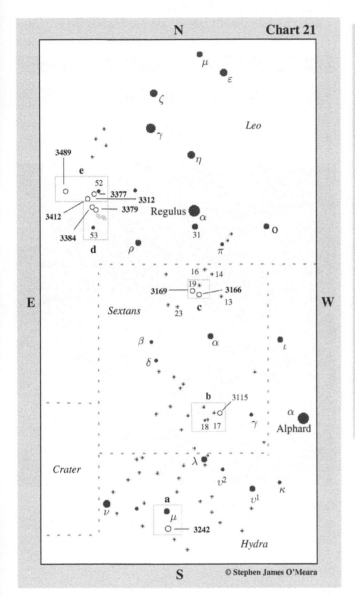

Chart 21

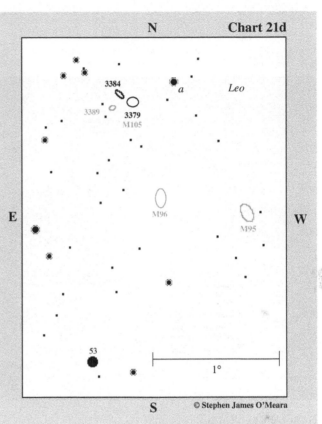

Chart 21d

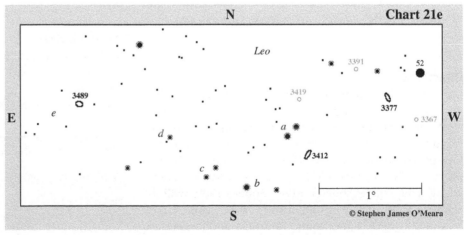

Chart 21e

SECOND NIGHT

1. NGC 3379 (H I-17) = M105

Type	Con	RA	Dec	Mag	Dim	Rating
Elliptical galaxy	Leo	10^h 47.8^m	+12° 35′	9.3	3.9′ × 3.9′	5

General description

NGC 3379 is a bright elliptical galaxy discovered by William Herschel almost 10° east, and slightly north, of Alpha (α) Leonis (Regulus). Actually, unknown to Herschel, Pierre Mechain had already discovered the object on March 24, 1781. But that fact remained hidden until 1947, when Canadian astronomer Helen Sawyer Hogg found a notation to it in Charles Messier's printed catalog. At Hogg's suggestion, the galaxy is now known as M105, though William Herschel first published its position. The galaxy is visible under a dark sky in 7×50 binoculars and can be seen huddled with two other Messier galaxies, M95 and M96.

Directions

Use Chart 21 to locate 1st-magnitude Alpha Leonis, then 4th-magnitude Rho (ρ) Leonis about 6½° to the southeast. Now use your unaided eyes or binoculars to find 5th-magnitude 53 Leonis. Center this star in your telescope at low power, then switch to Chart 21d. From 53 Leonis, move slowly and carefully less than 1½° north–northwest where you will encounter M96. Brighter M95 will be about 40′ to the west–southwest, while your target, NGC 3379 (M105), lies about 50′ to the north–northeast.

The Quick View

At 23× in the 4-inch, NGC 3379 is a well-defined, 4′-wide, circle of light that gradually brightens to a starlike center. At 72×, with attention, some knots can be seen decorating the galaxy's otherwise uniform halo.

Stop. Do not move the telescope. Your next target is nearby!

2. NGC 3384 (H I-18)

Type	Con	RA	Dec	Mag	Dim	Rating
Barred spiral galaxy	Leo	10^h 48.3^m	+12° 38′	9.9	5.5′ × 2.9′	4

General description

NGC 3384 is a bright and obvious companion galaxy to NGC 3379.

Directions

Use Chart 21d. NGC 3384 is only about 7′ northeast of NGC 3379.

The quick view

At 23× in the 4-inch, NGC 3384 has a round core and dim extensions that taper off to sharp points. At 72×, the galaxy's core becomes very bright, though the lens remains ill defined.

Stop. Do not move the telescope.

3. NGC 3377 (H II-99)

Type	Con	RA	Dec	Mag	Dim	Rating
Elliptical galaxy	Leo	10^h 47.7^m	+13° 59′	10.4	4.1′ × 2.6′	3

General description

NGC 3377 is a small and moderately dim elliptical galaxy less than $1\frac{1}{2}°$ due north of NGC 3379.

Directions

If you have a polar aligned equatorial mount, try centering NGC 3379 in your telescope, then swinging $1\frac{1}{2}°$ to the north to NGC 3377. Otherwise, you can use Chart 21 to locate 5th-magnitude 52 Leonis, which is about $1\frac{3}{4}°$ north–northwest of NGC 3379. NGC 3377 is less than 25′ southeast of NGC 52 Leonis. Chart 21e shows the field in more detail.

The quick view

At 23× in the 4-inch, NGC 3377 is 2′-wide circle of light that gets gradually brighter toward the middle. With averted vision it swells to a 4′×2′ ellipse, oriented northeast to southwest. At 72×, the galaxy is a bit more intense and appears somewhat spindle shaped with averted vision. Little else is apparent.

Stop. Do not move the telescope. Your next target is nearby!

General description

NGC 3412 is a small and somewhat dim barred spiral galaxy about 1° southeast of NGC 3377. While it is easy to see in large telescopes, the galaxy could be a challenge for small-telescope users from suburban locations.

Directions

Using Chart 21e, from NGC 3377 move 1° to the east–southeast, where you'll find a pair of 8th-magnitude stars (*a*), separated by about 5′ and oriented northwest to southeast. NGC 3412 is about 18′ southwest of Pair *a*.

The quick view

At 23×, NGC 3412 is a small (2′) ghostly glow that looks irregularly round with averted vision. At 72× and higher magnifications, the galaxy's core intensifies, and its form can be seen drawn out from northwest to southeast. In larger telescopes, the galaxy appears more condensed and intense, so it is an easier catch.

Stop. Do not move the telescope. Your next target is nearby!

4. NGC 3412 (H I-27)

Type	Con	RA	Dec	Mag	Dim	Rating
Barred spiral galaxy	Leo	10^h 50.9^m	+13° 25′	10.5	3.3′× 2.0′	3

5. NGC 3489 (H II-101)

Type	Con	RA	Dec	Mag	Dim	Rating
Mixed spiral galaxy	Leo	11^h 00.3^m	+13° 54′	10.3	3.2′× 2.0′	3

NGC 3412, so you will have to star-hop to it. There are few bright stars in the region, so you must sweep slowly and carefully.

Directions

Using Chart 21e, from NGC 3412, move 40′ southeast to 7.5-magnitude Star b, which has an 8.5-magnitude companion less than 20′ to the west–southwest. Now hop about 25′ east–northeast to a solitary pair of 9th-magnitude stars (c), which are separated by about 6′ and oriented northwest to southeast. A roughly 30′ hop northeast will bring you to 8.5-magnitude Star d, which has an 11th-magnitude companion about 2′ to the southeast. NGC 3489 is less than 1° further to the northeast, inside the northwest corner of a roughly 20′-wide triangle of three 9.5- to 10.5-magnitude stars (e). Again, sweep slowly and carefully to it.

The quick view

At 23× in the 4-inch, NGC 3489 is a small (2′) circular glow of nearly uniform light. A magnification of 72× reveals a very bright nucleus surrounded by a small circular halo that, with averted vision, expands to a dim lens 3′ in extent and oriented east–northeast to west–southwest.

General description

NGC 3489 is a small and somewhat dim spiral galaxy with a bright core. It lies less than $2\frac{1}{2}$° east–northeast of

Star charts for third night

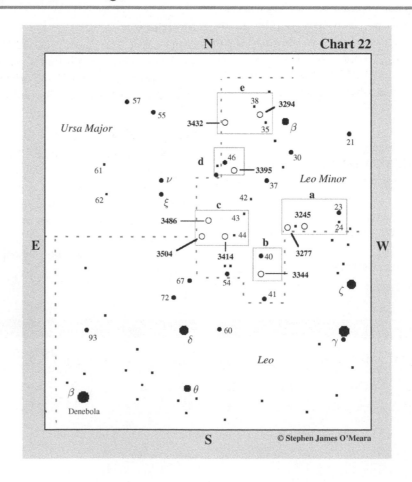

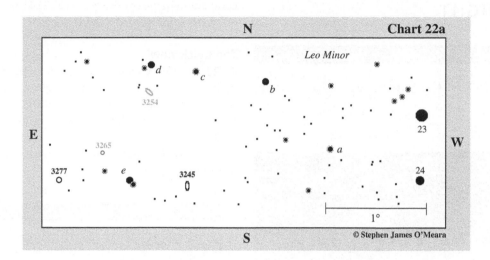

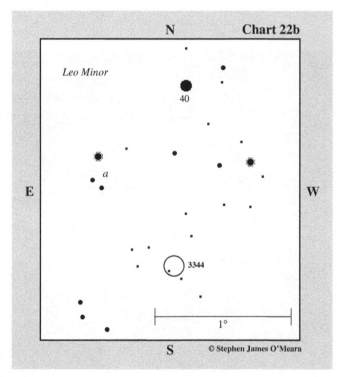

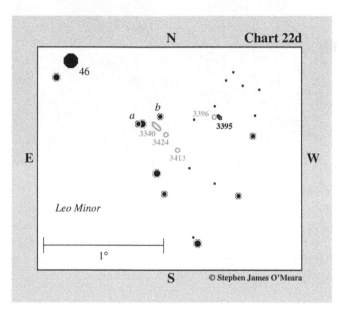

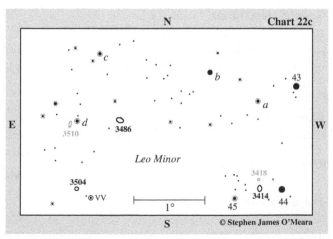

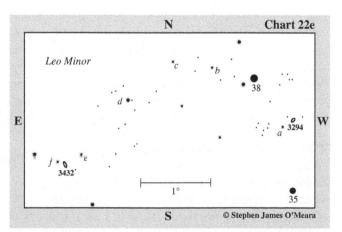

THIRD NIGHT

1. NGC 3245 (H I-86)

Type	Con	RA	Dec	Mag	Dim	Rating
Spiral galaxy	Leo Minor	10^h 27.3^m	+28° 30′	10.8	2.9′ × 2.0′	3

4′

General description

NGC 3245 is a small though somewhat bright spiral galaxy about $5\frac{1}{2}°$ northeast of 3rd-magnitude Zeta (ζ) Leonis in the Sickle of Leo. It may be a challenge to see from suburban locations. Be prepared to star-hop to its field, then use moderate magnification and averted vision.

Directions

Use Chart 22 to locate Zeta Leonis. Next, use your unaided eyes or binoculars to look about 6° north for 5.5-magnitude 23 Leonis Minoris; the star is easily identified because 6.5-magnitude 24 Leonis Minoris is about 40′ to its south. You want to center 23 Leonis Minoris in your telescope at low power, then switch to Chart 22a. From 23 Leonis Minoris, move 1° southeast to 7.5-magnitude Star a; note that Star a forms the eastern apex of a near equilateral triangle with 23 and 24 Leonis Minoris. Now move a little less than 1° to the northeast, where you'll find 6.5-magnitude Star b. Now sweep about 40′ east to 8th-magnitude Star c. About 25′ further to the east, and a little north, is 7th-magnitude Star d. (A 12th-magnitude galaxy, NGC 3254, lies about 15′ south–southeast of Star d.) Now make a slow and careful sweep $1\frac{1}{4}°$ south–southeast to Double Star e, which is composed of a 7th-magnitude primary and an 8.5-magnitude

companion to the southwest. NGC 3245 is a little more than 30′ west–southwest of Double Star e.

The quick view

At 23× in the 4-inch, NGC 3245 is moderately obvious with averted vision as a round, 2′ glow, like a small comet. At 72×, the galaxy is a 2′-long ellipse, oriented north to south, with a circular core that brightens to a starlike nucleus.

Stop. Do not move the telescope. Your next target is nearby!

2. NGC 3277 (H II-359)

Type	Con	RA	Dec	Mag	Dim	Rating
Spiral galaxy	Leo Minor	10^h 32.9^m	+28° 31′	11.7	2.2′ × 2.0′	2

4′

General description

NGC 3277 is a very small and faint spiral galaxy less than $1\frac{1}{2}°$ east of NGC 3245. Small-telescope users should be ready to use averted vision and moderate magnification to see this object.

Directions

Using Chart 22a, from NGC 3245, return to Double Star e. NGC 3277 is 40′ west of that double, about 6′ northeast of a 10.5-magnitude star.

The quick view

At 23× in the 4-inch under a dark sky, only the core of NGC 3277 is visible (just barely) with averted vision. At 72×, the galaxy is a 1′-wide circular glow with a bright core that seems mottled with averted vision. With averted vision and some attention, the galaxy has a tight but dim outer halo that turns the circular core into a tiny, 2′-long ellipse, which is oriented north–northeast to south–southwest.

3. NGC 3344 (H I-81)

Type	Con	RA	Dec	Mag	Dim	Rating
Mixed spiral galaxy	Leo Minor	10^h 43.5^m	+24° 55′	9.3	6.7′ × 6.3′	4

General description
NGC 3344 is the the largest and brightest galaxy in Leo Minor. Under a dark sky, the galaxy is bright and obvious in the smallest of telescopes.

Directions
Use Chart 22 to locate 2.6-magnitude Delta (δ) Leonis in the Lion's hindquarters. Just 6° to the northwest is 4th-magnitude 54 Leonis, which marks the eastern apex of a 3°-wide, near-equilateral triangle with 6th-magnitude 41 Leonis Minoris and slightly fainter 40 Leonis Minoris. (Use binoculars to confirm this configuration, if necessary.) Center 40 Leonis Minoris in your telescope at low power and switch to Chart 22b. You can get to your target by making two slow and careful 1° sweeps. From 40 Leonis Minoris, move 1° southeast to a 10′-long arc of three 8th-magnitude suns (a), which is oriented north to south. NGC 3344 is less than 1° southwest of Arc a.

The quick view
At 23×, NGC 3344 is a bright and obvious 5′-wide circular glow about 5′ west and slightly north of a magnitude 10.5 star, which sits on the very flank of the galaxy's halo. A roughly 12.5-magnitude star lies roughly midway between it and the galaxy's bright nucleus. Beware: at high power, a 14th-magnitude star lies even closer to the nucleus to the southeast. The galaxy's arms form a large halo of uniform milky light that envelopes all three of these suns. At 72×, the galaxy begins to lose its "milky" smooth luster and begins to look curdled.

Stop. Do not move the telescope.

4. NGC 3414 (H II-362)

Type	Con	RA	Dec	Mag	Dim	Rating
Lenticular galaxy	Leo Minor	10^h 51.3^m	+27° 59′	11.0	3.0′ × 2.7′	2

General description
NGC 3414 is a small and very faint lenticular galaxy near 44 Leonis Minoris. Small-telescope users will need to know exactly where to look to see it with averted vision. Think, "stellarlike."

Directions
Use Chart 22 to locate 6th-magnitude 44 Leonis Minoris, which is a little more than 2° northeast of 40 Leonis Minoris (the star you used to find NGC 3344). NGC 3414 is less than 20′ west of 44 Leonis Minoris. Use Chart 22c to identify its location.

The quick view
At 23× in the 4-inch, NGC 3414 is a very faint pip of fuzzy light that requires averted vision to see. At 72×, the galaxy is a concentrated spot of light (2′-wide) with a tiny stellar nucleus.

Stop. Do not move the telescope. Your next target is nearby!

Stop. Do not move the telescope. Your next target is nearby!

5. NGC 3486 (H I-87)

Type	Con	RA	Dec	Mag	Dim	Rating
Mixed spiral galaxy	Leo Minor	11^{h} 00.4^{m}	+28° 58′	10.5	6.6′× 4.7′	4

General description

NGC 3486 is a fairly bright spiral galaxy about $1\frac{3}{4}°$ northeast of NGC 3414. It is obvious even at low power in a small telescope under a dark sky.

Directions

Using Chart 22c, from NGC 3414, return to 44 Leonis Minoris. Now use your unaided eyes or binoculars to find 43 Leonis Minoris, about $1\frac{1}{2}°$ to the north–northwest. You want to center 43 Leonis Minoris in your telescope with low power, then switch to Chart 22c. From 43 Leonis Minoris, move 35′ southeast to solitary 8th-magnitude Star a. Next, move about 45′ northeast to 7th-magnitude Star b. From Star b, you want to make a slow and careful sweep $1\frac{1}{2}°$ east–northeast to equally bright Star c, which has a 9th-magnitude companion about 10′ to the southeast. Now look less than 1° to the south–southeast for 7.5-magnitude Star d; a very dim galaxy (of 13th magnitude), lies about 6′ to the southeast. NGC 3486 is 35′ east, and slightly north of Star d.

The quick view

At 23×, NGC 3486 is visible as an obvious 4′×2′ glow with a central brightening. At 72×, the slightly oblong disk appears delicately mottled.

6. NGC 3504 (H I-88)

Type	Con	RA	Dec	Mag	Dim	Rating
Mixed spiral galaxy	Leo Minor	11^{h} 03.2^{m}	+27° 58′	10.9	2.3′× 2.3′	3

General description

NGC 3504 is a very small but reasonably bright spiral galaxy about $1\frac{1}{4}°$ southeast of NGC 3486. In small telescopes it is very dim at low power under a dark sky but brightens significantly with magnification.

Directions

Using Chart 22c, from NGC 3486, return to Star d. NGC 3504 is a little less than 1° due south, about 15′ northeast of the semi-regular variable star VV Leonis Minoris.

The quick view

At 23× in the 4-inch, NGC 3504 is only a dim, 1′-wide glow. But at 72×, the galaxy brightens into a 2′-long elliptical disk, oriented northwest to southeast. Its core is quite bright and its nucleus stellarlike. The surrounding halo is somewhat dim and ill defined.

7. NGC 3395 (H I-116)

Type	Con	RA	Dec	Mag	Dim	Rating
Mixed spiral galaxy	Leo Minor	10^{h} 49.8^{m}	+32° 59′	12.1	1.6′× 0.9′	2

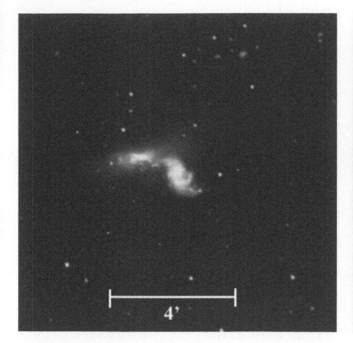

General description

NGC 3395 is a small and dim galaxy less than 1° from 46 Leonis Minoris. It is interacting with NGC 3396. In small telescopes at low power, the two galaxies appear as a single object. The surrounding field has several other galaxies, so be careful in your search. Be prepared to use moderate magnification to see your target well.

Directions

Use Chart 22 to locate 5th-magnitude 46 Leonis Minoris, which is about 6° north and slightly east of 44 Leonis Minoris, which you used to find NGC 3414. Center 46 Leonis Minoris in your telescope at low power, then switch to Chart 22d. From 46 Leonis Minoris, move a little less than 50′ southwest to 7.5-magnitude Star a, which has a 9th-magnitude companion immediately to the east. Now move about 10′ west–northwest to 9th-magnitude Star b. NGC 3395 is only 30′ west of Star b.

The quick view

At 23× in the 4-inch, the NGC 3395 and NGC 3396 blend to become a very low-surface-brightness glow, about 2′ in diameter, oriented east to west. At 72×, NGC 3995 and NGC 3396 are visible as two slightly oval 1′-wide glows touching. NGC 3395 is the westernmost of the two glows.

General description

NGC 3294 is a small and very dim galaxy 50′ southwest of 38 Leonis Minoris. It will be very difficult to see in a small telescope with any light pollution. It is best seen with moderate magnifications. Small-telescope users will need to know exactly where to look and use averted vision to see it.

Directions

Use Chart 22 to locate 4th-magnitude Beta (β) Leonis Minoris, which is about $5\frac{1}{2}$° northwest of 46 Leonis Minoris, or about 5° south–southeast of the southernmost toe in Ursa Major's left foot. Once you locate Beta Leonis Minoris, use your unaided eyes or binoculars to find 6th-magnitude 38 Leonis Minoris about $2\frac{1}{2}$° to the northeast. Center this star in your telescope at low power, then switch to Chart 22e. From 38 Leonis Minoris, move a little less than 50′ southwest to 8th-magnitude Star a. NGC 3294 is only about 11′ northwest of Star a, and about 5′ south of an 11th-magnitude star.

The quick view

At 23× in the 4-inch under very dark skies, NGC 3294 is a very dim and small (1′) round glow that is barely brighter than the background sky. At 72×, the galaxy is a bit more defined and looks like a 2′-wide comet with a diffuse, inner, and dimmer, outer coma and no nucleus.

Stop. Do not move the telescope. Your next target is nearby!

8.   NGC 3294 (H I-164)

Type	Con	RA	Dec	Mag	Dim	Rating
Spiral galaxy	Leo Minor	10ʰ 36.3ᵐ	+37° 20′	11.8	3.5′× 1.7′	2

9.   NGC 3432 (H I-172)

Type	Con	RA	Dec	Mag	Dim	Rating
Barred spiral galaxy	Leo Minor	10ʰ 52.5ᵐ	+36° 37′	11.2	6.9′× 1.9′	4

4'

General description

NGC 3432 is a very nice edge-on galaxy about 3° southeast of 38 Leonis Minoris. You will need to make a careful star-hop to it. So take your time and be patient. It is a very nice and striking galaxy even in a small telescope under a dark sky.

Directions

Using Chart 22e, from NGC 3294, return to 38 Leonis Minoris, then, using low power, move about 40′ east–northeast to 9th-magnitude Star *b*. Now move 30′ east–northeast to similarly bright Star *c*. Next make a 50′ swing southeast, to 7.5-magnitude Star *d*. A careful 1° sweep further to the southeast will bring you to 9th-magnitude Star *e*. NGC 3432 lies about 15′ southeast of Star *e* and 5′ west–southwest of 9th-magnitude Star f.

The quick view

At 23×, NGC 3432 is a sharp 5′-long glow, oriented northeast to southwest, with a bright core that tapers away from the center on either end (especially when seen with with averted vision). At 72×, the galaxy is an elongated spindle with a prominent star punctuating the galaxy's southwestern tip.

Star charts for fourth night

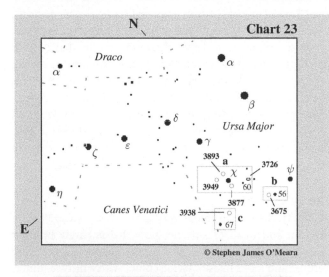

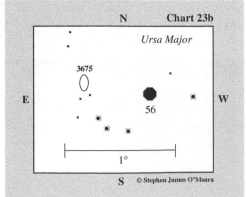

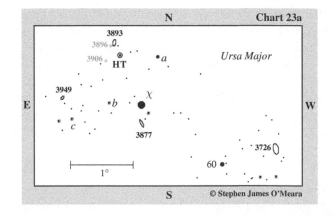

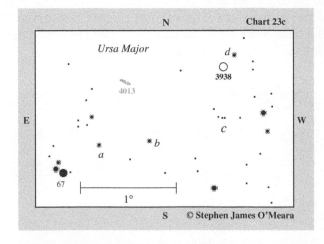

FOURTH NIGHT

1. NGC 3877 (H I-201)

Type	Con	RA	Dec	Mag	Dim	Rating
Spiral galaxy	Ursa Major	11ʰ 46.1ᵐ	+47° 30′	11.0	5.1′× 1.1′	2

General description

NGC 3877 is a very small and dim galaxy immediately south of 3.7-magnitude Chi (χ) Ursae Majoris. In small telescopes it appears very stellar and can be seen only when you know exactly where to look with averted vision. It is best seen at moderate magnifications.

Directions

Use Chart 23 to locate Chi Ursae Majoris, which is only 6° south–southwest of 2nd-magnitude Gamma (γ) Ursae Majoris in the Big Dipper's bowl. To see NGC 3877, you need only center Chi Ursae Majoris in your telescope and look 17′ to the south. If necessary, use Chart 23a to identify the field.

The quick view

I could not see NGC 3877 at 23× in the 4 inch until after I had seen it at 72×. Even then, it was only a fleck of fuzziness with averted vision – and that's from a very dark sky. At 72×, only the galaxy's nucleus and round central lens is visible. Larger scopes will show it as a dim 4′-long needle of light, oriented northeast to southwest.

  Stop. Do not move the telescope. Your next target is nearby!

2. NGC 3893 (H II-738)

Type	Con	RA	Dec	Mag	Dim	Rating
Mixed spiral galaxy	Ursa Major	11ʰ 48.6ᵐ	+48° 43′	10.5	4.2′× 2.3′	4

General description

NGC 3893 is a somewhat bright and obvious spiral galaxy about 1° northeast of Chi (χ) Ursae Majoris.

Directions

Using Chart 23a, from NGC 3877, return to Chi Ursae Majoris. Now move about 45′ northwest to 7th-magnitude Star a. NGC 3893 is a little more than 40′ east–northeast of Star a and about 10′ northeast of the variable star HT Ursae Majoris.

The quick view

At 23×, NGC 3893 is a small (2′) glow, uniform in brightness, just 3′ northeast of an 11th-magnitude star. At 72×, the galaxy transforms into a larger (4′×2′) spindle with a soft oval core and fainter extensions. There is no sharp stellar core, just a soft glow. A 12.5-magnitude star lies on its northwest flank. Larger telescopes should reveal the galaxy's highly mottled texture and S-shaped spiral arms.

  Stop. Do not move the telescope. Your next target is nearby!

3. NGC 3949 (H I-202)

Type	Con	RA	Dec	Mag	Dim	Rating
Spiral galaxy	Ursa Major	11ʰ 53.7ᵐ	+47° 52′	11.1	2.6′× 1.6′	2

General description

NGC 3949 is a very small and very faint spiral galaxy less than $1\frac{1}{2}°$ east of Chi (χ) Ursae Majoris. This galaxy will be extremely difficult to see from skies with any kind of light pollution, especially for small-telescope users.

Directions

Using Chart 23a, from NGC 3893, return to Chi Ursae Majoris. Now, with low power, move a little more than 30′ east to 9th-magnitude Star *b*, which marks the northern apex of an 8′-wide equilateral triangle with two 10.5-magnitude stars. Now move about 40′ southeast to 8.5-magnitude Star *c*. NGC 3949 is less than 20′ northeast of Star *c*.

The quick view

At 23× in the 4-inch, under very dark skies, NGC 3949 is a very dim circular glow, about 1.5′ in diameter – and that's with keen averted vision. At 72×, the galaxy is a very delicate oval glow, oriented northwest to southeast, with a very soft core of light.

Stop. Do not move the telescope. Your next target is nearby!

General description

NGC 3726 is a relatively large, low-surface-brightness galaxy about $2\frac{1}{4}°$ west–southwest of Chi (χ) Ursae Majoris. Its appearance is very much like that of a dim comet just beginning to shine.

Directions

Using Chart 23a, from 3949, return to Chi Ursae Majoris. Now use your unaided eyes or binoculars to find 6th-magnitude 60 Ursae Majoris about $1\frac{1}{2}°$ to the southwest. Center 60 Ursae Majoris in your telescope at low power, then move a little less than 1° west–northwest to your target, NGC 3726.

The quick view

At 23×, NGC 3726 is a 4′ comet like glow with a weakly condensed inner lens surrounded by a dim circular halo. At 72×, the galaxy's core, with averted vision, brightens a bit, though it remains, overall, a large amorphous glow. An 11th-magnitude star is projected against its northern flank.

4. NGC 3726 (H II-730)

Type	Con	RA	Dec	Mag	Dim	Rating
Mixed spiral galaxy	Ursa Major	11^h 33.3^m	+47° 02′	10.4	5.6′ × 3.8′	3

5. NGC 3675 (H I-194)

Type	Con	RA	Dec	Mag	Dim	Rating
Spiral galaxy	Ursa Major	11^h 26.1^m	+43° 35′	10.2	6.2′ × 3.2′	4

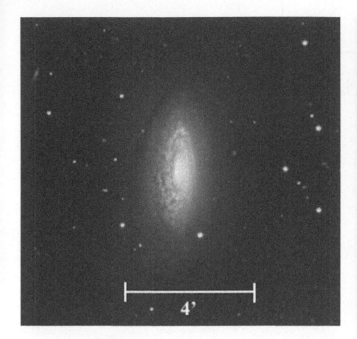

General description

NGC 3675 is a fairly bright spiral galaxy near 5th-magnitude 56 Ursae Majoris. With its bright core, the galaxy is a good object for small telescopes and a prominent object in larger telescopes.

Directions

Use Chart 23 to locate 3rd-magnitude Psi (ψ) Ursae Majoris, which is a little more than 7° southwest of Chi Ursae Majoris. From Psi Ursae Majoris, use your unaided eyes or binoculars to find 5th-magnitude 56 Ursae Majoris, only $2\frac{1}{2}°$ further to the east–southeast. NGC 3675 is only 35′ east–northeast of 56 Ursae Majoris. Use Chart 23b to identify the object's exact location.

The quick view

At 23×, NGC 3675 is a fairly bright, 4′-long ellipse, oriented north–south, that swells with averted vision. The inner core is very bright and the halo is quite diffuse. A magnification of 72× reveals a bright inner lens, about 2′ across, surrounded by a mottled elliptical halo twice that extent. With some concentration an S-shaped spiral structure can be discerned. A very nice object for small-telescope users.

6. NGC 3938 (H I-203)

Type	Con	RA	Dec	Mag	Dim	Rating
Spiral galaxy	Ursa Major	11ʰ 52.8ᵐ	+44° 07′	10.4	4.9′ × 4.7′	3

General description

NGC 3938 is a small but somewhat bright spiral galaxy near 5th-magnitude 67 Ursae Majoris. You will need to make a careful star-hop to it. So move slowly and be patient as you confirm various stars. Fortunately, the galaxy has a bright core, so small-telescope users should have no problem identifying the galaxy.

Directions

Use Chart 23 to locate 67 Ursae Majoris, which is $5\frac{1}{2}°$ southeast of Chi Ursae Majoris. You want to center this star in your telescope at low power, then switch to Chart 23c. Note: 67 Ursae Majoris is easy to identify because it is a fabulous multiple star whose brightest components comprise part of a 15′-long asterism that somewhat resembles the W of Cassiopeia with unevenly bright stars. From 67 Ursae Majoris, move about 30′ northwest to 8.5-magnitude Star a. Next, hop 30′ due west to solitary 9th-magnitude Star b. Now make a very slow and careful sweep 50′ northwest to a 5′-wide row of three roughly 9.5- and 10th-magnitude stars (c), oriented east to west. NGC 3938 is 30′ due north of Row c, just about 10′ southeast of 9th-magnitude Star d.

The quick view

At 23×, NGC 3938 is a relatively bright 4′-wide circular glow wedged between two stars. At 72×, the galaxy is a highly concentrated bright mass with a bright inner core and outer halo.

Star charts for fifth night

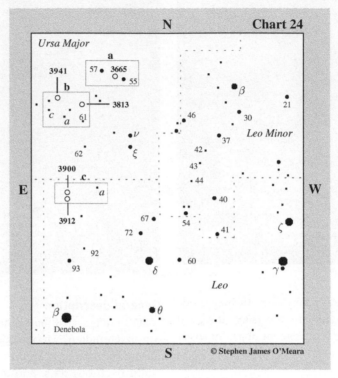

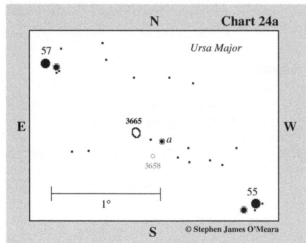

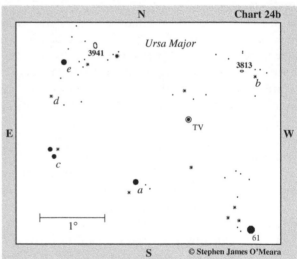

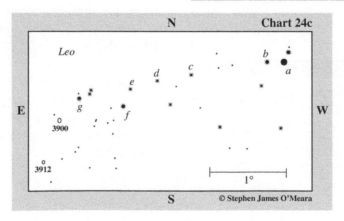

FIFTH NIGHT

1. NGC 3665 (H I-219)

Type	Con	RA	Dec	Mag	Dim	Rating
Spiral galaxy	Ursa Major	11^h 24.7^m	+38° 46′	10.8	3.5′× 3.1′	3

General description

NGC 3665 is a small and relatively faint spiral galaxy midway between the 5th-magnitude stars 55 and 57 Ursae Majoris. In small telescopes it is best seen under a dark sky at moderate magnification.

Directions

Use Chart 24 to locate 55 Ursae Majoris, which is 5° north–northeast of 3.5-magnitude Nu (ν) Ursae Majoris. Center this star in your telescope at low power, then switch to Chart 24a. From 55 Ursae Majoris, move 1° northeast to 9th-magnitude Star a. NGC 3665 is only about 12′ further to the northeast.

The quick view

At 23×, NGC 3665 is visible with averted vision as a moderate glow 1.5′ in diameter. Nearby field stars give the illusion that the galaxy is elongated. At 72×, the galaxy appears soft and elegant, with a slight brightening toward the core. The outer halo is tight and round. With averted vision, the halo extends slightly to the north–northeast and south–southwest.

2. NGC 3813 (H I-94)

Type	Con	RA	Dec	Mag	Dim	Rating
Spiral galaxy	Ursa Major	11^h 41.3^m	+36° 33′	11.7	1.9′× 1.1′	2

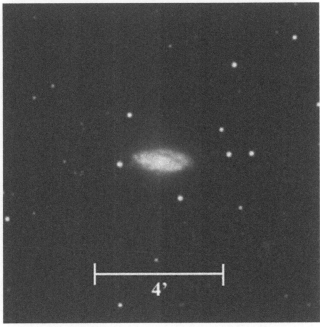

General description

NGC 3813 is a very small and very dim galaxy about $2\frac{1}{2}$° north of 5th-magnitude 61 Ursae Majoris. Small-telescope users will need to star-hop very carefully to its position and use moderate magnification to see it. Any light pollution will obliterate it.

Directions

Use Chart 24 to locate 61 Ursae Majoris, which is about $4\frac{1}{2}$° east–northeast of 3.5-magnitude Nu (ν) Ursae Majoris. From 61 Ursae Majoris use your unaided eyes or binoculars to find 6th-magnitude Star a. Center Star a in your telescope at low power, then switch to Chart 24b. From Star a, move $1\frac{1}{4}$° northwest to 7th-magnitude TV Ursae Majoris. A little more than 1° west–northwest is 8th-magnitude Star b. NGC 3813 is less than 15′ east–northeast of Star b.

The quick view

At 23×, NGC 3813 is extremely faint. You have to know exactly where to look to see it with averted vision. I suggest that once you get Star b centered in your telescope, try switching to moderate magnification. At 72× in the 4-inch, the galaxy is visible as a roughly 2′-wide ellipse with a slight central brightening. Larger telescopes will show the galaxy's halo to be more extended east to west, trapped in a triangle of three roughly 14th-magnitude stars.

Stop. Do not move your telescope.

3. NGC 3941 (H I-173)

Type	Con	RA	Dec	Mag	Dim	Rating
Barred spiral galaxy	Ursa Major	11^h 52.9^m	+36° 59′	10.3	3.7′ × 2.6′	4

General description

NGC 3941 is a relatively small but bright barred spiral galaxy less than 4° northeast of 61 Ursae Majoris. The galaxy has a high surface brightness, making it a good target for small-telescope users.

Directions

Using Chart 24b, from NGC 3813 return to Star *a*. Now make a slow and careful sweep a little less than 1$\frac{1}{2}$° northeast to a pair of 7th-magnitude stars (*c*), separated by about 5′ and oriented northeast to southwest. Now make another careful sweep about 50′ north to 8th-magnitude Star *d*. Next move about 30′ north–northwest to 6.5-magnitude Star *e*. NGC 3941 is about 20′ northwest of Star *e*.

The quick view

At 23×, NGC 3941 is a very condensed 2′-wide circular glow; it's relatively bright and quite easy to see. At 72×, the galaxy is a very intense and round mass with a starlike nucleus. Larger scopes will show an outer envelope oriented north–northeast to south–southwest.

General description

NGC 3900 is a small and dim spiral galaxy about 10° northeast of 2nd-magnitude Delta (δ) Leonis. It lies in a very desolate field, one extremely devoid of bright stars, so you will need to take some time to star-hop to it.

Directions

Use Chart 24 to locate 6th-magnitude Star *a* in Leo, which is about 5$\frac{1}{2}$° southeast of 4th-magnitude Xi (ξ) Ursae Majoris. Star *a* is the brightest star in the area. Center Star *a* in your telescope at low power, then switch to Chart 24c. Star *a* should be easy to identify, because it is the brightest of three or four stars huddled together. From Star *a*, move about 12′ east to 7.5-magnitude Star *b*. Now make a slow sweep 1° east–southeast to 8.5-magnitude Star *c*. Note that Star *c*, together with stars *d* and *e*, form a graceful 50′ arc oriented west–northwest to east–southeast. Center Star *e* in your scope, then drop about 12′ south–southeast to 7.5-magnitude Star *f*. Now make a 35′ hop east–northeast to similarly bright Star *g*. NGC 3900 is about 25′ southeast of Star *g*.

The quick view

At 23×, NGC 3900 is extremely difficult to see; it is not convincingly visible with averted vision. At 72×, the galaxy does materialize as a dim, hyperfine glow between (and slightly west of) two dim stars oriented northeast and southwest. With averted vision the galaxy appears slightly elongated north to south.

Stop. Do not move the telescope. Your next target is nearby!

4. NGC 3900 (H I-82)

Type	Con	RA	Dec	Mag	Dim	Rating
Spiral galaxy	Leo	11^h 49.2^m	+27° 01′	11.3	2.9′ × 1.5′	1.5

5. NGC 3912 (H II-342)

Type	Con	RA	Dec	Mag	Dim	Rating
Mixed spiral galaxy	Leo	11^h 50.1^m	+26° 29′	12.4	1.6′ × 0.9′	1

General description

NGC 3912 is an extremely small and dim galaxy near NGC 3900.

Field note

This is one of the most difficult objects to see in the Herschel 400 catalog, especially for small telescope users. A dark sky is required, as is high power and patience. Be prepared to spend some time validating your sighting.

Directions

Use Chart 24c. NGC 3912 is about 35′ south–southeast of NGC 3900.

The quick view

NGC 3912 is not visible at 23× in my 4-inch under very dark skies. The surrounding field is a vast and lonely void. At 72×, the galaxy is a tiny fuzzy "star" about 2′ or 3′ south–southeast of a roughly 13.5-magnitude star. These two objects almost blend together. If you're using a small telescope, just resign yourself to seeing only the galaxy's fuzzy stellar nucleus. It will give itself away with averted vision, because it swells ever so slightly whereas the neighboring star will not. Be very patient and hyperventilate to feed oxygen to your brain and eyes. This galaxy can be a challenge to those using 10-inch telescopes from a decent suburban sky. So be patient and give this one time. If you fail, check off this object for a return visit on another night and move on. Do not dwell.

Star charts for sixth night

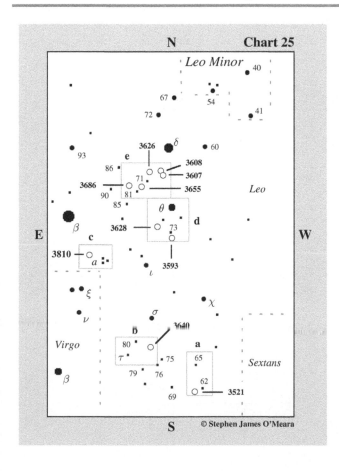

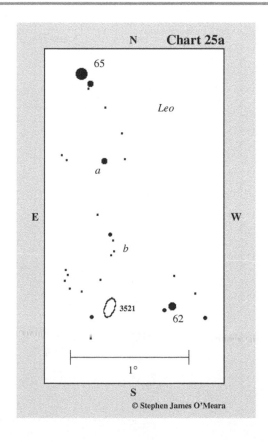

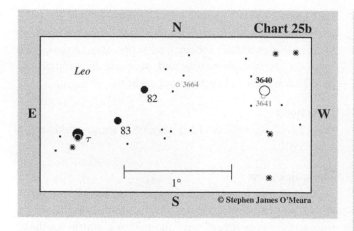

Chart 25b

N

Leo

E

W

3664

3640

82

3641

83

τ

1°

S

© Stephen James O'Meara

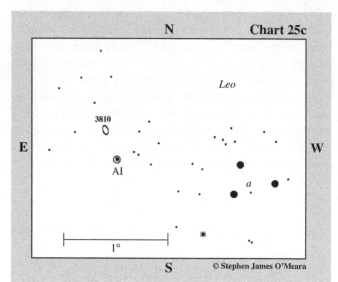

Chart 25c

N

Leo

E

W

3810

AI

a

1°

S

© Stephen James O'Meara

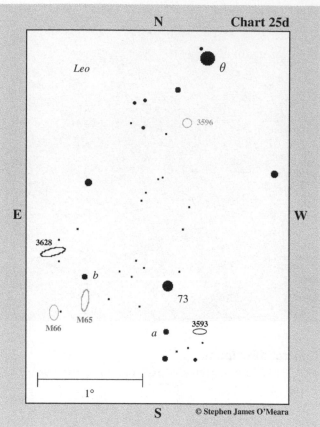

Chart 25d

N

Leo

θ

E

W

3596

3628

b

73

M65

3593

M66

a

1°

S

© Stephen James O'Meara

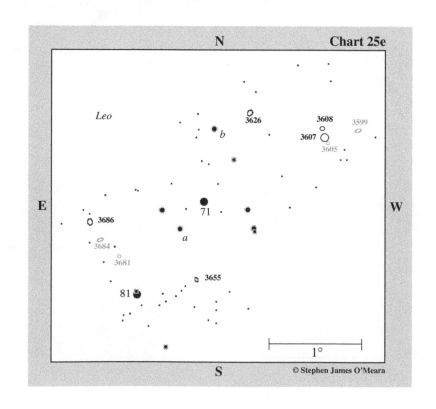

Chart 25e

N

Leo

3626

3608   3599

b

3607

3605

E

W

71

3686

a

3684

3681

3655

81

1°

S

© Stephen James O'Meara

SIXTH NIGHT

1. NGC 3521 (H I-13)

Type	Con	RA	Dec	Mag	Dim	Rating
Mixed spiral galaxy	Leo	11^h 05.8^m	−00° 02′	9.1	11.7′ × 6.5′	4

4′

General description

NGC 3521 is a bright and relatively large spiral galaxy near 6th-magnitude 62 Leonis, which is deep in the southern recesses of Leo. From a dark-sky site, the galaxy is visible in 7×50 binoculars. The smallest of telescopes will show it as a faint splotch of light 10′ northwest of an 8th-magnitude star.

Directions

Use Chart 25 to locate the right rear foot of Leo. It is composed of a 3°-wide polygon of five 5th- and 6th-magnitude stars: 62, 65, 69, 75, 76, and 79 Leonis – about 15° due south of 3rd-magnitude Theta (θ) Leonis. Use your unaided eyes or binoculars to star-hop to the polygon. Start at Theta Leonis and look 5½° southeast for 4th-magnitude Iota (ι) Leonis. Another hop 4½° south–southwest brings you to 4th-magnitude Sigma (σ) Leonis. Now look 4° further to the south–southwest for 5th-magnitude 75 Leonis, the northern tip of the polygon. You'll know if you have 75 Leonis because it has a 6th-magnitude companion about 30′ to the southeast. Just 2½° west of 75 Leonis is 5.5-magnitude 65 Leonis. You can find the galaxy from here in one of two ways. (1) Using

your unaided eyes or binoculars, look 2° to the southwest for 6th-magnitude 62 Leonis; NGC 3521 lies a little more than 30′ east and slightly south of 62 Leonis. You can also find it by (2) centering 65 Leonis in your telescope at low power, then switching to Chart 25a. Note that 65 Leonis has a 7th-magnitude companion about 6′ to the southwest. About 45′ to the south–southwest is 6th-magnitude Star a. A 40′ hop south will bring you to a nice 10′-long arc of four roughly 10th-magnitude suns (b), which is oriented north to south. Your target is only 35′ south of Arc b.

The quick view

At 23×, the galaxy looks like a spindle of light (oriented northwest to southeast) in a rich field of dim suns. It is a pleasing sight, bright and obvious. With averted vision, it appears cometlike, being a simple oval within an oval, with a gradual brightening toward the center. At 72×, the galaxy is a more complex sight, though all the details are delicate. The nucleus appears sharp and bright, and an oval coma surrounds it. The southeast side of the galaxy is also sharply defined and has what appears to be an arm reaching out from the center, like a bar. The inner disk looks warped and mottled, but these latter features are subtle.

2. NGC 3640 (H II-33)

Type	Con	RA	Dec	Mag	Dim	Rating
Elliptical galaxy	Leo	11^h 21.1^m	+03° 14′	10.4	4.6′ × 4.1′	4

4′

General description

NGC 3640 is a small but fairly bright elliptical galaxy $2\frac{3}{4}°$ due south of 4th-magnitude Sigma (σ) Leonis. It is very condensed, so it should be a good object for suburban skies in small telescopes.

Directions

Use Chart 25 to locate Sigma Leonis. Then use your unaided eyes or binoculars to locate 5th-magnitude Tau (τ) Leonis $3\frac{1}{2}°$ to the southeast. Center Tau Leonis in your telescope at low power then switch to Chart 25b. (Tau Leonis is easy to identify because it has a magnitude 7.5 companion immediately to the south, and another 7.5-magnitude star about 7' to the southeast.) From Tau Leonis, move 20' northwest to 6th-magnitude 83 Leonis. Next, make another short 25' hop northwest to 6th-magnitude 82 Leonis. NGC 3640 is a little more than 1° west and a tad south of 82 Leonis.

The quick view

At 23×, NGC 3640 is extremely difficult to see and not convincingly visible at this power. At 72×, the galaxy is a dim, hyperfine glow between, and slightly west of, two dim stars oriented northeast and southwest. With averted vision the galaxy appears slightly elongated north to south.

General description

NGC 3810 is a small and somewhat dim galaxy for small-telescope users, but a bright object for larger telescopes. It lies about 4° east–northeast of 4th-magnitude Iota (ι) Leonis. Any light pollution will certainly wash it out in a small telescope.

Directions

Use Chart 25 to locate Iota Leonis, which is a little more than 2° southeast of 3rd-magnitude Theta (θ) Leonis. Now use binoculars to look for a 20'-wide near-equilateral triangle (a) composed of three 7th-magnitude stars. Center Triangle a in your telescope at low power, then switch to Chart 25c. From the northernmost star in Triangle a, make a slow and careful sweep 1° east to 9th-magnitude AI Leonis. NGC 3810 is less than 20' north–northeast of AI Leonis.

The quick view

At 23×, NGC 3810 is a soft oval glow, about 2' in length, oriented northeast to southwest. At 72×, the galaxy is a 2'-wide amorphous smudge of pale light with a ghostlike appearance. With averted vision and some concentration, a tiny brightening can be seen at the center. Larger telescopes will show the core to be more concentrated with a somewhat triangular shape with averted vision.

3. NGC 3810 (H I-21)

Type	Con	RA	Dec	Mag	Dim	Rating
Spiral galaxy	Leo	11^h 41.0^m	+11° 28'	10.8	3.8' × 2.6'	3

4. NGC 3593 (H I-29)

Type	Con	RA	Dec	Mag	Dim	Rating
Spiral galaxy	Leo	11^h 14.6^m	+12° 49'	10.9	5.3' × 2.2'	3

General description

NGC 3593 is a small and rather faint spiral galaxy near 5th-magnitude 73 Leonis. Small-telescope users will see it best at moderate powers.

Directions

Use Chart 25 to locate 3rd-magnitude Theta (θ) Leonis. Now use your unaided eyes or binoculars to locate 73 Leonis a little more than 2° to the south–southeast. Center 73 Leonis in your telescope at low power, then, switching to Chart 25d, look about 30′ southwest for 7th-magnitude Star *a*. NGC 3593 is 20′ due west of Star *a*.

The quick view

At 23×, NGC 3593 is a very small (1.5′) diffuse circular glow. At 72×, the galaxy swells to twice its size, transforming into an elliptical glow, oriented east to west, that gradually, then suddenly, gets brighter in the middle. Large telescopes may show the galaxy's needle-like dust lane just north of the galaxy's major axis.

Stop. Do not move the telescope. Your next target is nearby!

General description

NGC 3628 is a fairly bright and large edge-on galaxy near the famous Messier galaxies M65 and M66.

Directions

Using Chart 25d, from NGC 3593, return to 73 Leonis. Less than 50′ east and slightly north of 73 Leonis is 7.5-magnitude Star *b*. M65 and M66 lie about 20′ south and southeast (respectively) of Star *b*, while NGC 3628 is about 20′ to the northeast.

The quick view

At 23× in the 4-inch, NGC 3628 is a very obvious and very beautiful needle of light immersed in an elliptical haze (5′×3′) oriented east-west. The ends of this ellipse are warped and broad. The dust lane is quite obvious with averted vision under a dark sky. At 72×, the dust lane can be followed throughout the body, its brightest parts are segmented, and its western end is definitely fainter than the eastern end. The galaxy's core looks beaded and, with great concentration, tube tapping, and gentle breathing, I can see the dust lane against the core with scalloped edges.

5. NGC 3628 (H V-8)

Type	Con	RA	Dec	Mag	Dim	Rating
Spiral galaxy	Leo	11ʰ 20.3ᵐ	+13° 35′	9.5	14.8′× 3.3′	4

6. NGC 3655 (H I-5)

Type	Con	RA	Dec	Mag	Dim	Rating
Spiral galaxy	Leo	11ʰ 22.9ᵐ	+16° 35′	11.7	1.5′× 0.9′	2

General description

NGC 3655 is an extremely small and dim spiral galaxy near 5.5-magnitude 81 Leonis. It will be extremely difficult to see in a small telescope with any light pollution. Think "stellar fuzzy."

Directions

Use Chart 25 to locate 5.6-magnitude 81 Leonis, which is less than 3° northeast of 3rd-magnitude Theta (θ) Leonis. Center 81 Leonis in your telescope at low power, then switch to Chart 25e. NGC 3655 is only 40′ northwest of 81 Leonis.

The quick view

At 23×, NGC 3655 is a tiny 1′-wide fleck of hyperfine light – and that's with averted vision under a very dark sky. To see it all, pinpoint its exact location, then avert your gaze. At 72×, the galaxy is a 2′-wide ellipse of dim, uniform light. Larger scopes may see some concentration of light at the core.

Stop. Do not move the telescope. Your next target is nearby!

General description

NGC 3686 is an extremely difficult barred spiral galaxy about 55′ northeast of 81 Leonis. Small-telescope users need to be under a dark sky, have to have extreme patience, and use averted vision. Take your time with this one. Any light pollution will erase this dim ornament from our skies.

Directions

Using Chart 25e, from NGC 3655, return to 81 Leonis. NGC 3686 is about 55′ northeast of 81 Leonis, just 2.5′ south of a roughly 11th-magnitude star.

The quick view

At 23×, NGC 3686 is an extremely faint glow. Just its 1.5′-wide core is visible in the 4-inch, and it takes great effort with averted vision to see it. The core is faint and ill-defined, just a butterfly's breath of light with no concentration. Larger scopes will see it twice as large and with a slight central condensation.

Stop. Do not move the telescope. Your next target is nearby!

7. NGC 3686 (H II-160)

Type	Con	RA	Dec	Mag	Dim	Rating
Barred spiral galaxy	Leo	11ʰ 27.7ᵐ	+17° 13′	11.3	2.8′× 2.3′	1

8. NGC 3626 (H II-52)

Type	Con	RA	Dec	Mag	Dim	Rating
Spiral galaxy	Leo	11ʰ 20.1ᵐ	+18° 21′	11.0	2.6′× 1.8′	2

General description

NGC 3626 is a small and dim spiral galaxy near 7th-magnitude 71 Leonis. Its light is somewhat concentrated, so it is easier to see than NGC 3686. Again, think, "stellar fuzzy."

Directions

Using Chart 25e, from NGC 3686, return to 81 Leonis. Now move 50′ northwest to 7.5-magnitude Star *a*. Next hop about 25′ northwest to 71 Leonis. A 45′ sweep to the north–northwest will bring you to 8th-magnitude Star *b*. NGC 3626 is less than 30′ northwest of Star *b*.

The quick view

At 23×, NGC 3626 is very small (∼1′) but condensed, so it appears as a fuzzy star with averted vision – but only when you know exactly where to look. The galaxy is better seen at 72×, when the 1′ core swells a bit and is surrounded by a thin outer halo.

Stop. Do not move the telescope. Your next target is nearby!

9.   NGC 3607 (H II-50)

Type	Con	RA	Dec	Mag	Dim	Rating
Spiral galaxy	Leo	11^h 16.9^m	+18° 03′	9.9	4.6′× 4.1′	4

General description

NGC 3607 is a fairly bright but small spiral galaxy about 50′ southwest of NGC 3626. It is a good object for small telescopes, given that it is small and condensed.

Directions

Use Chart 25e. NGC 3607 is about 50′ southwest of NGC 3626.

The Quick View

At 23×, NGC 3607 is a fairly bright and concentrated fuzzy spot, 1′ in diameter. This is but the intense central region of an object three times that size. At 72×, the intense core is surrounded by a dim glow that extends another 1′ of arc. Overall, the galaxy appears featureless and circular. Larger telescopes will show it slightly elongated northwest to southeast.

Stop. Do not move the telescope. Your next target is nearby!

10.   NGC 3608 (H II-51)

Type	Con	RA	Dec	Mag	Dim	Rating
Elliptical galaxy	Leo	11^h 17.0^m	+18° 09′	10.8	2.7′× 2.3′	3

General description

NGC 3608 is a small but fairly obvious elliptical galaxy about 6′ north–northeast of NGC 3607.

Directions

Use Chart 25e. NGC 3608 is only 6′ north–northeast of NGC 3607.

The quick view

At 23×, NGC 3608 is a fairly obvious 1′-wide glow that appears slightly smaller and dimmer than NGC 3607. At 72×, the galaxy displays an intense starlike core immersed in a tight and bright core surrounded by a dim outer halo of light.

Star charts for seventh night

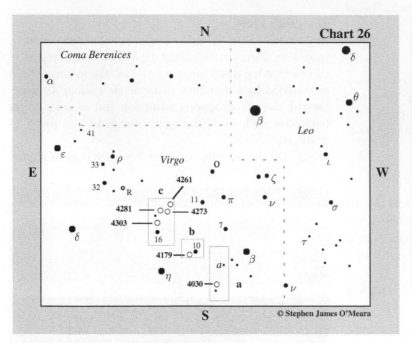

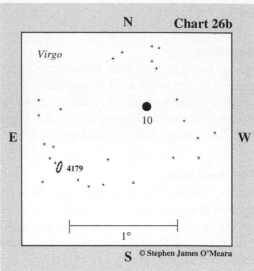

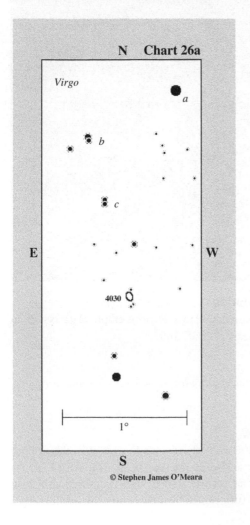

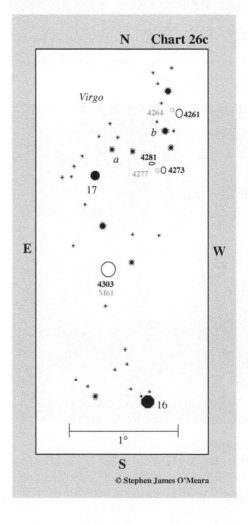

SEVENTH NIGHT

1. NGC 4030 (H 1–121)

Type	Con	RA	Dec	Mag	Dim	Rating
Spiral galaxy	Virgo	12^h 00.4^m	−01° 06′	10.6	3.8′ × 2.9′	3

4'

General description
NGC 4030 is a small but somewhat obvious spiral galaxy about 4$\frac{3}{4}$° southeast of 3.6-magnitude Beta (β) Virginis. It is nestled between two 10th-magnitude stars and is fairly condensed, making it a reasonable target for small-telescope users.

Directions
Use Chart 26 to locate Beta Virginis, which is about 12° south, and slightly east of 2nd-magnitude Beta Leonis. Now use your binoculars to locate 6.5-magnitude Star a, which marks the eastern tip of a 1$\frac{1}{4}$°-wide triangle of 6.5 and 7th-magnitude stars 1$\frac{3}{4}$° southeast of Beta Virginis. Once you locate Star a, center it in your telescope at low power, then switch to Chart 26a. From Star a, move about 50′ southeast to the 7.5-magnitude pair of stars b. Next hop 30′ south–southwest to the roughly 8th-magnitude pair of stars c. NGC 4030 is about 50′ south–southwest of pair c nestled in a dim triangle of suns.

The quick view
At 23×, NGC 4030 and its two neighboring 10th-magnitude stars to the south–southwest form a tiny cluster that is obvious to the eye as a singular spot of interest. At 72×, the galaxy is a 3′-wide glow that brightens suddenly to a roughly 1′-wide core. The surrounding halo is moderately, and uniformly, bright.

2. NGC 4179 (H I-9)

Type	Con	RA	Dec	Mag	Dim	Rating
Lenticular galaxy	Virgo	12^h 12.9^m	+01° 18′	11.0	3.9′ × 1.1′	2.5

4'

General description
NGC 4179 is small, thin, and fairly dim lenticular galaxy near 6th-magnitude 10 Virginis. Small-telescope users should be prepared to use moderate magnification to see it well. It will virtually be impossible at low power with any light pollution.

Directions
Use Chart 26 to locate 10 Virginis, which is 4$\frac{1}{2}$° east of 3.6-magnitude Beta (β) Virginis. Use binoculars if needed. The star is also north of the midpoint between Beta and Eta (η) Virginis. Center 10 Virginis in your telescope at low power, then switch to Chart 26b. NGC 4179 is 1° southeast of 10 Virginis.

The quick view
At 23× in the 4-inch under a very dark sky, NGC 4179 is a pip of fuzzy light, like a fuzzy star. This is just the galaxy's sharp nuclear region. At 72×, the galaxy is a very tiny (2′-long) needle of light with a sharp, starlike core.

Small-telescope users will need to use averted vision to see the galaxy as elongated northwest to southeast. Through larger telescopes, the galaxy's core is quite bright and the galaxy is quite the attractive edge-on object.

3. NGC 4303 (H I-139) = M61

Type	Con	RA	Dec	Mag	Dim	Rating
Mixed spiral galaxy	Virgo	12^h 21.9^m	+04° 28′	9.7	6.0′ × 5.9′	4

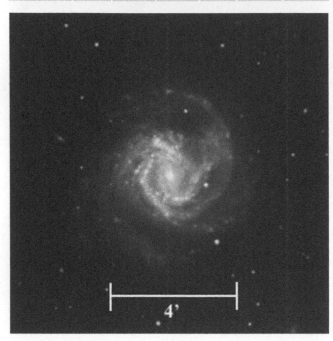

4′

General description
NGC 4303 (Messier 61) is a bright galaxy between the stars 16 and 17 Virginis. Barnabus Oriani first discovered it in 1779. Messier had mistaken the object for the comet of 1779 on three nights. William Herschel also encountered this object during his famous sweeps of the heavens. It is one of the most pleasing open-face galaxies in Messier's list and is visible under a dark sky in 7×50 binoculars.

Directions
Use Chart 26 to locate 5th-magnitude 16 Virginis, which is 4° due north of 4th-magnitude Eta (η) Virginis. Center 16 Virginis in your telescope at low power, then switch to Chart 26c. NGC 4303 is about 1$\frac{1}{4}$° to the north–northeast. Sweep slowly!

The quick view
At 23×, the galaxy is small and somewhat inconspicuous. It is tucked away in the crook of a nice wishbone asterism of 9th-magnitude stars. Once you spot it, try using about 130×,

which will show the galaxy as a 4′-wide circular glow with a bright nucleus inside a mottled, diamond-shaped inner lens. With time and averted vision, the galaxy's warped spiral appendages can be glimpsed.

Stop. Do not move the telescope. Your next target is nearby!

4. NGC 4281 (H II-573)

Type	Con	RA	Dec	Mag	Dim	Rating
Lenticular galaxy	Virgo	12^h 20.4^m	+05° 23′	11.3	2.5′ × 1.3′	2

4′

General description
NGC 4281 is a small, low-surface-brightness lenticular galaxy near 17 Virginis. It is in the same field of view as NGC 4273, our next Herschel object.

Directions
Using Chart 26c, from NGC 4303 (M61), move 50′ north–northeast to 6.4-magnitude 17 Virginis. Now hop 20′ northwest to a pair of 9th-magnitude stars (a), separated by 10′ and oriented roughly east to west. NGC 4281 is only about 12′ southwest of the westernmost star in Pair a.

The quick view
At 23×, NGC NGC 4281 is a small, 2′-wide ellipse of dim, uniform light. No concentration can be seen. At 72×, the galaxy is a bit more condensed with a brightening toward the center like a star.

Stop. Do not move the telescope. Your next target is nearby!

5. NGC 4273 (H II-569)

Type	Con	RA	Dec	Mag	Dim	Rating
Barred spiral galaxy	Virgo	12^h 19.9^m	+05° 21′	11.9	2.3′ × 1.1′	2

6. NGC 4261 (H II-139)

Type	Con	RA	Dec	Mag	Dim	Rating
Lenticular galaxy	Virgo	12^h 19.4^m	+05° 49′	10.4	3.5′ × 3.1′	2

General description

NGC 4273 is a small and dim barred-spiral galaxy near NGC 4281.

Directions

Using Chart 26c, from NGC 4281, move 9′ southwest to NGC 4273.

The quick view

At 23×, NGC 4273 is a small (1′) amorphous glow. At 72×, the galaxy has a bright center that, with averted vision, is surrounded by an irregular halo of dim light that swells the galaxy to perhaps 2′. Larger telescopes should be able to reveal the 13.5-magnitude galaxy NGC 4277, which lies 2′ to its east.

Stop. Do not move the telescope. Your next target is nearby!

General description

NGC 4261 is a small but very concentrated galaxy almost 1° northwest of 6.4-magnitude 17 Virginis.

Directions

Using Chart 26c. From NGC 4273, move 20′ north to 7.5-magnitude Star b. NGC 4261 is only about 12′ northwest of Star b.

The quick view

At 23×, NGC 4261 is a highly concentrated, 2′-wide glow. At 72×, the galaxy gradually, then suddenly, gets much brighter in the middle to a swollen starlike nucleus. Larger scopes may show tiny starlike NGC 4264 about 3.5′ east–northeast of NGC 4261.

5 · May

Star charts for first night

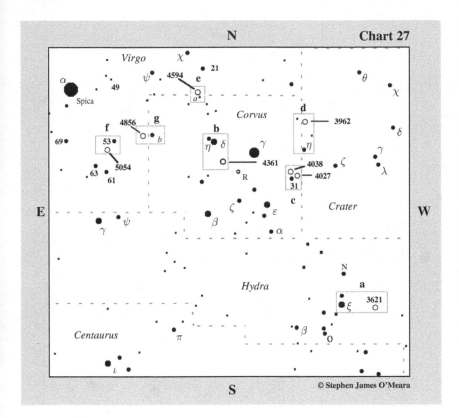

Chart 27

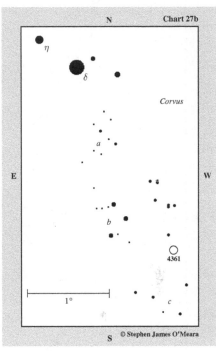

Chart 27b

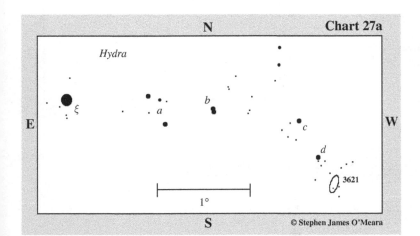

Chart 27a

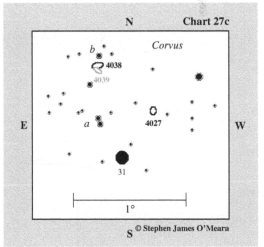

Chart 27c

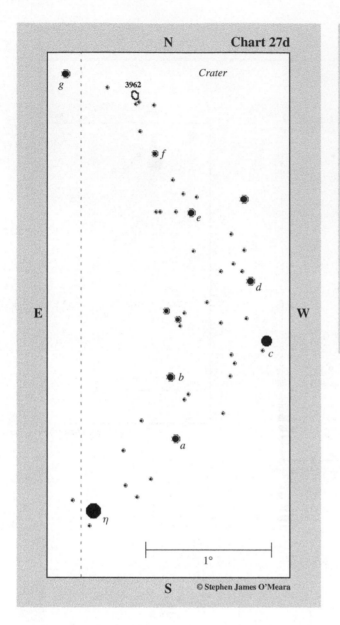

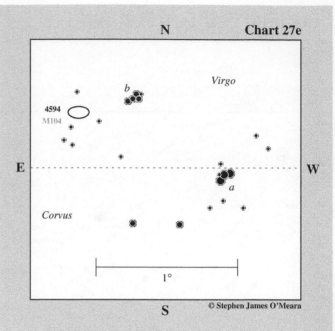

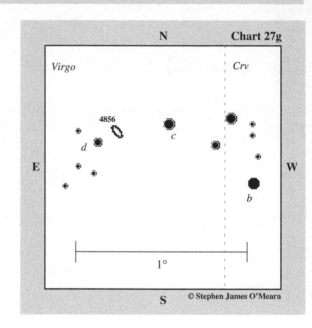

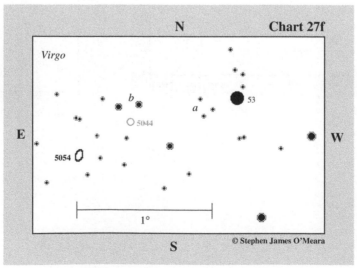

FIRST NIGHT

1. NGC 3621 (H I-241)

Type	Con	RA	Dec	Mag	Dim	Rating
Spiral galaxy	Hydra	11ʰ 18.3ᵐ	−32° 49′	8.5	14.9′ × 7.4′	4

6′

General description

NGC 3621 is a bright spiral galaxy $3\frac{1}{4}°$ west and slightly south of 3.5-magnitude Xi (ξ) Hydrae. From dark-sky sites, especially at southern locals, it is visible in 7 × 50 binoculars as a soft, peach-fuzz glow, even with the crescent Moon up. The smallest of telescopes will show NGC 3621 as a dim, blotted haze; the challenge in this star-poor region is to know exactly where to look for it.

Directions

Use Chart 27 to find Xi Hydrae. To spy it, first locate 3rd-magnitude Epsilon (ε) Corvi, the southwestern star in the famous Corvus Keystone. About 2° south and slightly west of Epsilon Corvi is 4th-magnitude Alpha (α) Corvi. Xi Hydrae is about 10° (a fist) to the southwest. Center Xi Hydrae in your telescope, then switch to Chart 27a. From Xi Hydrae, make a careful sweep about 1° west–southwest, where you should encounter a pair of 8th-magnitude stars (a) 20′ apart and oriented northeast to southwest. Now move about 40′ west to a fine double star (b) with an 8.0-magnitude primary and 9.8-magnitude secondary to the south–southwest. Less than 1° further to the west, and slightly south, is solitary 7th-magnitude Star c. Next move about $\frac{1}{2}°$ southwest to 8th-magnitude Star d. NGC 3621 is about $\frac{1}{2}°$ further to the southwest.

The quick view

In the 4-inch at 23×, NGC 3621 is a bright ellipse with three dim stars forming a cap on the galaxy's southeastern edge. In fact, the galaxy looks more like an elongated globular cluster just starting to be resolved. The galaxy is divided into two parts: a bright inner region (measuring about 5′ × 3′) and a gentle outer halo, which looks like a breath of moist air on a cold morning. That halo will vanish in small apertures with any increase in power. At 72×, the bright inner lens is fantastic. With any concentration it splinters into a veritable ornament of dim lights. The northeastern side is darker than the southwestern side, which is indicative of the dark and dappled veins of dust that riddle the galaxy in photographs.

2. NGC 4361 (H I-65)

Type	Con	RA	Dec	Mag	Dim	Rating
Planetary nebula	Corvus	12ʰ 24.5ᵐ	−18° 47′	10.2	1.9′ × 1.9′	4

4′

General description

NGC 4361 is one of the finest non-Messier objects of its class. It lies in the north central part of the Corvus Keystone (also known as the Sail), and marks the southern apex of a near-right triangle with Delta (δ) Crv and Gamma (γ) Corvi. Under a dark sky, it may be glimpsed, with difficulty, in 7 × 50 binoculars. It can be seen without trouble in any telescope.

Directions

Use Chart 27 to center Delta Corvi (a beautiful double star) in your telescope, then switch to Chart 27b. From Delta Corvi, look a little less than 1° southwest for a pair of 8th-magnitude stars, which are part of a little line of 8th- and 9th-magnitude suns (a). Another 1° sweep to the south will bring you to a

20′-wide isosceles triangle of 7th-magnitude suns (*b*). NGC 4361 is about 50′ southwest of Triangle *b*, about 30′ north of a 40′-wide Y-shaped asterism of 8th-magnitude suns (*c*).

The quick view

At 23× in the 4-inch, the nebula is easily spotted as a small, round, glow, about 2′ wide. But with a good boost in power (I found 182× worked best in my small telescope), the nebula appears as an almost uniform glow around a 13th-magnitude star.

3. NGC 4027 (H II-296)

Type	Con	RA	Dec	Mag	Dim	Rating
Barred spiral galaxy	Corvus	11ʰ 59.5ᵐ	−19° 16′	11.2	3.8′ × 2.3′	2

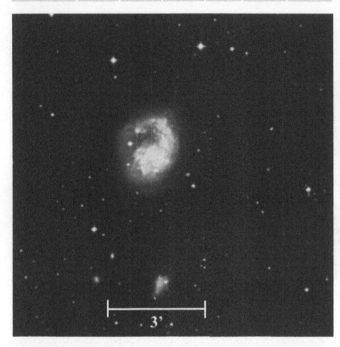

General description

NGC 4027 is a small, diffuse, and dim galaxy near 5th-magnitude 31 Crateris – which, today, lies in Corvus. Small-telescope users will need averted vision to make it swell to prominence.

Directions

Use Chart 27 to locate 31 Crateris, which is a little more than 4° southwest of 2.6-magnitude Gamma (γ) Corvi. From a dark sky, the star is visible to the unaided eye, but use binoculars if necessary. Center 31 Crateris in your telescope at low power, then switch to Chart 27c. NGC 4027 is only 30′ northwest of that star.

The quick view

At 23× in the 4-inch under a dark sky, NGC 4027 was not obvious at first. I had to relax my gaze then use averted vision to see it's diminutive 2′-wide circular form nestled between two 11th-magnitude stars. With averted vision, the

galaxy appears slightly elongated east–northeast to west–southwest. At 72×, the galaxy's roundish core is surrounded by a diffuse and irregular shell of dim light.

Stop. Do not move the telescope. Your next target is nearby!

4. NGC 4038 (H IV-28)

Type	Con	RA	Dec	Mag	Dim	Rating
Irregular barred spiral galaxy	Corvus	12ʰ 01.9ᵐ	−18° 52′	10.5	11.2′ × 5.9′	3.5

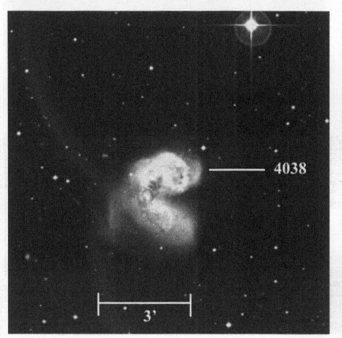

General description

NGC 4038 is the brightest member of the famous Ringtail Galaxy pair. It lies about $\frac{3}{4}$° north and slightly east of 31 Crateris and is just at the verge of visibility in 7×50 binoculars from a dark sky – but that is with extreme effort. While it is somewhat bright and obvious from a dark sky, seeing the galaxy may be a challenge for small-telescope users who have to deal with a bit of light pollution low in the sky.

Directions

Use Chart 27c. NGC 4038 is about 50′ north and slightly east of 31 Crateris. You can star-hop to it if the galaxy is not immediately obvious at low power. From 31 Crateris, move 20′ north–northeast to a pair of roughly 9th-magnitude stars (*a*), which are 5′ apart and oriented north–northeast to south–southwest. Now look for 9th-magnitude Star *b* a little more than 30′ due north of Pair *a*. NGC 4038 is about 5′ south and slightly east of Star *b*.

The quick view

At 23× in the 4-inch, NGC 4038 looks elongated, with a dark and tenuous wedge to the west. That 1′-wide wedge is the minute gulf that separates NGC 4038 from its interacting companion NGC 4039 to the south. Together, the two galaxies look like an apostrophe or the lower Greek letter nu (ν). At 72×, NGC 4038, your target, looks like a clumpy wad of wet tissue.

Stop. Do not move the telescope.

5. NGC 3962 (H I-67)

Type	Con	RA	Dec	Mag	Dim	Rating
Elliptical galaxy	Crater	11^h 54.7^m	−13° 58′	10.7	2.6′× 2.2′	2.5

3′

General description

NGC 3962 is a very small and somewhat dim galaxy a little more than 3° north of 5th-magnitude Eta (η) Crateris. Finding it requires a decent star-hop, so take your time and be patient; your target will be but a tiny fleck of light when seen at low power, so you must be certain of your field.

Directions

Use Chart 27 to locate Eta Crateris, which is 3° north–northwest of 31 Crateris. Use binoculars to confirm this star, if necessary. Center Eta Crateris in your telescope at low power then switch to Chart 27d. From Eta Crateris, move 50′ northwest to 7.5-magnitude Star a. Now move about 30′ north and slightly east to similarly bright Star b. Another 50′ sweep to the northwest will bring you to 6.5-magnitude Star c. Next, move 30′ north–northeast to 7.5-magnitude Star d. A little more than 40′ to the northeast is 8th-magnitude Star e. Now make a 30′ hop further to the northeast to 9th-magnitude Star f. NGC 3962 is about 30′ north–northeast of Star f, or about 35′ west–southwest of 7th-magnitude Star g.

The quick view

At 23× in the 4-inch, NGC 3962 is extremely close to two 11th-magnitude stars, so the area appears as a tight and concentrated 4′-wide glow that should grab your attention. Once you see the glow, use averted vision; the galaxy will appear as a tiny dab of light 1′ across, looking much like a dim fuzzy star. At 72×, the galaxy's core is moderately concentrated and round. A slightly elliptical outer halo, oriented north to south, stands out with averted vision. The view does not change much with larger telescopes.

6. NGC 4594 (H I-43) = M104

Type	Con	RA	Dec	Mag	Dim	Rating
Spiral galaxy	Virgo	12^h 40.0^m	−11° 37′	8.0	7.1′× 4.4′	4

3′

General description

NGC 4594 is a William Herschel discovery that was renamed M104 after Canadian astronomer Helen Sawyer Hogg unearthed in 1947 a letter from Pierre Mechain to J. Bernoulli at Berlin in about 1783. In that letter, Mechain mentions the discovery of four nebulae, the first of which is NGC 4594. More popularly known as the Sombrero galaxy, this giant nearly edge-on spiral system is about 5½° northeast of Delta (δ) Corvi, just across the border in Virgo. Under a dark sky, it can be picked up in a sweep with 7×50 binoculars and is a marvel in telescopes of all sizes.

Directions

Use Chart 27 to locate Delta Corvi, the northeast corner of the Sail. Note that Eta (η) Corvi is only 30′ to the northeast. Now use your unaided eyes or binoculars to look 3° north and slightly east of Delta and Eta Corvi for a 2°-long inverted Y-shaped asterism comprising four 5th- and 6th-magnitude stars. Center the Y's northeasternmost star (*a*) in your telescope at low power; there is no mistaking Star *a*, because, through a telescope, this single "star" becomes three 7th-magnitude suns forming a 7′-wide arc oriented north–northwest to south–southeast. Now switch to Chart 27e. NGC 4594 is a little more than 1° northeast of Arc *a*, about 25′ east–southeast of a roughly 4′-wide sideways Y asterism (*b*) of 9th-magnitude suns.

The quick view

At 23× in the 4-inch, NGC 4594 is a well concentrated oval glow, 4′ wide and oriented east to west. At 72×, the galaxy has a brilliant core, which seems to illuminate the surrounding oval from within, like a distant bonfire seen through a thick fog. A sharp line running just south of the galaxy's major axis marks the location of the galaxy's prominent dust lane, which is easily seen in larger telescopes.

7. NGC 5054 (H II-513)

Type	Con	RA	Dec	Mag	Dim	Rating
Spiral galaxy	Virgo	13ʰ 17.0ᵐ	−16° 38′	10.9	4.8′ × 2.8′	2.5

3′

General description

NGC 5054 is a small and faint galaxy a little less than 6° southwest of 1st-magnitude Alpha (α) Virginis (Spica), near 5th-magnitude 53 Virginis. The object is not well concentrated, so it may be difficult to see in small telescopes from suburban locations, which may only allow you to see the galaxy's core.

Field note
NGC 5044, a galaxy equally as bright as NGC 5054, is within 30′ of your target! So be sure to take the time to identify the correct object.

Directions

Use Chart 27 to locate Alpha Virginis. Now use your unaided eyes or binoculars to identify 53 Virginis 6° to the southwest. Center 53 Virginis in your telescope at low power then switch to Chart 27f. From 53 Virginis, move 15′ east–southeast to a 5′-wide triangle (*a*) of 10th- and 11th-magnitude stars. Now move 30′ west to a pair of 8th- and 9th-magnitude stars (*b*). NGC 5044 (not your target) is about 7′ south of the pair's midpoint. NGC 5054 (your target) is a little less than 30′ southeast of the easternmost star in Pair *b*.

The quick view

At 23× in the 4-inch, NGC 5054 is a diffuse and dim glow, about 2′ wide, which, with averted vision, has a slight bit of central condensation. At 72×, the object is irregularly round with a very slight brightening toward the center. Averted vision and some concentration brings out some mottling, but overall the galaxy is a moderately diffuse glow. Larger telescopes may show a roughly 13.5-magnitude star about 1.5′ east–northeast of the galaxy's core.

Stop. Do not move the telescope.

8. NGC 4856 (H I-68)

Type	Con	RA	Dec	Mag	Dim	Rating
Barred spiral galaxy	Virgo	12ʰ 59.3ᵐ	−15° 02′	10.5	3.1′ × 0.9′	3

3′

General description

NGC 4856 is a tiny but somewhat obvious galaxy about $3\frac{1}{4}°$ northwest of 53 Virginis, near a 6.5-magnitude star. Although small, the galaxy is nicely concentrated and elongated, making it relatively easy to spot, especially with averted vision.

Directions

Use Chart 27 to locate 6.5-magnitude Star b in Corvus, which is about $3\frac{1}{4}°$ northwest of 53 Virginis. The star is easy to identify in binoculars because it marks the southwest tip of a 35'-wide and squat pyramid of four 6.5- to 8th-magnitude stars. Once you identify this pyramid, switch to Chart 27g. Center Star c (the northeasternmost star in the pyramid) in your telescope at low power. NGC 4856 is only 20' east–southeast of Star c, about 7' northwest of 9.5-magnitude Star d.

The quick view

At 23×, NGC 4856 is a small, condensed, elongated glow, about 4' in extent and oriented northeast to southwest. It is easily spied with averted vision; it can also be seen with a direct gaze. At 72×, the galaxy has a starlike nucleus in a small (1') core of light that gradually tapers into a 4'-long lens. Very nice for such a small object.

Star charts for second night

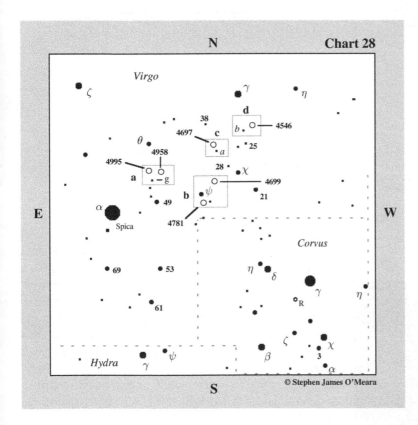

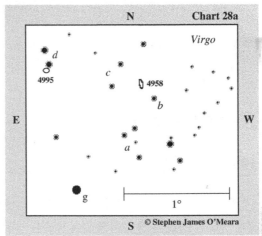

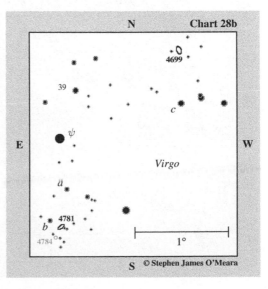

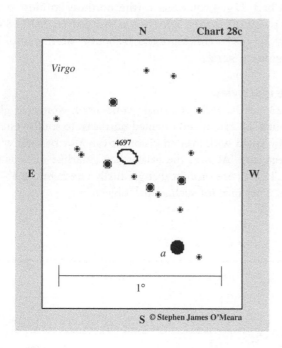

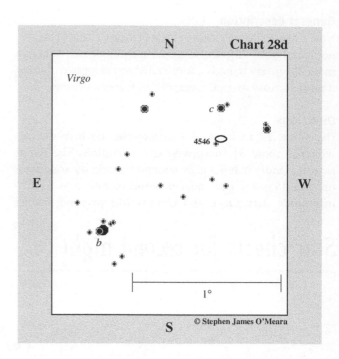

SECOND NIGHT

1. NGC 4958 (H I-130)

Type	Con	RA	Dec	Mag	Dim	Rating
Lenticular galaxy	Virgo	13^h 05.8^m	−08° 01′	10.7	3.6′ × 1.4′	3

General description

NGC 4958 is a very small but reasonably bright galaxy about 6° northwest of 1st-magnitude Alpha (α) Virginis (Spica).

Directions

Use Chart 28 to locate Alpha Virginis (Spica), then, using your unaided eyes or binoculars, find 5th-magnitude 49 Virginis 4° to the west–northwest. Note that 49 Virginis is the southernmost star in a $1\frac{3}{4}$° arc of 5th- and 6th-magnitude stars, oriented north to south, that ends with 5th-magnitude g Virginis. Once you confirm this arc, center g Virginis in your telescope at low power, then Switch to Chart 28a. From g Virginis, move 45′ northwest to a 15′-wide Y-shaped asterism (a) comprising four 9.5- to 11th-magnitude suns. Now move just 25′ further to the northwest, where you'll find 9th-magnitude Star b. NGC 4958 is only 10′ northeast of Star b.

The quick view

At 23× in the 4-inch, NGC 4958 is a small but well-concentrated circular glow, 2′ wide. At 72×, the galaxy remains round, but the core increases dramatically to a starlike center. Larger telescopes will show that core to be lens-shaped (oriented north–northeast to south–southwest). Under dark skies, the lens-shaped core is seated in a dim, tapered halo of feeble light that extends to nearly 4′.

Stop. Do not move the telescope. Your next target is nearby!

2. NGC 4995 (H I-42)

Type	Con	RA	Dec	Mag	Dim	Rating
Mixed spiral galaxy	Virgo	13^h 09.7^m	−07° 50′	11.1	2.5′ × 1.8′	2

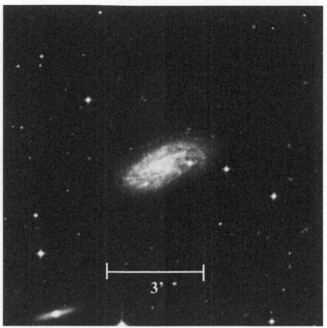

General description

NGC 4995 is a small and dim spiral galaxy 1° east–northeast of NGC 4958. It is extremely close to an 8th-magnitude star, which will help you identify its location.

Directions

Using Chart 28a, from NGC 4958, move 15′ east–northeast to a pair of 10th-magnitude stars (c). Now move 45′ further east–northeast to a pair of 8th-magnitude stars (d), which are oriented northeast to southwest. NGC 4995 is 3′ south–southeast of the southernmost star in Pair d.

The quick view

At 23×, NGC 4995 is a very dim object that looks like an asymmetrical foggy star. At 72×, the galaxy is more defined, showing a faint, round core, about 1′ wide, in an elongated haze that's 2′ wide and oriented east to west. The generally hazy disk has some irregularities that can be glimpsed with averted vision.

Stop. Do not move the telescope.

General description

NGC 4781 is a small and dim galaxy near 4.8-magnitude Psi (ψ) Virginis. Small-telescope users will need to rely on averted vision and have some patience. It will be best seen with moderate magnifications.

Directions

Use Chart 28 to locate Psi Virginis, which is about 3$\frac{1}{2}$° west–northwest of 49 Virginis. Now switch to Chart 28b. Note that NGC 4781 is 1° due south of Psi Virginis. You can make a short star-hop to it, though. From Psi Virginis move about 30′ south and slightly west to 9th-magnitude Star a. Now make a roughly 20′ hop southeast to 8.5-magnitude Star b. NGC 4781 is about 10′ southwest of Star b.

The quick view

At 23×, NGC 4781 is a moderately faint glow of low surface brightness; it is best seen with averted vision. At 72×, the galaxy is difficult to fathom because two stars lie at its western flank. The galaxy's core is very round and the surrounding lens only slightly elongated west–southwest to east–southeast. Larger scopes should be able to resolve the 12.5-magnitude star immediately west of the galaxy's nucleus (see the photograph above).

Stop. Do not move the telescope. Your next target is nearby!

3. NGC 4781 (II I-134)

Type	Con	RA	Dec	Mag	Dim	Rating
Barred spiral galaxy	Virgo	12^h 54.4^m	−10° 32′	11.1	2.9′ × 1.3′	2

4. NGC 4699 (H I-129)

Type	Con	RA	Dec	Mag	Dim	Rating
Mixed spiral galaxy	Virgo	12^h 49.0^m	−08° 40′	9.5	3.1′ × 2.5′	4

General description

NGC 4699 is a small but very bright spiral galaxy about $1\frac{1}{2}^{\circ}$ northwest of Psi Virginis. Shining at magnitude 9.5, NGC 4699 equals M90 in brightness but it is about three times smaller. Under a dark sky, the galaxy can be seen in 7×50 binoculars as a star with a tiny halo. It looks much the same way in small telescopes, but only a tad brighter.

Directions

Using Chart 28b, from NGC 4781, return to Psi Virginis. Now look for 8th-magnitude 39 Virginis 30′ to the north–northwest. NGC 4699 is about $1\frac{1}{4}^{\circ}$ northwest of 39 Virginis and 30′ north of 7.5-magnitude Star *c*.

The quick view

At 23× in the 4-inch, NGC 4699 shows a very intense and starlike core in a small but intense lens about 1′ in extent. This inner lens lies in a larger elliptical envelope that is 3′ in extent and oriented northeast to southwest. At 72×, the galaxy is highly mottled. The intense core is definitely elongated, indicative of a tiny bar that runs along the major axis of the central ellipse. The southern flank is slightly brighter than the northwestern one and looks curdled. I call NGC 4699 the Vinyl LP Galaxy because it looks like an old 78 LP record held at a slight angle under the moonlight.

Stop. Do not move the telescope.

General description

NGC 4697 is a bright and large elliptical galaxy a little more than 3° northeast of 4.7-magnitude Chi (χ) Virginis. The galaxy is well concentrated and can be seen in 7×50 binoculars. It is easy to see in telescopes of all sizes.

Directions

Use Chart 28 to locate Chi Virginis, which is 4° northwest of Psi (ψ) Virginis. Now use binoculars to identify 7th-magnitude 28 Virginis just 30′ to the northeast. Next use your binoculars to locate 7th-magnitude Star *a* $1\frac{3}{4}^{\circ}$ further to the northeast; it is the brightest star in that region. Center Star *a* in your telescope at low power, then switch to Chart 28c. NGC 4697 is a little more than 30′ northeast of Star *a*.

The quick view

At 23×, NGC 4697 has a dense, yet fuzzy, core of light surrounded by a fainter circular glow about 3′ in diameter. Averted vision reveals the galaxy's slightly oval shape, which is oriented northeast to southwest. At 72×, the galaxy's core is very concentrated and prominent. The core is not stellar, but appears as an intense orb of light that, with averted vision, morphs into a lens with tapered edges.

Stop. Do not move the telescope.

5. NGC 4697 (H I-39)

Type	Con	RA	Dec	Mag	Dim	Rating
Elliptical galaxy	Virgo	12ʰ 48.6ᵐ	−05° 48′	9.0	7.1′× 5.4′	4

6. NGC 4546 (H I-160)

Type	Con	RA	Dec	Mag	Dim	Rating
Barred spiral galaxy	Virgo	12ʰ 35.5ᵐ	−03° 48′	10.3	3.2′× 1.4′	3.5

General description

NGC 4546 is a very small but reasonably bright barred spiral galaxy about $2\frac{1}{2}°$ southwest of Gamma (γ) Virginis, and 2°

northwest of 6th-magnitude 25 Virginis. You will need a careful star-hop to this galaxy so be patient.

Directions

Use Chart 28 and binoculars to locate 25 Virginis, which is about $2\frac{1}{4}°$ north–northwest of Chi Virginis. You want to find 7th-magnitude Star b, which is $1\frac{1}{2}°$ to the north–northeast, and center it in your telescope. Now switch to Chart 28d. There is little mistaking Star b, since it is it is the brightest star in a tight grouping of a half dozen stars. NGC 4546 is 1° northwest of Star b, and 15′ due south of 8th-magnitude Star c.

The quick view

At 23×, NGC 4546 is a tiny fuzzy star just northwest of a 10th-magnitude star. Together the two comprise a fuzzy pair – one that looks like the famous double star M40 in Ursa Major. But averted vision will show that the fuzziness here is due to the galaxy. At 72×, the galaxy is extremely small, about 1′ in extent, and slightly oval. With averted vision, a dense central condensation can be see. Larger scopes will show the galaxy expand into a lens 3′ wide, oriented east–northeast to west–southwest.

Star charts for third night

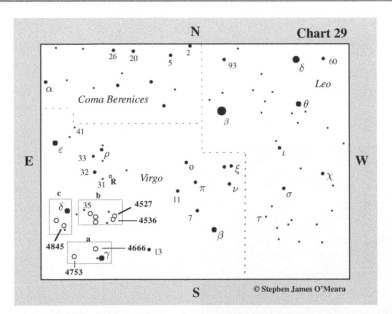

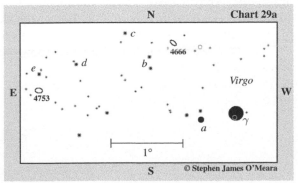

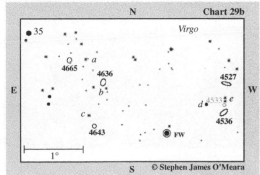

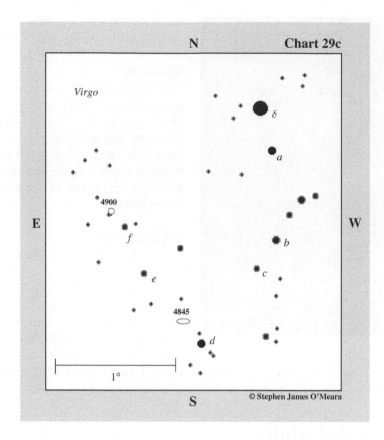

N   Chart 29c

Virgo

4900

E   W

f

e

4845

d

1°

S

© Stephen James O'Meara

THIRD NIGHT

1. NGC 4666 (H I-15)

Type	Con	RA	Dec	Mag	Dim	Rating
Spiral galaxy	Virgo	12ʰ 45.1ᵐ	−00° 28′	10.7	4.1′ × 1.3′	3.5

4′

General description

NGC 4666 is a small but moderately bright elongated galaxy less than $1\frac{1}{2}°$ from Gamma (γ) Virginis. Small-telescope users may need averted vision to spy it at low power. It is best to find the surrounding star field, use moderate power, then use averted vision.

Directions

Use Chart 29 to locate Gamma Virginis. Center the star in your telescope at low power, then switch to Chart 29a. From Gamma Virginis, move 30′ east–southeast to 6th-magnitude Star *a*. Now make a gentle 1° sweep to the northeast, to a pair of 8th-magnitude stars (*b*). NGC 4666 is a little less than 30′ northwest of Pair *b*.

The quick view

At 23×, NGC 4666 is a small but very nice sight in a dark sky. It is a 4′-long needle of light, oriented northeast to southwest, that is best seen with averted vision. At 72×, the galaxy has a bright elongated core, perhaps about 1′ in extent. With averted vision, the northeast sector seems a little brighter than the southwest sector. Larger telescopes may show the lens with a mottled texture.

Stop. Do not move the telescope. Your next target is nearby!

2. NGC 4753 (H I-16)

Type	Con	RA	Dec	Mag	Dim	Rating
Peculiar galaxy	Virgo	12^h 52.4^m	−01° 12′	9.9	4.1′ × 2.3′	4

4′

General description
NGC 4753 is a reasonably bright and obvious galaxy 2$\frac{3}{4}$° east–northeast of Gamma (γ) Virginis. Its core is well concentrated, making it a good target to hunt from suburban locations.

Directions
Using Chart 29a, from NGC 4666, return to Pair b. Now move 30′ northeast to 8th-magnitude Star c. Now drop 45′ southeast to 8.5-magnitude Star d, which has a 10.5-magnitude companion about 4′ to the northeast. A short hop 25′ further to the southeast will bring you to 7.5-magnitude Star e. NGC 4753 is only about 15′ south–southeast of Star e.

The quick view
At 23× in the 4-inch, NGC 4753 is a bright, multilayered oval glow about 3′ wide. The galaxy gets gradually, then suddenly, much brighter in the middle. At 72×, NGC 4753 displays a tiny diffuse nuclear region immediately surrounded by a bright and mottled inner lens. A dim halo of light, oriented east to west, surrounds these features.

3. NGC 4665 (H I-142)

Type	Con	RA	Dec	Mag	Dim	Rating
Barred spiral galaxy	Virgo	12^h 45.1^m	+03° 03′	10.5	4.1′ × 4.1′	3.5

4′

General description
NGC 4665 is a very small and round galaxy about 2$\frac{3}{4}$° east–southeast of 3rd-magnitude Delta (δ) Virginis and a mere 1.5′ northeast of an 11th-magnitude star. Think "small" and "fuzzy pair" (like M40), in your search.

Directions
Use Chart 29 to locate Delta Virginis, which is about 6° northeast of Gamma Virginis. Once you locate Delta Virginis, use binoculars to find 6.5-magnitude 35 Virginis, which is a little less than 2° to the east and slightly north. Center 35 Virginis in your telescope at low power, then switch to Chart 29b. NGC 4665 is about 50′ southwest of 35 Virginis, about 18′ east of 9th-magnitude Star a and 1.5′ northeast of an 11th-magnitude star.

The quick view
At 23×, NGC 4665 is a well condensed, 2′-wide circular glow 1.5′ northeast of an 11th-magnitude star. The pair is conspicuous because, seen together, they look like a larger extended object oriented northeast to southwest. At 72×, the galaxy is clearly separated from the star and appears as a very bright patch with an intense core that intensifies inward to a starlike center. Larger telescopes may reveal the dim bar that extends north and south from the round core.

Stop. Do not move the telescope. Your next target is nearby!

4. NGC 4636 (H II-38)

Type	Con	RA	Dec	Mag	Dim	Rating
Elliptical galaxy	Virgo	12^h 42.8^m	+02° 41′	9.5	7.1′× 5.2′	4

General description

NGC 4636 is a bright and conspicuous elliptical galaxy 1$\frac{1}{2}$° southwest of Delta Virginis and 40′ southwest of NGC 4665. It is a nice object for small-telescope users.

Directions

Use Chart 29b. NGC 4636 is only 40′ southwest of NGC 4665 and about 5′ northwest of a pair of 8th-magnitude stars (b).

The quick view

At 23×, NGC 4636 is a bright oval glow about 4′ in extent. It is visible with direct vision and swells slightly with averted vision. At 72×, the galaxy is a fine oval – eye-shaped – with a uniformly bright outer halo and a condensed circular core.

  Stop. Do not move the telescope. Your next target is nearby!

General description

NGC 4643 is a very small barred spiral galaxy about 3$\frac{1}{2}$° southwest of Delta Virginis and a little more than 40′ south–southeast of NGC 4636. Think, "tiny, fuzzy star."

Directions

Using Chart 29b, from NGC 4636, move 5′ southeast to Pair b. Now hop a little less than 30′ to 8.5-magnitude Star c. NGC 4643 is about 12′ south–southwest of Star c.

The quick view

At 23×, NGC 4643 is a small (2′-wide) ellipse about 3′ northwest of a roughly 12th-magnitude star. At 72×, the galaxy is clearly separated from its stellar "companion" and appears as a relatively circular 1′ glow with a bright starlike core.

Field note

NGC 4643 may only appear elongated at low power owing to the proximity of the 12th-magnitude star whose light combines with the galaxy's, creating an illusion of length.

  Stop. Do not move the telescope. Your next target is nearby!

5. NGC 4643 (H I-10)

Type	Con	RA	Dec	Mag	Dim	Rating
Barred spiral galaxy	Virgo	12^h 43.3^m	+01° 59′	10.8	3.0′× 3.0′	3

6. NGC 4536 (H V-2)

Type	Con	RA	Dec	Mag	Dim	Rating
Mixed spiral galaxy	Virgo	12^h 34.5^m	+02° 11′	10.6	6.4′× 2.6′	3

General description

NGC 4536 is a moderately large though somewhat difficult spiral galaxy about 1° northwest of 6th-magnitude FW Virginis. Light pollution will greatly affect its visibility in small telescopes.

Directions

Use Chart 29b. From NGC 4643, make a slow and careful sweep about $1\frac{1}{4}°$ west and slightly south to 6th-magnitude FW Virginis; the star varies by only 0.1 magnitude so you do not have to worry about it "disappearing" from view. FW Virginis has an 8.5-magnitude companion about 8′ to the north–northwest. From FW Virginis, make a slow and careful sweep about 50′ northwest to 7.5-magnitude Star *d*. NGC 4536 is a little more than 10′ southwest of Star *d*.

The quick view

At 23×, NGC 4536 is a 4′-long ellipse, oriented northwest to southeast, with little or no concentration. At 72×, the galaxy is more difficult to see because its diffuse light spreads out. Still, with averted vision, a starlike core can be seen inside a small oval lens that gradually fades outward into a broad elliptical halo.

Stop. Do not move the telescope. Your next target is nearby!

General description

NGC 4527 is a moderately bright spiral galaxy 30′ north and a little west of NGC 4536. It is slightly more difficult to see in small telescopes than NGC 4536. Use averted vision. It is best seen at low power. Light pollution will greatly affect its visibility.

Directions

Use Chart 29b. NGC 4527 is only 30′ north and slightly west of NGC 4536 and just 12′ north–north-west of a pair of 9th-magnitude stars (*e*), oriented north to south.

The quick view

At 23×, NGC 4527 is a moderately bright and slender ellipse 3′ in length, oriented northeast to southwest, that swells a bit with averted vision. The galaxy has a translucent quality to it. At 72×, NGC 4527 becomes a difficult object to see, in that it almost disappears, forcing you to use averted vision. With averted vision and some concentration a small nuclear region can be seen in a tiny central lens.

7. NGC 4527 (H II-37)

Type	Con	RA	Dec	Mag	Dim	Rating
Mixed spiral galaxy	Virgo	12^{h} 34.1^{m}	+02° 39′	10.5	6.0′× 2.1′	2.5

8. NGC 4845 (H II-536)

Type	Con	RA	Dec	Mag	Dim	Rating
Spiral galaxy	Virgo	12^{h} 58.0^{m}	+01° 35′	11.2	4.8′× 1.2′	1.5

General description

NGC 4845 is a very small and dim lens-shaped galaxy 2° south–southeast of Delta (δ) Virginis. In small telescopes it may not be visible at low power. It is best viewed with averted vision and moderate magnifications. Be prepared to make a careful star-hop and to spend some time in confirming this object.

Directions

Use Chart 29 to find Delta Virginis. Center that star in your telescope at low power, then switch to Chart 29c. From Delta Virginis, move about 25' south–southwest to 7th-magnitude Star a. Now move about 45' south and slightly west to 8th-magnitude Star b. A shorter 20' hop to the southeast will bring you to 9th-magnitude Star c. Next make a 45' sweep further to the southeast to 7th-magnitude star d. NGC 4845 is only 10' northeast of Star d.

The quick view

NGC 4845 was not visible at 23× in my 4-inch under a very dark sky. At 72×, the galaxy could just be seen as a 1'-long needle of light trapped between two roughly 12th-magnitude stars oriented north–northeast to south–southwest. You might want to try increasing the power to at least 100×, which will help separate the galaxy from the nearby stars. With averted vision, a tiny lens with a dim central concentration could be suspected.

Stop. Do not move the telescope. Your next target is nearby!

9. NGC 4900 (H I-143)

Type	Con	RA	Dec	Mag	Dim	Rating
Barred spiral galaxy	Virgo	13ʰ 00.6ᵐ	+02° 30'	11.4	2.3' × 2.3'	2.5

General description

NGC 4900 is a very small and dim galaxy $1\frac{1}{2}°$ southeast of Delta Virginis. You will need to make a slow and careful star-hop to it from NGC 4845, so be patient.

Directions

Using Chart 29c, from NGC 4845 move 30' northeast to 9th-magnitude Star e. A roughly 25' hop further to the northeast will bring you to similarly bright Star f. NGC 4900 is about 10' northeast of Star f and just about 1' northwest of a roughly 11th-magnitude star.

The quick view

At 23×, NGC 4900 is simply a tiny 2'-wide diffuse glow next to an 11th-magnitude star. At 72×, the galaxy is an intriguing sight, since the 11th-magnitude star sits on its southeast flank, mimicking a supernova. Also, with averted vision, the disk appears to form a ring around the nucleus. Otherwise, it is just a dim haze, like an asymmetrical fog around a star.

Star charts for fourth night

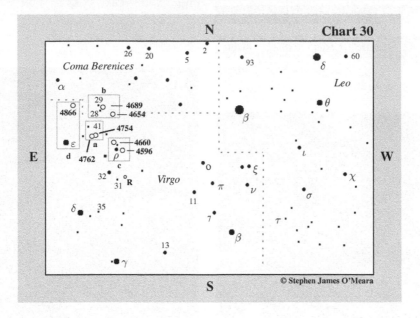

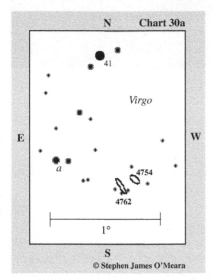

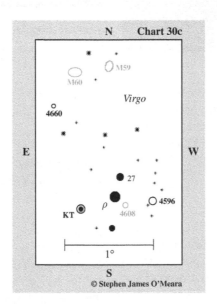

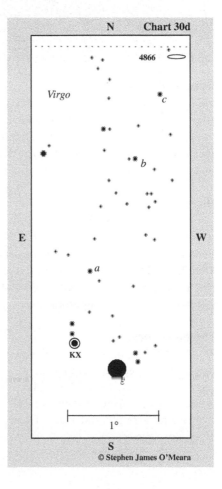

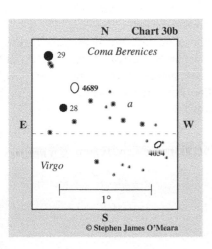

FOURTH NIGHT

1. NGC 4762 (H II-75)

Type	Con	RA	Dec	Mag	Dim	Rating
Barred spiral galaxy	Virgo	12^h 52.9^m	+11° 14′	10.3	9.1′ × 2.2′	3.5

General description

NGC 4762 is a small, stunningly thin, yet reasonably bright, edge-on galaxy about 2° west and slightly north of 3rd-magnitude Epsilon (ε) Virginis. Together with NGC 4754 (see below), it forms a beautiful pair that can be appreciated in a small telescope, especially from a dark sky.

Directions

Use Chart 30 to locate Epsilon Virginis. Now use binoculars to find 6.5-magnitude 41 Virginis 2½° to the northwest. Center this star in your telescope, then switch to Chart 30a. From 41 Virginis, make a careful sweep 1° southeast to 7th-magnitude Star *a*. NGC 4762 is 40′ southwest of Star *a*, between two 9.5-magnitude stars.

The quick view

At 23×, NGC 4762 is a sharp needle of light, 4′ long and oriented northeast to southwest. At this power, it is best seen with averted vision. The galaxy becomes much more obvious at 72×. Averted vision reveals a mottled or beaded core inside a bright and tapered lens that gradually fades away from the center.

Stop. Do not move the telescope. Your next target is nearby!

2. NGC 4754 (H I-25)

Type	Con	RA	Dec	Mag	Dim	Rating
Barred spiral galaxy	Virgo	12^h 52.3^m	+11° 19′	10.6	4.6′ × 2.6′	3.5

General description

NGC 4754 is a small but condensed galaxy near NGC 4762.

Directions

Use Chart 30a. NGC 4754 is only 10′ northwest of NGC 4762.

The quick view

At 23×, NGC 4754 is is a small galaxy northwest of two 11th- and 12th-magnitude stars. It has a bright circular core (about 2′ across), which can be seen with direct vision, though its overall form is best seen with averted vision. At 72×, NGC 4754 remains bright and round. The glow becomes gradually, then suddenly, much brighter in the middle, though not to a starlike nucleus. Larger telescopes will show its dim and extended (4.6′-long) outer envelope, which is also oriented northeast to southwest.

Stop. Do not move the telescope.

3. NGC 4689 (H II-128)

Type	Con	RA	Dec	Mag	Dim	Rating
Spiral galaxy	Coma Berenices	12^h 47.8^m	+13° 46′	10.9	3.7′ × 3.2′	3

General description

NGC 4689 is a small and somewhat dim galaxy about 2° northwest of 41 Virginis, near the 6th-magnitude stars 28 and 29 Comae Berenices. Light pollution will greatly affect the visibility of this object in small telescopes.

Directions

From NGC 4754, return to 41 Virginis, then use Chart 30 to locate 28 and 29 Comae Berenices about 2° to the northwest; this may require binoculars. You want to center 28 Comae Berenices in your telescope at low power, then switch to Chart 30b. NGC 4689 is only about 15′ northwest of 28 Comae Berenices.

The quick view

At 23×, NGC 4689 is a moderately small circular glow, about 3′ wide, that is best seen with averted vision. With concentration, a small central brightening can be detected. At 72×, the galaxy is an amorphous round glow with a bit of mottling toward the center. With averted vision the circular disk becomes irregularly round.

Stop. Do not move the telescope. Your next target is nearby!

General description

NGC 4654 is a small spiral galaxy a little more than 1° southwest of 28 Comae Berenices and NGC 4689, just over the border in Virgo. Averted vision helps greatly in seeing its elongated form.

Directions

Use Chart 30b. From NGC 4689, follow the 40′-long and lazy Z-shaped asterism of four 8.5-magnitude stars (a) to the southwest. NGC 4654 is less than 20′ southwest of the southwestern-most star in Asterism a and less than 3′ east-southeast of a 10th-magnitude star.

The quick view

At 23×, NGC 4654 is a small 2′-wide circular glow that transforms into a 3′-wide ellipse with averted vision. The galaxy is oriented northwest to southeast and is flanked to the north by two roughly 12th-magnitude stars. At 72×, the galaxy looks nicely elongated. The disk is uniformly bright until it gradually and then suddenly brightens toward the middle to a starlike nucleus. Larger telescopes may show the inner lens mottled.

4. NGC 4654 (H II-126)

Type	Con	RA	Dec	Mag	Dim	Rating
Mixed spiral galaxy	Virgo	12ʰ 44.0ᵐ	+13° 08′	10.5	4.9′ × 2.7′	3

5. NGC 4596 (H I-24)

Type	Con	RA	Dec	Mag	Dim	Rating
Barred spiral galaxy	Virgo	12ʰ 39.9ᵐ	+10° 11′	10.4	4.6′ × 4.1′	3.5

General description

NGC 4596 is a small but concentrated barred-spiral galaxy near 5th-magnitude Rho (ρ) Virginis.

Directions

Use Chart 30 to locate Rho Virginis, which is about 5° west–southwest of Epsilon Virginis. (Use binoculars, if necessary). Rho Virginis is easy to identify because it is the brightest star at the center of a roughly 30′-wide upside-down Y-shaped asterism with three 6.5- to 7th-magnitude stars. Center Rho Virginis in your telescope at low power, then switch to Chart 30c. NGC 4596 is only 30′ west of Rho Virginis. Do not confuse it with dimmer NGC 4608, which is only 10′ southwest of Rho Virginis.

The quick view

At 23×, NGC 4596 is little more than a small round glow, 2′ wide and fairly uniform. It is immediately north–northwest of an 11th-magnitude star. At 72×, the galaxy displays a small 1′-wide core of bright light surrounded by a circular dimmer halo of uniform light. With averted vision, that halo appears to taper to the east–northeast and south–southwest; here are the galaxy's bright bars. Larger telescopes may show the galaxy's larger circular disk extend almost 5′ across.

Stop. Do not move the telescope. Your next target is nearby!

General description

NGC 4660 is a very small, almost stellar galaxy about $1\frac{1}{4}°$ northeast of Rho Virginis and about 25′ southeast of the bright Messier galaxy, M60. Think, "tiny fuzzy star."

Directions

Use Chart 30c. From NGC 4596, return to Rho Virginis, then hop a little more than 10′ northwest to 6.5-magnitude 27 Virginis. You'll find M60 a little more than 1° to the north–northeast. (Note that equally conspicuous M59 is about 25′ northwest of M60.) From M60, make a very slow and careful move about 25′ southeast to NGC 4660.

The quick view

At 23×, NGC 4660 is a tiny "fuzzy star" appearing about as bright as a star of 12th magnitude. It is literally a star surrounded by a thin and faint collar of light that is best seen with averted vision. At 72×, the galaxy remains very small, about 1′ in diameter; the view is essentially of the galaxy's bright star like nucleus surrounded by a tightly wound lens, which, when seen with averted vision, is oriented, more or less, west to east.

6. NGC 4660 (H II-71)						
Type	Con	RA	Dec	Mag	Dim	Rating
Elliptical galaxy	Virgo	12^h 44.5^m	+11° 11′	11.2	2.4′ × 2.1′	2.5

7. NGC 4866 (H I-162)						
Type	Con	RA	Dec	Mag	Dim	Rating
Spiral galaxy	Virgo	12^h 59.5^m	+14° 10′	11.2	5.5′ × 1.2′	3.5

General description

NGC 4866 is a small and faint galaxy about $3\frac{1}{4}°$ north–northwest of 3rd-magnitude Epsilon (ε) Virginis. Although it is dim, the galaxy is condensed enough so that it is fairly obvious with averted vision in a small telescope

under a dark sky. Light pollution will affect the visibility of this interesting nearly edge-on spiral galaxy.

Directions

If you have a rich-field telescope, you might want to first try sweeping $2\frac{1}{2}°$ due east of 29 Virginis to pick up the galaxy. Otherwise, use Chart 30 to find Epsilon Virginis. Center that star in your telescope at low power, then switch to Chart 30d. From Epsilon Virginis, move 30′ east–northeast to the 7.5-magnitude variable star KX Virginis. Next move 45′ north–northwest to 8th-magnitude Star a. Now make a slow and careful $1\frac{1}{4}°$ sweep north–northwest to 8th-magnitude Star b, which has a 9.5-magnitude companion to the east. Next move a little more than 40′ north–northwest to 8th-magnitude Star c. NGC 4866 is a little more than 20′ further to the north–northwest.

The quick view

In the 4-inch at 23× under a dark sky, NGC 4866 is a nice 5′-long oval oriented east to west. It is best seen with averted vision and has an odd visual quality to it: if you do not look at it just right, it can vanish from view; look at it just right, and it swells lengthwise into view. At 72×, the galaxy is a neat and narrow ellipse of uniform light with a bright and obvious starlike core. The galaxy takes magnification well.

Star charts for fifth night

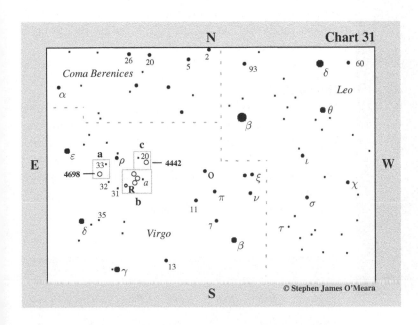

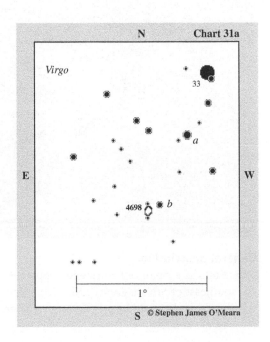

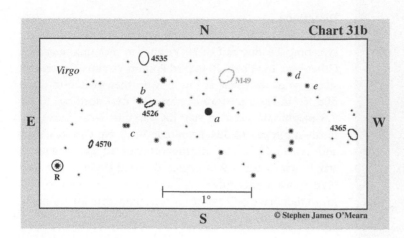

Chart 31b
N
Virgo
4535
M49
b
4526
d
e
a
c
4570
4365
R
E
W
S
1°
© Stephen James O'Meara

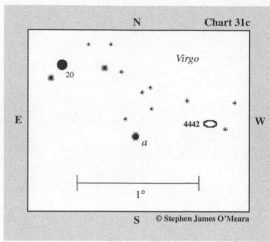

Chart 31c
N
Virgo
20
E
4442
W
a
1°
S
© Stephen James O'Meara

FIFTH NIGHT

1. NGC 4698 (H I-8)

Type	Con	RA	Dec	Mag	Dim	Rating
Barred spiral galaxy	Virgo	12^h 48.4^m	+08° 29′	10.6	3.2′× 1.7′	3.5

4′

General description

NGC 4698 is a small but somewhat apparent galaxy about 2$\frac{1}{2}$° southeast of 5th-magnitude Rho (ρ) Virginis. Its core is very bright and should be a good target for small-telescope users in the suburbs.

Directions

Use Chart 31 to locate Rho Virginis, which is about 5° west–southwest of Epsilon (ε) Virginis. (Use binoculars, if necessary.) As mentioned above, Rho Virginis is easy to identify because it is the brightest star at the center of a roughly 30′-wide upside-down Y-shaped asterism with three 6.5- to 7th-magnitude stars. You want to use your unaided eye or binoculars to locate 5.7-magnitude 33 Virginis about 1$\frac{1}{2}$° to the southeast, then switch to Chart 31a. From 33 Virginis, move 30′ south–southeast to 7th-magnitude Star a. Now hop 35′ further to the south–southeast to 8th-magnitude Star b. NGC 4698 is only 5′ east–southeast of Star b, trapped between two roughly 11th-magnitude stars.

The quick view

At 23×, NGC 4698 is a very nice elongated glow, measuring about 1.5′ in length with a starlike core. The galaxy stretches lengthwise with averted vision. At 72×, the galaxy's core is very bright and sharp. It sits inside a small cocoon of fuzzy light whose slightly tapered edges seem more pronounced than its interior.

Stop. Do not move the telescope.

2. NGC 4526 (H I-31 = H I-38)

Type	Con	RA	Dec	Mag	Dim	Rating
Mixed lenticular galaxy	Virgo	12^h 34.0^m	+07° 42′	9.9	7.4′× 2.7′	4

4'

General description

NGC 4526 is a fairly bright and obvious galaxy about 3°
southwest of Rho Virginis. It is visible under a dark sky in
7×50 binoculars with averted vision and is quite noticeable
in the smallest of telescopes. It is very well condensed,
making it a nice object for telescopes of all sizes.

Directions

Use Chart 31. Note that Rho and 33 Virginis are part of a
roughly $3\frac{1}{2}°$-wide oval of similarly bright stars. The other
stars in the oval, moving from east to west, are 5th-magnitude
32 Virginis, 6th-magnitude 31 Virginis, R Virginis (a long-
period variable star whose brightness fluctuates from
6.0-magnitude to 12.1-magnitude every 146 days), 6th-
magnitude Star a, and 6th-magnitude 20 Virginis. You
want to take the time to locate Star a with your unaided eyes
or binoculars, center it in your telescope at low power, then
switch to Chart 31b. First, make sure that the 8th-magnitude
galaxy M49 is 35' northwest of Star a. Now look a little more
than 30' east and slightly north for a roughly 18'-wide
triangle (b) of 7th- and 7.5-magnitude stars. NGC 4526 lies
midway between the two 7th-magnitude stars (separated by
15') forming the triangle's base.

The quick view

At 23× in the 4-inch, the galaxy is a very condensed fuzzy
navel, about 2' in length, nestled between two stud-like
7th-magnitude stars. With concentration, the bright core
has a starlike nucleus. The surrounding bulge at low power
is very round. A magnification of 72× enhances the central
lens, causing the dim elliptical outer regions to look like
two spheres, one on each side of the central lens. The view
is reminiscent of Galileo's first impression of Saturn with

its nearly edge-on rings in 1610. A 12.5-magnitude star
lies just a few minutes due south of the nucleus.

Stop. Do not move the telescope. Your next target is
nearby!

3.   NGC 4535 (H II-500)

Type	Con	RA	Dec	Mag	Dim	Rating
Mixed spiral galaxy	Virgo	12^h 34.3^m	+08° 12'	10.0	7.1'× 6.4'	3.5

4'

General description

NGC 4535 is a reasonably bright galaxy near NGC 4526.
Unlike NGC 4526, it is a somewhat sizable amorphous
glow, which may make it difficult to see in a small telescope
if light pollution is a problem.

Directions

Use Chart 31b. NGC 4535 is only 30' north of NGC 4526.

The quick view

At 23×, NGC 4535 is simply a large (5') ill-defined, uniform glow. At 72×, the core of the galaxy seems to condense ever so slightly toward the middle. No structure is
visible. It is like a featureless comet with no central condensation. Larger telescopes will reveal a starlike nucleus
within the broad inner core. They might also reveal an
irregular waviness to the halo and some 13th-magnitude
and fainter mottlings.

Stop. Do not move the telescope. Your next target is
nearby!

4. NGC 4570 (H I-32)

Type	Con	RA	Dec	Mag	Dim	Rating
Lenticular galaxy	Virgo	12ʰ 36.9ᵐ	+07° 15′	10.9	4.3′× 1.3′	2

General description
NGC 4570 is a very small and dim galaxy with a bright nucleus about 3° south–southwest of Rho Virginis, near NGC 4526. Small-telescope users should think, "dim, fuzzy star."

Directions
Using Chart 31b, from NGC 4535, return to NGC 4526. Now move 20′ southeast to a close pair of 8.5- and 9th-magnitude stars (c). NGC 4570 is just 30′ further to the southeast in a field devoid of bright stars. Note that it is also midway between Pair c and R Virginis.

The quick view
At 23×, NGC 4750 is a faint mishmash of dim stars and fuzz but only with averted vision. At 72×, the galaxy displays a sharp starlike nucleus in a small and bright central condensation. Together these two features appear like a fuzzy star with averted vision. With concentration, faint, fuzzy extensions can be seen to the northwest and southeast.

Stop. Do not move the telescope. Your next target is nearby!

5. NGC 4365 (H I-30)

Type	Con	RA	Dec	Mag	Dim	Rating
Elliptical galaxy	Virgo	12ʰ 24.5ᵐ	+07° 19′	9.6	5.6′× 4.6′	3.5

General description
NGC 4365 is a bright and sizable elliptical galaxy 5° south-west of Rho Virginis and about 1¾° southwest of M49. Overall the galaxy is bright and easy to see from a dark-sky site. Although the galaxy is moderately condensed, small-telescope users living under light-polluted skies may lose its diffuse outer envelope.

Directions
Using Chart 31b, from NGC 4570, return to NGC 4526. Now move back to Star a, which is about 40′ to the west and slightly south. Next hop a little more than 30′ northwest to M49. From M49, move about 30′ west to 8th-magnitude Star d. Now move 18′ southwest to similarly bright Star e. NGC 4365 is less than 50′ further to the southwest.

The quick view
At 23×, NGC 4365 is a bright, large (∼4′) and diffuse glow tucked inside the southwest corner of a 25′-wide trapezoid of 10th-magnitude stars. With averted vision there's a sizable inner core, about 2′-wide, that gets gradually, then suddenly, brighter toward the middle. At 72×, a dim but sharp nucleus can be seen in a circular inner core. The core is surrounded by an elliptical disk oriented northeast to southwest.

6. NGC 4442 (H II-156)

Type	Con	RA	Dec	Mag	Dim	Rating
Barred lenticular galaxy	Virgo	12ʰ 28.1ᵐ	+09° 48′	10.4	4.6′× 1.9′	3.5

4'

General description

NGC 4442 is a very small but condensed galaxy about $3\frac{1}{2}°$ west–southwest of Rho Virginis, near 6th-magnitude 20 Virginis. It is elongated and has a bright starlike core.

Directions

From Chart 31, using the unaided eye or binoculars, locate 6th-magnitude 20 Virginis, which is 2° west of Rho Virginis. Center 20 Virginis in your telescope at low power, then switch to Chart 31c. From 20 Virginis, move 50' southwest to 7th-magnitude Star *a*. NGC 4442 is a little less than 40' west–northwest of Star *a*.

The quick view

At 23×, NGC 4442 is a "fuzzy star," in a 2'-long ellipse of dim light, that becomes more prominent with averted vision. At 72×, the galaxy, which is oriented east to west, appears much the same as it did at 23×, only it is much more apparent. With averted vision, a tiny stellarlike core can be seen in a tiny (1') inner lens.

Field note

Memorize the location of 20 Virginis before you retire. Your next night of observations will, once again, begin with this star.

Star charts for sixth night

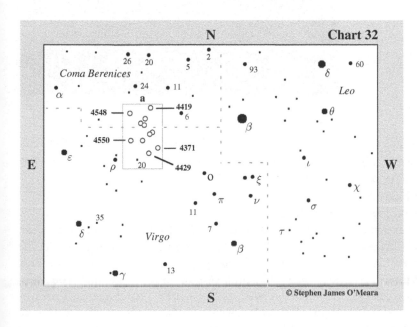

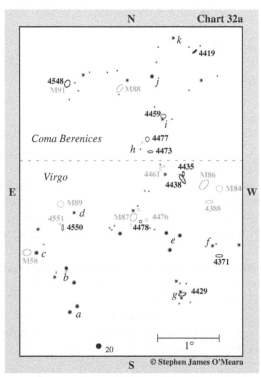

SIXTH NIGHT

1.   NGC 4550 (H I-36)

Type	Con	RA	Dec	Mag	Dim	Rating
Barred spiral galaxy	Virgo	12ʰ 35.5ᵐ	+12° 13′	11.7	3.3′× 1.0′	1.5

General description
NGC 4550 is an extremely small and dim galaxy 2° northeast of 6th-magnitude 20 Virginis, near M89. Dark skies are a must and small-telescope users should be prepared to use moderate magnification.

Directions
Use Chart 32 to locate Rho (ρ) Virginis, which is about 5° west–southwest of Epsilon (ε) Virginis. (Use binoculars, if necessary.) Rho Virginis is easy to identify because it is the brightest star at the center of a roughly 30′-wide upside-down Y-shaped asterism with three 6.5- to 7th-magnitude stars. Now use your unaided eye or binoculars to find 6th-magnitude 20 Virginis 2° to the west. Center 20 Virginis in your telescope at low power, then switch to Chart 32a. From 20 Virginis, move about 40′ northeast to a pair of 7.5-magnitude stars (a). Next move about 30′ north–northeast to a 20′-wide triangle of three 7.5- to 8th-magnitude suns (b). M58, a 9.6-magnitude galaxy, is about 45′ to the northeast, about 7′ east of 8th-magnitude Star c. Now carefully sweep 50′ northwest of Star c to 9.7-magnitude M89. NGC 4550 is only 20′ south of M89.

The quick view
NGC 4550 is not readily visible at 23× in the 4-inch under very dark skies. At 72×, the galaxy is simply a 1′-long spindle of dim light, oriented north to south. Round NGC 4551 is just as conspicuous and just as faint. Larger telescopes do not show much more than a brightening toward the center of NGC 4550.

Stop. Do not move the telescope. Your next target is nearby!

2.   NGC 4478 (H II-124)

Type	Con	RA	Dec	Mag	Dim	Rating
Elliptical galaxy	Virgo	12ʰ 30.3ᵐ	+12° 20′	11.4	1.7′× 1.4′	2.5

General description
NGC 4478 is a very small and dim galaxy near 8.6-magnitude M87. Small-telescope users should use moderate magnification to see it well.

Directions
Using Chart 32a, from NGC 4550, return to M89. Next, make a small hop 12′ southwest to 9th-magnitude Star d. Now move 1° due west to M87. NGC 4478 is a little less than 10′ southwest of M87. Note that you must differentiate NGC 4478 from its 12th-magnitude companion NGC 4476, which is about 4.5′ to the northwest. NGC 4478 is the brighter and more obvious of the two, so, if you see only one object, you have found your target.

The quick view

At 23×, NGC 4478 is very small, like a star, and very faint. It is best seen with averted vision, even under a dark sky. At 72×, the galaxy is more robust and surprisingly easy to see as a 1'-wide glow with a bright core.

Stop. Do not move the telescope. Your next target is nearby!

3. NGC 4371 (H I-22)

Type	Con	RA	Dec	Mag	Dim	Rating
Barred spiral galaxy	Virgo	12^h 24.9^m	+11° 42'	10.8	4.6'× 2.2'	2.5

General description

NGC 4371 is a moderately small and dim galaxy a little more than 1½° southwest of M87. It is best seen in small telescopes with moderate magnification.

Directions

Using Chart 32a, from M87, move about 45' southwest to a 20'-wide equilateral triangle of three 8th-magnitude stars (e). NGC 4371 is 50' west–southwest of Triangle e, about 15' south–southwest of a pair of 9th- and 10th-magnitude stars (f).

The quick view

At 23×, NGC 4371 is a small (1.5') oval glow of uniform brightness, which is best seen with averted vision. The galaxy is much more obvious at 72×, which reveals a pinpoint nucleus in a 2'-wide lens of uniform light, oriented east to west.

Stop. Do not move the telescope. Your next target is nearby!

4. NGC 4429 (H II-65)

Type	Con	RA	Dec	Mag	Dim	Rating
Spiral galaxy	Virgo	12^h 27.4^m	+11° 07'	10.0	5.6'× 2.6'	3.5

General description

NGC 4429 is a moderately small but reasonably bright galaxy 1½° south–southwest of M87 and 50' southeast of NGC 4371. It is very condensed so it makes a target for small-telescope users.

Directions

Using Chart 32a, from NGC 4371, move 50' southeast, where you will see NGC 4429 bracketed by a pair of 9th-magnitude stars (g), oriented north to south and separated by about 7'.

The quick view

At 23×, NGC 4429 is a small 2'-wide glow bracketed between two 9th-magnitude stars. The galaxy is closer to the northern star in Pair g. At 72×, the galaxy has a highly condensed and elongated core, oriented west–northwest to east–southeast. The core sits inside a moderately bright and tapered lens that measures about 3' in length. Some mottling can be seen with averted vision along the galaxy's major axis.

Stop. Do not move the telescope.

5. NGC 4438 (H I-28,2)

Type	Con	RA	Dec	Mag	Dim	Rating
Spiral galaxy	Virgo	12^h 27.8^m	+13° 01′	10.2	8.9′ × 3.6′	4

6. NGC 4435 (H I-28,1)

Type	Con	RA	Dec	Mag	Dim	Rating
Barred spiral galaxy	Virgo	12^h 27.7^m	+13° 05′	10.8	3.2′ × 2.0′	3.5

General description

NGC 4438 is a rather bright and obvious galaxy near 9th-magnitude M86. It is paired with smaller and dimmer NGC 4435 (see below) immediately to the northwest.

Directions

Using Chart 32a, from NGC 4429, move 50′ north–northeast to Triangle *e*. M86 is 1° to the northwest. Confirm that M86 is flanked to the west–southwest by equally bright M84, only about 18′ away. NGC 4438, your target, is 20′ east–northeast of M86.

The quick view

At 23×, NGC 4438 is immediately obvious as a roughly 2′-long lens of reasonably bright light. (Do not be fooled by the dimensions listed above, which include the galaxy's dim and warped extensions, as shown in the accompanying photograph.) At 72×, the galaxy's core becomes bright and condensed but not to a starlike nucleus. Also, the dim and warped extensions can be suspected. With the extensions, the galaxy appears 4′ long with averted vision.

Stop. Do not move the telescope. Your next target is nearby!

General description

NGC 4435 is a small but reasonably bright galaxy paired with NGC 4438.

Directions

Use Chart 32a. NGC 4435 is about 4′ north and slightly west of NGC 4438's core.

The quick view

At 23×, NGC 4435 becomes visible with averted vision after NGC 4438 is spotted. It is mostly round. At 72×, NGC 4435 suddenly, and surprisingly, becomes more obvious than neighboring NGC 4438. That's because the core of NGC 4435 is more condensed than NGC 4438's. With averted vision, NGC 4435 lengthens into a 1′-long lens of light.

Stop. Do not move the telescope. Your next target is nearby!

7. NGC 4473 (H II-114)

Type	Con	RA	Dec	Mag	Dim	Rating
Elliptical galaxy	Coma Berenices	12^h 29.8^m	+13° 26′	10.2	3.7′ × 2.4′	4

General description

NGC 4473 is a moderately bright and obvious galaxy 1° northeast of M86, near NGC 4438.

Directions

Use Chart 32a. NGC 4473 is 40' northeast of NGC 4438 and just 10' west–southwest of 10th-magnitude Star h. It shares the field with equally obvious NGC 4477 (see below), which lies about 12' north and slightly west.

The quick view

At 23×, NGC 4473 is a moderately conspicuous, 2'-wide, lens of uniform light. At 72×, the galaxy's core appears rounded with a tiny stellar nucleus.

Stop. Do not move the telescope. Your next target is nearby!

General description

NGC 4477 is a moderately bright and obvious galaxy that shares the field with equally bright and obvious NGC 4473. NGC 4477 is the more northerly of the pair.

Directions

Use Chart 32a. NGC 4477 lies only about 12' north and slightly west of NGC 4473.

The quick view

At 23×, NGC 4477, like NGC 4573, is moderately bright and obvious. At 72×, NGC 4477 appears larger and more textured than NGC 4473. With averted vision, the galaxy has a bright core surrounded by a mottled and slightly irregular envelope.

Stop. Do not move the telescope. Your next target is nearby!

8. NGC 4477 (H II-115)

Type	Con	RA	Dec	Mag	Dim	Rating
Barred spiral galaxy	Coma Berenices	12^h 30.0^m	+13° 38'	10.4	3.9'× 3.6'	4

9. NGC 4459 (H I-169)

Type	Con	RA	Dec	Mag	Dim	Rating
Spiral galaxy	Coma Berenices	12^h 29.0^m	+13° 59'	10.4	3.5'× 2.8'	3.5

General description

NGC 4459 is a small but concentrated glow that kisses a 9.5-magnitude star to the south–southeast. It is best seen at moderate powers in small telescopes.

Directions

Use Chart 32a. NGC 4459 is only 25′ north–northeast of NGC 4477, immediately north–northwest of 9.5-magnitude Star i.

The quick view

At 23×, NGC 4459 is a small and tight core of light that swells to prominence with averted vision. Its proximity to Star i makes it difficult to detect with direct vision. At 72×, the galaxy is just a tiny 1′-wide glow of uniform light that very suddenly brightens to a tight but diffuse center.

Stop. Do not move the telescope. Your next target is nearby!

General description

NGC 4548 (M91) was one of Messier's "missing" objects until 1969 when William C. Williams of Fort Worth, Texas, proposed it was William Herschel's H II-120 (NGC 4548). It is a moderately bright and large object near similarly bright M88.

Directions

Using Chart 32a, from NGC 4459 move 40′ north–northeast to 7.5-magnitude Star j. Now move to 9.8-magnitude M88 only 35′ to the southeast. NGC 4548 (M91) is 50′ east and slightly north of M88.

The quick view

At 23×, NGC 4548 is a very modest glow, definitely out of round – a broad spiral shape with a bright center. At 72×, the galaxy begins to display complexities, such as a central bar, knots, and hazy arcs.

Stop. Do not move the telescope. Your next target is nearby!

10.   NGC 4548 (H II-120) = M91

Type	Con	RA	Dec	Mag	Dim	Rating
Barred spiral galaxy	Coma Berenices	12ʰ 35.4ᵐ	+14° 30′	10.2	5.0′ × 4.1′	4

11.   NGC 4419 (H II-113)

Type	Con	RA	Dec	Mag	Dim	Rating
Barred spiral galaxy	Coma Berenices	12ʰ 26.9ᵐ	+15° 03′	11.2	2.8′ × 0.9′	2

General description

NGC 4419 is a small and faint barred spiral galaxy a little less than $1\frac{1}{2}°$ northwest of 9.6-magnitude M88 in Coma Berenices. Small-telescope users should be prepared to use moderate magnification to see it well.

Directions

Using Chart 32a, from NGC 4548, return to M88, then to Star j. Now hop 45′ north–northwest to 9th-magnitude Star k. NGC 4419 is a little more than 20′ southwest of Star k, about 6′ southeast of a 10th-magnitude star.

The quick view

At 23×, NGC 4419 is a small and faint elongated glow, oriented northwest to southeast; it extends about 2′ in length, as seen with averted vision. At 72×, the galaxy is more distinct. It has a sharply defined elongated core surrounded by a thin halo of diffuse light. It looks like a snapshot of a faint meteor with a fuzzy aura.

Star charts for seventh night

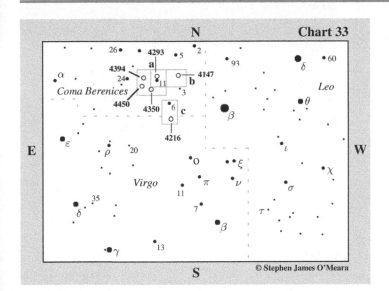

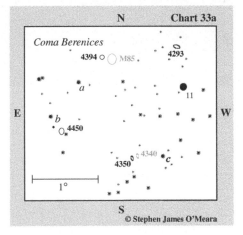

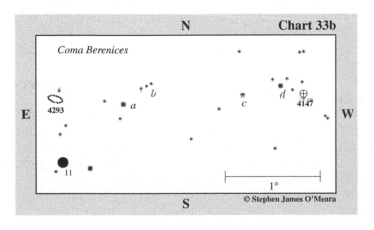

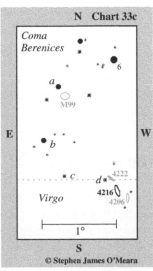

Seventh night

1.   NGC 4293 (H V-5)

Type	Con	RA	Dec	Mag	Dim	Rating
Barred spiral galaxy	Coma Berenices	12^h 21.2^m	+18° 23′	10.4	5.3′ × 3.1′	3.5

General description
NGC 4293 is a reasonably bright and elongated galaxy near 5th-magnitude 11 Comae Berenices. It is fairly condensed and should be a nice sight even in small telescopes, especially at moderate magnification.

Directions
Use Chart 33 to find 11 Comae Berenices, which is 9° northwest of Rho (ρ) Virginis. It also forms the northwest apex of a 5°-wide isosceles triangle with the similarly bright stars 24 and 6 Comae Berenices. Center 11 Comae Berenices in your telescope at low power then switch to Chart 33a. NGC 4293 is only 35′ north and slightly east of 11 Comae Berenices.

The quick view
At 23×, NGC 4293 is simply a nice elongated glow, about 3′-long. It is oriented east–northeast to west–southwest and shines with a uniform brightness. At 72×, the galaxy is a very interesting sight. It suddenly has a bright inner core that, with averted vision, fragments into three sections – a nucleus flanked on either side of the galaxy's major axis by two equally bright knots. With averted vision, the galaxy also swells to 5′ in length. That added length comprises the

galaxy's surrounding halo, which, with concentration under a dark sky, appears warped.

Stop. Do not move the telescope. Your next target is nearby!

2.   NGC 4394 (H II-55)

Type	Con	RA	Dec	Mag	Dim	Rating
Barred spiral galaxy	Coma Berenices	12^h 25.9^m	+18° 13′	10.9	3.3′ × 3.1′	3

General description
NGC 4394 is a moderately sized, though somewhat dim galaxy near 9.1-magnitude M85. It is best seen in small telescopes at moderate magnifications.

Directions
Using Chart 33a, from NGC 4293 make a slow and gentle 1° sweep east–southeast to M85 – a powerfully glowing oval mass. NGC 4394 is only about 8′ to the east–northeast of M85.

The quick view
At 23×, NGC 4394 is a faint, round, diffuse glow, about 3′ across. A brightening can be seen toward the center with averted vision. At 72×, the galaxy is more obvious, appearing quite large, almost half the size as M85. Its core is bright to a starlike nucleus. The core is also elongated northwest to southeast. The surrounding disk is diffuse and irregular, a mishmash of both light and dark patches.

Stop. Do not move the telescope. Your next target is nearby!

3. NGC 4450 (H II-56)

Type	Con	RA	Dec	Mag	Dim	Rating
Spiral galaxy	Coma Berenices	12^h 28.5^m	+17° 05′	10.1	5.0′× 3.4′	4

General description

NGC 4450 is a small but fairly bright spiral galaxy a little less than 1$\frac{1}{2}$° southeast of M85. It is nicely compact, making it a good target for small-telescope users.

Directions

Using Chart 33a, from NGC 4394 and M85, move about 35′ southeast to 7.5-magnitude Star *a*. Now move 40′ southeast to equally bright Star *b*. NGC 4450 is about 20′ southwest of Star *b*, and 4′ northeast of a 9.5-magnitude star.

The quick view

At 23×, NGC 4450 is a bright oval glow about 3′ in extent. It's major axis is oriented north to south, and it gradually, then suddenly, gets brighter toward the middle. At 72×, the galaxy is very apparent. It has a bright circular core with a starlike nucleus. The core is surrounded by an eye-shaped inner lens and a diffuse outer halo. Larger telescopes may show the galaxy's spiral structure.

General description

NGC 4350 is a small but concentrated lenticular galaxy almost 1$\frac{1}{2}$° southeast of 11 Comae Berenices. It is best seen at moderate magnifications in a small telescope.

Directions

Using Chart 33a, from NGC 4450, return to 5th-magnitude 11 Comae Berenices and center it in your telescope at low power. Now make a slow and careful sweep, just a little more than 1° to the south–southeast, where you will find 7th-magnitude Star *c*. NGC 4350 is a little less than 30′ east, and slightly south, of Star *c*. It is also about 6′ east–southeast of the 11th-magnitude galaxy NGC 4340.

The quick view

At 23×, NGC 4350 is a very small (2′) ellipse of uniform light, oriented northeast to southwest. At 72×, the galaxy's light is very concentrated. Neighboring NGC 4340 can also be seen; although it is larger, it appears smaller and fainter in small telescopes because only the core is visible.

Stop. Do not move the telescope.

4. NGC 4350 (H II-86)

Type	Con	RA	Dec	Mag	Dim	Rating
Lenticular galaxy	Coma Berenices	12^h 24.0^m	+16° 42′	11.0	2.5′× 1.0′	3.5

5. NGC 4147 (H I-19)

Type	Con	RA	Dec	Mag	Diam	Rating
Globular cluster	Coma Berenices	12^h 10.1^m	+18° 32′	10.4	4′	3

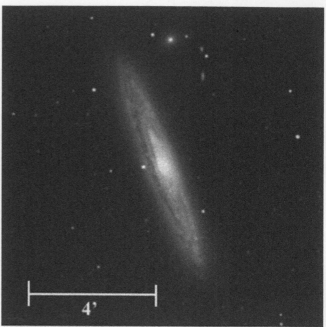

General description

NGC 4147 is a small, tight, and faint globular cluster a little more than $2\frac{1}{2}°$ west–northwest of 11 Comae Berenices. You will need to make a slow and careful star-hop to it, since it is in a field somewhat void of bright stars.

Directions

Return to 11 Comae Berenices then switch to Chart 33b. From 11 Comae Berenices move a little more than 50′ northwest to 8th-magnitude Star *a*. Next move about 20′ further to the northwest to a roughly 8′-long line of three 11th-magnitude stars (*b*), oriented northwest to southeast. Now make a slow and careful 1° sweep west, and ever-so-slightly south, to 9th-magnitude Star *c*. A short 25′ hop to the west–northwest will bring you to 8th-magnitude Star *d*. NGC 4147 is a little less than 15′ southwest of Star *d*.

The quick view

At 23×, NGC 4147 is a tiny 1′-wide circular glow, almost starlike; it swells slightly with averted vision. At 72×, it is a cometlike glow with a soft inner core surrounded by a diffuse and irregularly round halo with no hint of resolution. With concentration, however, some mottling can be suspected.

6. NGC 4216 (H I-35)

Type	Con	RA	Dec	Mag	Dim	Rating
Mixed spiral galaxy	Virgo	12ʰ 15.9ᵐ	+13° 09′	10.0	7.9′ ×1.7′	4

General description

NGC 4216 is a small but bright and highly inclined galaxy near 5th-magnitude 6 Comae Berenices, but just over the border in Virgo. It is obvious in a small telescope under a dark sky.

Directions

Use Chart 33 to locate 5th-magnitude 6 Comae Berenices. Center this star in your telescope at low power, then switch to Chart 33c. From 6 Comae Berenices, move 50′ southeast to the 10th-magnitude galaxy M99, which is only 10′ southwest of 6.5-magnitude Star *a*. Now hop 40′ south–southeast to 7th-magnitude Star *b*. A little more than 30′ to the southwest will bring you to 9.5-magnitude Star *c*. A roughly 35′ hop west, and slightly south, will bring you to similarly bright Star *d*. NGC 4216 lies about 15′ southwest of Star *d*.

The quick view

At 23×, NGC 4216 is a bright and beautiful, 5′-long silver streak, oriented north–northeast to south–southwest. With averted vision the edges sharpen to a knife edge. The northern half of the lens also looks slightly brighter than the southern half. At 72×, the galaxy shows a bright core – one that is tack sharp and extremely luminous. That core seems to burn through the surrounding soft and smooth spindle, the ends of which gradually taper to a point, or so it seems. At 101×, the brilliant nucleus lies at the center of a white egg-shaped core beyond which extends the oblique and dynamic lens. Observers have spied the galaxy's dust lane clearly in 12-inch and larger telescopes.

6 · June

Star charts for first night

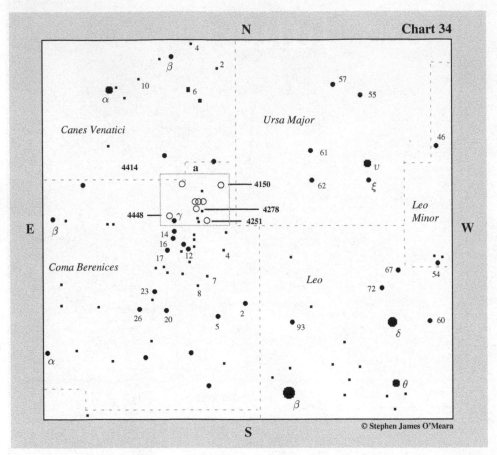

Chart 34

N

Canes Venatici

Ursa Major

4414

a

4150

4448

4278

4251

γ

Leo Minor

E

β

W

14
16
17
12
4

Coma Berenices

7

8

23

26 20

5

2

93

Leo

67

54

72

60

δ

α

θ

β

S

© Stephen James O'Meara

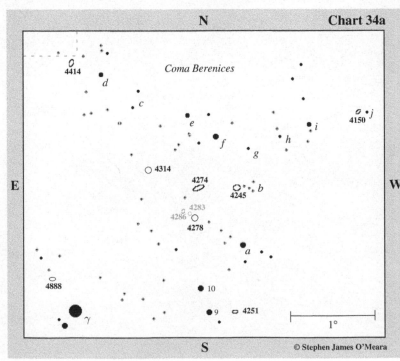

Chart 34a

N

4414

Coma Berenices

d

c

e

j
4150

i

f

h

g

4314

4274

b

4245

4283

4286

4278

a

E

W

4888

10

γ

9 4251

1°

S

© Stephen James O'Meara

FIRST NIGHT

1. NGC 4448 (H I-91)

Type	Con	RA	Dec	Mag	Dim	Rating
Barred spiral galaxy	Coma Berenices	12^h 28.2^m	+28° 37′	11.1	3.7′× 1.4′	3

4′

General description

NGC 4448 is a small and dim galaxy near 4th-magnitude Gamma (γ) Comae Berenices. It has a starlike core which should not be too difficult to see in a small telescope, especially at moderate magnification.

Directions

Use Chart 34 to locate Gamma Comae Berenices, the northernmost star in the inverted Y asterism that forms the heart of the Coma Berenices Star Cluster. Center Gamma Comae Berenices in your telescope, then switch to Chart 34a. NGC 4448 is a little less than 30′ northeast of that star.

The quick view

In the 4-inch at 23×, NGC 4448 is a small (2′) spindle of faint fuzzy light with a starlike core. The spindle is elongated east to west, though it is the nucleus that stands out first. At 72×, the galaxy is a faint, uniform ellipse of light with a dim inner lens that gradually gets brighter toward the center to a fuzzy nucleus; the nucleus does not appear as sharp as it did at low power.

Stop. Do not move the telescope. Your next target is nearby.

2. NGC 4251 (H I-89)

Type	Con	RA	Dec	Mag	Dim	Rating
Barred lenticular galaxy?	Coma Berenices	12^h 18.1^m	+28° 10′	10.7	3.7′× 2.1′	3

4′

General description

NGC 4251 is a small but relatively conspicuous lenticular galaxy near 6th-magnitude 9 Comae Berenices. Its core is nicely condensed and should be a reasonable target for small-telescope users. Think, "elongated, fuzzy star."

Directions

Refer to Chart 34a. Use your unaided eyes or binoculars to locate the 6th-magnitude stars 9 and 10 Comae Berenices, which are about 1$\frac{1}{2}$° west of Gamma Comae Berenices. Center 9 Comae Berenices in your telescope at low power. NGC 4251 is less than 20′ west of that star.

The quick view

At 23×, NGC 4251 is a 2′-long ellipse, oriented, like NGC 4448, east to west, with a bright core. At 72×, the galaxy shows a bright central pip inside a round and bright core that gradually fades and tapers along its major axis.

Stop. Do not move the telescope. Your next target is nearby.

3. NGC 4278 (H I-90)

Type	Con	RA	Dec	Mag	Dim	Rating
Elliptical galaxy	Coma Berenices	12^h 20.1^m	+29° 17'	10.2	3.5'× 3.5'	3.5

General description
NGC 4278 is a small but obvious elliptical galaxy about $1\frac{3}{4}°$ northwest of Gamma Comae Berenices. It appears highly concentrated and round in small telescopes.

Directions
Using Chart 34a, from NGC 4251, return to 9 Comae Berenices. Note that 10 Comae Berenices should be 20' to the northeast. Center 10 Comae Berenices in your telescope, then move about 40' northwest, to 6th-magnitude Star α. NGC 4278 is 40' northeast of Star α. (You can follow a line of three roughly 10.5-magnitude stars, which lead to it from Star α.)

Field note
NGC 4278 is easy to identify because it has a small (1') and condensed 12th-magnitude companion (NGC 4283) 3.5' to the northeast; both galaxies are visible in small telescopes, though your target will be the brighter and more obvious one in the pair.

The quick view
At 23×, NGC 4278 is a 2' round glow that gradually, then suddenly brightens toward the center. At 72×, the galaxy's

round core stands out well, and is surrounded by a dimmer halo of light. Larger telescopes should show the halo as a slight ellipse oriented northeast to southwest.

Stop. Do not move the telescope. Your next target is nearby!

4. NGC 4274 (H I-75)

Type	Con	RA	Dec	Mag	Dim	Rating
Barred spiral galaxy	Coma Berenices	12^h 19.8^m	+29° 37'	10.4	6.7'× 2.5'	3

General description
NGC 4274 is a somewhat large and elongated galaxy, a little more than 2° northwest of Gamma Comae Berenices, near NGC 4278. It is not as condensed as the latter object and may require darker skies to see in a small telescope.

Directions
Use Chart 34a. NGC 4274 is only 20' north and ever-so-slightly west of NGC 4278.

Field note
Take your time in moving to the north, making sure you do not stray to the west or east, where your next two targets lie.

The quick view
At 23×, NGC is a 4'-long ellipse of diffuse light, oriented west–northwest to east–southeast, with a

slightly brighter inner lens. At 72×, the galaxy's inner lens is more defined. With averted vision that lens gradually brightens toward the center, though not to a stellar nucleus.

Stop. Do not move the telescope. Your next target is nearby!

5. NGC 4245 (H I-74)

Type	Con	RA	Dec	Mag	Dim	Rating
Barred lenticular galaxy	Coma Berenices	12ʰ 17.6ᵐ	+29° 36′	11.4	3.2′ × 3.0′	2

4'

General description
NGC 4245 is a very small and dim lenticular galaxy west of NGC 4274. It is a faint but condensed fuzzy patch in small telescopes. Be prepared to make a careful sweep to it with moderate magnification.

Directions
Use Chart 34a. NGC 4245 is 30′ west of NGC 4274, immediately east and a tad south of a 6′-wide pyramid of four roughly 10.5-magnitude stars (b).

The quick view
At 23× in the 4-inch, NGC 4274 is little more than a 12th-magnitude star with a slight fuzz, as seen with averted vision. At 72×, the view is much the same.

Stop. Do not move the telescope. Your next target is nearby!

6. NGC 4314 (H I-76)

Type	Con	RA	Dec	Mag	Dim	Rating
Barred spiral galaxy	Coma Berenices	12ʰ 22.6ᵐ	+29° 53′	10.6	4.2′ × 4.1′	2.5

4'

General description
NGC 4314 is a somewhat small, low-surface galaxy northeast of NGC 4274. It has a condensed core surrounded by a large diffuse halo. Its size is much more on a par with NGC 4274 than with NGC 4245.

Directions
Use Chart 34a. NGC 4314 is 40′ northeast of NGC 4274.

The quick view
At 23×, NGC 4314 is a 3′-wide oval, oriented northwest to southeast, with a bright starlike core surrounded by a large and diffuse halo. At 72×, the halo has, with averted vision, a sharp ring-like structure. Larger scopes may see this barred spiral galaxy's two sweeping arms, which form a letter S around the central oval.

Stop. Do not move the telescope. Your next target is nearby!

7. NGC 4414 (H I-77)

Type	Con	RA	Dec	Mag	Dim	Rating
Spiral galaxy	Coma Berenices	12ʰ 26.4ᵐ	+31° 13′	10.1	4.4′ × 3.0′	4

General description

NGC 4414 is a moderately small but very conspicuous lens-shaped spiral galaxy about 3° north and very slightly west of Gamma Comae Berenices. It has a relatively high surface brightness and should be a good target for small-telescope users under suburban skies.

Directions

Using Chart 34a, from NGC 4314 move slowly and carefully 45′ north to a pair of 9th-magnitude suns (c), which are separated by about 12′ and oriented north–northwest to south–southeast. From Pair c, move 30′ northeast to 7.5-magnitude Star d. NGC 4414 is about 25′ northeast of Star d.

The quick view

At 23×, NGC 4414 is moderately large (4′) and surprisingly bright. Most prominent (even with direct vision) is its bright starlike core, which is surrounded by a fuzzy elliptical halo oriented northwest to southeast. At 72×, the galaxy is very beautiful, appearing as a large tapered lens with a very prominent nucleus.

General description

NGC 4150 is a very small and dim lenticular galaxy that, fortunately, has a condensed core, which is a reasonable target for small-telescope users when seen with moderate magnification.

Directions

Use Chart 34a. You will need to make a slow and patient star-hop to this galaxy. Start by backtracking. From NGC 4414, return to Star d, then Pair c. Now move 40′ southwest to 8th-magnitude Star e. About 20′ further to the southwest is 6.5-magnitude Star f. Continuing to the southwest, hop 25′ to solitary 9th-magnitude Star g. A little more than 20′ to the northwest is 8.5-magnitude Star h. And another 20′ hop to the northwest brings you to 7.5-magnitude Star i. NGC 4150 is a little less than 40′ to the northwest, just 6′ east of 9.5-magnitude Star j.

The quick view

At 23×, NGC 4150 is a small (1′) and condensed galaxy that looks much like a slightly out-of-focus star. Averted vision is required to see it at this power in a small telescope. At 72×, the galaxy appears round and reasonably bright. The galaxy's halo is thin and grows suddenly brighter toward the center, though not to a starlike nucleus. Larger telescopes will show the disk elongated northwest to southeast.

8. NGC 4150 (H I-73)

Type	Con	RA	Dec	Mag	Dim	Rating
Lenticular galaxy	Coma Berenices	12ʰ 10.6ᵐ	+30° 24′	11.6	2.1′ × 1.5′	2.5

Star charts for second night

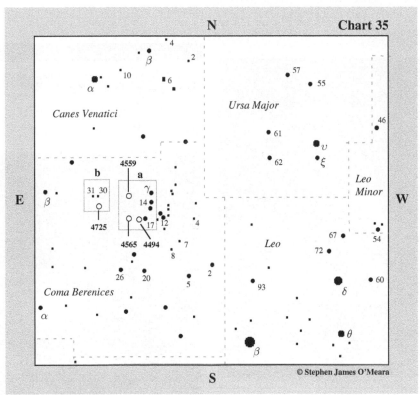

Chart 35

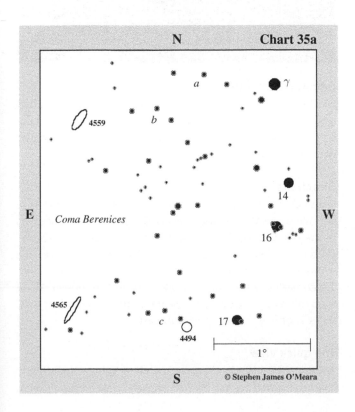

Chart 35a

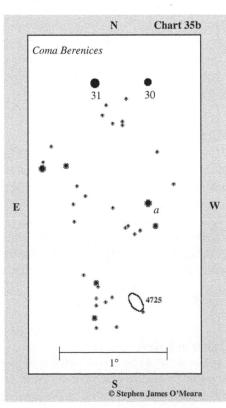

Chart 35b

SECOND NIGHT

1. NGC 4559 (H I-92)

Type	Con	RA	Dec	Mag	Dim	Rating
Mixed spiral galaxy	Coma Berenices	12^h 36.0^m	+27° 58′	10.0	11.3′ × 5.0′	4

General description

NGC 4559 is a large and bright spiral galaxy 2° east–southeast of 4th-magnitude Gamma (γ) Comae Berenices. The galaxy is visible in 7 × 50 binoculars (with effort) from a dark-sky site. Telescopically, it is a fine and obvious oval glow.

Directions

Use Chart 35 to locate Gamma Comae Berenices, the northernmost star in the famous inverted Y asterism that forms the heart of the Coma Berenices Star Cluster. Center Gamma Comae Berenices in your telescope at low power, then switch to Chart 35a. From Gamma Comae Berenices, move 40′ east, and slightly north, where you will find a roughly 35′-long arc (a) of three 9.5-magnitude suns. From Arc a, move 35′ southeast to 25′-long arc b, which comprises three 9th- and 10th-magnitude stars. NGC 4559 is 45′ east and slightly south of Arc b.

The quick view

At 23×, NGC 4559 is a 10′-wide oval glow, oriented northwest to southeast, that very suddenly becomes brighter in the middle. Three 12.5- to 13.5-magnitude stars form an arc

on the galaxy's southeastern flank. At 72×, the galaxy's core comprises a fuzzy round spot with an intense starlike nucleus. With averted vision, little spikes of light flare to the northwest and, with effort, a dim spiral arm can be faintly seen to the southeast.

Stop. Do not move the telescope. Your next target is nearby.

2. NGC 4494 (H I-83)

Type	Con	RA	Dec	Mag	Dim	Rating
Elliptical galaxy	Coma Berenices	12^h 31.4^m	+25° 47′	9.8	4.6′ × 4.4′	3

General description

NGC 4494 is a small and somewhat bright elliptical galaxy near 6.6-magnitude 17 Comae Berenices. In small telescopes, it is best viewed at moderate magnifications under dark skies.

Directions

Using Chart 35a, from NGC 4559, return to Gamma Comae Berenices. Now use your unaided eyes or binoculars to locate 17 Comae Berenices, which is about 2$\frac{1}{2}$° south–southeast of Gamma Comae Berenices. It also marks the southern tip of the inverted Y asterism. You want to center 17 Comae Berenices in your telescope at low power; there is no mistaking the star, because it is a stunning double. NGC 4494 is only $\frac{1}{2}$° to the east–southeast.

The quick view

At 23×, NGC 4494 is a 2′ round patch of light with a slightly brighter central region. At 72×, the galaxy is better defined, appearing as a circular glow, like a dim comet with a diffuse and slightly brighter central condensation.

Stop. Do not move the telescope. Your next target is nearby!

3. NGC 4565 (H V-24)

Type	Con	RA	Dec	Mag	Dim	Rating
Spiral galaxy?	Coma Berenices	12^h 36.3^m	+25° 59'	9.6	16.2' × 2.3'	4

General description
NGC 4565 is one of the most famous and sought-after edge-on spiral galaxies in the northern night sky. While an absolute marvel in large telescopes, it is a delicate sight in small telescopes – one that is best appreciated at moderate magnifications.

Directions
Using Chart 35a, from NGC 4494 move 15' northwest to a 20'-wide arc (c) of three 8th- to 9th-magnitude stars, oriented east–northeast to west–southwest. NGC 4565 is less than 1° due east of Arc c. Sweep slowly and carefully to it.

The quick view
At 23×, NGC 4565 is a 15'-long wafer of light, oriented northwest to southeast, with a small but swollen center. At 72×, the galaxy's central hub is more apparent. A 13.5-magnitude star shines just northeast of it. With averted vision and some concentration, a dust lane can followed along the galaxy's major axis.

Stop. Do not move the telescope.

4. NGC 4725 (H I-84)

Type	Con	RA	Dec	Mag	Dim	Rating
Barred spiral galaxy	Coma Berenices	12^h 50.4^m	+25° 30'	9.2	10.5' × 8.1'	4

General description
NGC 4725 is a beautiful galaxy about 2° south–southeast of 30 Comae Berenices, which is about 4° east–southeast of Gamma Comae Berenices. The only galaxy in Coma Berenices brighter than it is M64, located only 4° to the southeast. NGC 4725 is visible in 7×50 binoculars under a dark sky, and it appears as a soft and slightly elliptical glow in the smallest of telescopes. It is a beautiful object in telescopes of all sizes.

Directions
Using Chart 35, return to 4th-magnitude Gamma Comae Berenices. Next use your unaided eyes or binoculars to locate 30 Comae Berenices about 4° to the east–southeast. Center 30 Comae Berenices in your telescope at low power, then switch to Chart 35b. From 30 Comae Berenices, make a slow and careful sweep 1° south–southeast to 7th-magnitude Star a. NGC 4725 is about 1° south and slightly east of Star a.

The quick view
At 23×, the galaxy is immediately obvious as a well-defined glow 20' west of a 20'-wide clustering of about a dozen 8th- to 12th-magnitude suns. The galaxy is also clearly elliptical, being oriented northeast to southwest; it appears very layered. Look for a bright starlike nucleus, nestled inside a

soft inner lens that rests inside a larger elliptical halo about 5′ across. At 72×, subtle details start to emerge. First, the inner lens breaks down into what appears to be a bar oriented roughly north–northeast to south–southwest inside a sharp but broken S-shaped ring. A fainter, amorphous glow surrounds this inner lens. The glow, however, does match the region of the galaxy's faint outer arms, which split and loop around the inner ring.

Star charts for third night

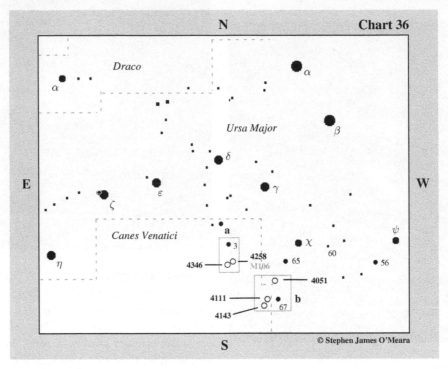

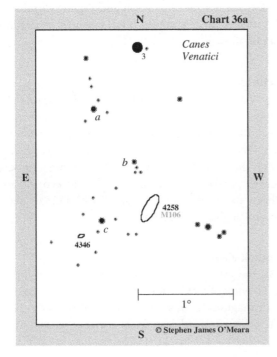

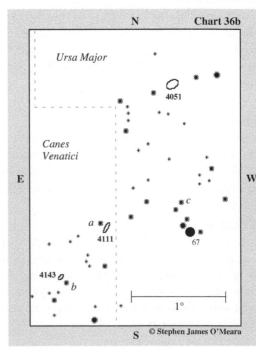

THIRD NIGHT

1. NGC 4258 (H V-43) = M106

Type	Con	RA	Dec	Mag	Dim	Rating
Mixed spiral galaxy	Canes Venatici	12^h 19.0^m	+47° 18′	8.4	20.0′ × 8.4′	4

4′

General description

NGC 4258, better known as M106, was discovered in 1781 by Pierre Mechain. But this fact escaped detection until 1947, when Canadian astronomer Helen Sawyer Hogg unearthed a letter by Mechain to Bernoulli in which he describes the discovery of four new nebulae in 1781 – one of which is William Herschel's V-43 (NGC 4258). The galaxy is large and bright and can be spied in 7 × 50 binoculars under a dark sky. It is a wonder in telescopes of all sizes.

Directions

Use Chart 36 to locate 5th-magnitude 3 Canum Venaticorum (use binoculars if necessary), which is a little less than 6° east–northeast of 3.7-magnitude Chi (χ) Ursae Majoris in the Great Bear's hind leg. Center 3 Canum Venaticorum in your telescope at low power, then switch to Chart 36a. From 3 Canum Venaticorum, move a little less than 50′ southeast to 7.5-magnitude Star a. Next, hop a little more than 40′ southwest to 8th-magnitude Star b; note that a tiny triangle of 10.5 magnitude stars lies 5′ to its south. NGC 4258 is 30′ south–southwest of Star b.

The quick view

At 23×, NGC 4258 is a bright 10′-wide oval of light, oriented north–northwest to south–southeast, that gradually, then suddenly, gets brighter toward the middle to a nucleus. With averted vision and some concentration, the northern half of the galaxy appears brighter than the southern half. Faint, S-shaped arms can also be spied extending beyond the bright oval. At 72×, inner arms can also be seen with an S-shaped curvature, and some dark mottling can be seen in the nuclear region.

Stop. Do not move the telescope. Your next target is nearby.

2. NGC 4346 (H I-210)

Type	Con	RA	Dec	Mag	Dim	Rating
Lenticular galaxy	Canes Venatici	12^h 23.5^m	+47° 00′	11.1	3.2′ × 1.4′	3

4′

General description

NGC 4346 is a small but bright lenticular galaxy near NGC 4258 (M106). Its core is very condensed, which will help users of small telescopes to see it well, especially with moderate magnifications.

Directions

Using Chart 36a, from NGC 4258, move 30′ east–southeast to 6.5-magnitude Star c. NGC 4346 is 20′ southeast of Star c.

The quick view

At 23×, NGC 4346 is small (2′) and round. With averted vision, the galaxy's core appears highly condensed and relatively bright and obvious. At 72×, the galaxy's core is very small and starlike. It is surrounded by a tight inner halo of light, which is at the core of a tapered ellipse of extremely faint light, oriented roughly east to west.

3. NGC 4111 (H I-195)

Type	Con	RA	Dec	Mag	Dim	Rating
Lenticular galaxy	Canes Venatici	12^h 07.1^m	+43° 04′	10.7	4.4′× 0.9′	3

4′

General description

NGC 4111 is a small but reasonably bright lenticular galaxy near 6th-magnitude 67 Ursae Majoris. In small telescopes it will appear as a fuzzy star. It is best seen at moderate magnifications.

Directions

Use Chart 36 (and binoculars, if necessary) to locate 67 Ursae Majoris, which is about 5½° southeast of 4th-magnitude Chi (χ) Ursae Majoris in the hind leg of the Great Bear. Center 67 Ursae Majoris in your telescope at low power, then switch to Chart 36b. Note that 67 Ursae Majoris is the brightest star in a 25′-wide W-shaped asterism, oriented northeast to southwest. NGC 4111 is 55′ due east of 67 Ursae Majoris, about 4′ southwest of 8.5-magnitude Star a.

The quick view

At 23×, and direct vision, NGC 4111 is very small (~1′) and very bright (like a star). With averted vision it swells to 2′, appearing as a star with a round and tenuous halo. At 72×, the galaxy has a very intense, quasi-stellar nucleus surrounded by a very bright and condensed inner core of light. Under a very dark sky with averted vision, the galaxy's tapered disk can be seen very faintly, being oriented northwest to southeast. Larger telescopes show this needle-sharp galaxy quite spectacularly.

Stop. Do not move the telescope. Your next target is nearby!

4. NGC 4143 (H IV-54)

Type	Con	RA	Dec	Mag	Dim	Rating
Mixed spiral galaxy	Canes Venatici	12^h 09.6^m	+42° 32′	10.7	2.9′× 1.9′	2.5

4′

General description

NGC 4143 is a small and somewhat dim galaxy near NGC 4111. It is best seen at moderate magnifications. The galaxy is not as condensed as NGC 4111, making it a bit more difficult to detect.

Directions

Use Chart 36b. NGC 4143 is a little more than 40′ southeast of NGC 4111, about 5′ northeast of 9th-magnitude Star b.

The quick view

At 23×, NGC 4143 is a dim, round spot, about 2′ in diameter. Averted vision is required to see it well in a small telescope, even under a dark sky. At 72×, the galaxy becomes a 3′-wide lens, oriented northwest to southeast, with a very small but bright starlike core. The surrounding disk is of uniform brightness.

Stop. Do not move the telescope. Your next target is nearby!

5. NGC 4051 (H IV-56)

Type	Con	RA	Dec	Mag	Dim	Rating
Mixed spiral galaxy	Ursa Major	12^h 03.2^m	+44° 32′	10.2	5.5′× 4.6′	4

General description

NGC 4051 is bright and obvious galaxy near 67 Ursae Majoris.

Directions

Using Chart 36b, from NGC 4143, return to 67 Ursae Majoris. Now center the northernmost star in the W-shaped asterism, 9.5-magnitude Star c. NGC 4051 is $1\frac{1}{4}°$ north and slightly west of that star.

The quick view

At 23×, NGC 4051 appears very bright and obvious. A first glance shows it to be a round spot, about 3′ in extent. Averted vision shows it to be more elongated, but this is an illusion created by an 11th-magnitude star immediately to its southwest. At 72×, the galaxy displays a very bright starlike nucleus surrounded by a round central lens in a slightly elliptical halo, oriented northwest to southeast. With averted vision, the disk appears irregular and mottled. It also swells to 5′. Larger scopes will show this mottled texture to be H II regions and dust lanes in the galaxy's S-shaped inner spiral arms.

Star charts for fourth night

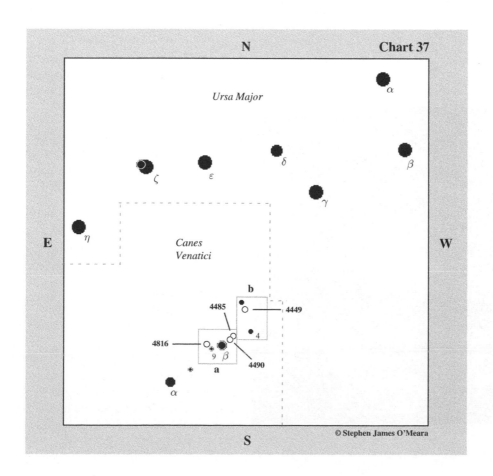

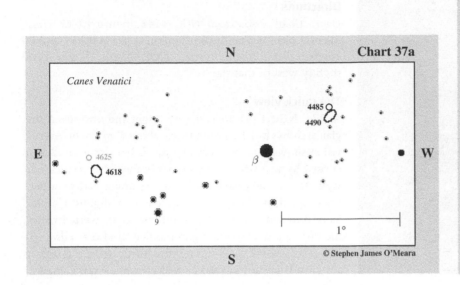

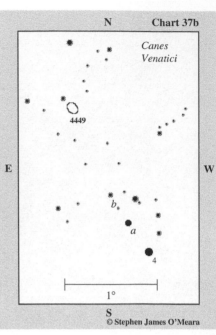

FOURTH NIGHT

1. NGC 4490 (H I-198)

Type	Con	RA	Dec	Mag	Dim	Rating
Barred spiral galaxy	Canes Venatici	12^h 30.6^m	+41° 39'	9.5	5.6' × 2.8'	4

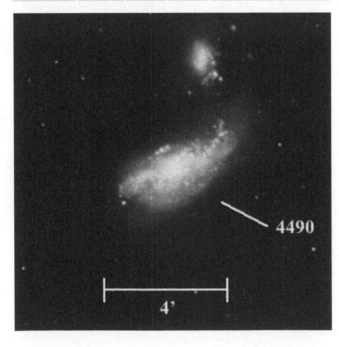

General description

NGC 4490, the famous Cocoon Galaxy, is a prominent barred spiral near 4th-magnitude Beta (β) Canum Venaticorum. It is an easy target for small-telescope users. In fact, from a dark sky, it is visible in 7 × 50 binoculars as a small dim glow. Small telescopes will show it as a large, though ill-defined haze. Otherwise, it is a very condensed and extremely distinct glow, so observers under suburban skies should have no problem with this one.

Field note

NGC 4490 forms an interacting pair with NGC 4485 (see below), which is two magnitudes fainter and only about 3.5' to the north. Together the pair constitute a curious interacting pair, which is also called Arp 269.

Directions

Use Chart 37 to locate Beta Canum Venaticorum. Center that star in your telescope at low power, then switch to Chart 37a. NGC 4490 is only 40' northwest of Beta Canum Venaticorum.

The quick view

At 23×, NGC 4490 is a 5'-long, flying-saucer-shaped lens of light, oriented northwest to southeast. With averted vision the galaxy looks mottled, and has faint extensions. At 72×, the galaxy is simply bright and beautiful. Look for a faint starlike nucleus in a bright round core that's embedded in a cocoon of slightly fainter light. With averted vision the galaxy's inner lens fragments into quarters, with enhancements at the four cardinal directions.

Stop. Do not move the telescope. Your next target is nearby.

2. NGC 4485 (H I-197)

Type	Con	RA	Dec	Mag	Dim	Rating
Barred irregular galaxy	Canes Venatici	12^h 30.5^m	+41° 42′	11.9	2.7′ × 2.3′	3

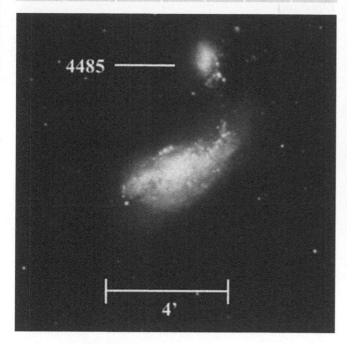

General description

NGC 4485 is a very small and dim irregular galaxy that forms an interacting pair with NGC 4485 (see above). It is a small but relatively bright object when seen at moderate magnifications.

Directions

Use Chart 37a. NGC 4485 is only about 3.5′ north of NGC 4490.

The quick view

At 23×, NGC 4485 can be seen with averted vision as a dim circular spot of light, about 2′ in diameter, like a slightly out-of-focus star. At 72×, NGC 4485 really stands out as a small but obvious spot of light. At 303×, NGC 4485 shows some minute structure, namely the otherwise circular spot becomes slightly elongated. And with much attention a bright core and a dim patch to the south can be seen.

Stop. Do not move the telescope. Your next target is nearby.

3. NGC 4618 (H I-178)

Type	Con	RA	Dec	Mag	Dim	Rating
Barred spiral galaxy	Canes Venatici	12^h 41.5^m	+41° 09′	10.8	4.1′ × 3.2′	3

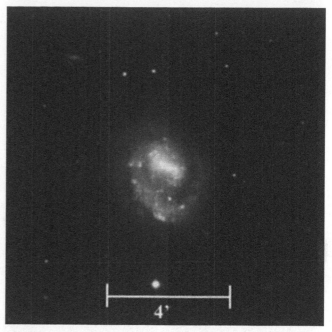

General description

NGC 4618 is a small but moderately bright barred spiral galaxy about $1\frac{1}{2}°$ east–southeast of Beta Canum Venaticorum, near 6th-magnitude 9 Canum Venaticorum. In a small telescope, it is best seen at moderate magnifications.

Directions

Using Chart 37a, from NGC 4485, return to Beta Canum Venaticorum. Now make a careful sweep 1° southeast to 6th-magnitude 9 Canum Venaticorum. NGC 4618 is about 35′ northeast of 9 Canum Venaticorum.

The quick view

At 23×, NGC 4618 is a moderately diffuse, irregular patch of light 6′ north of an 11th-magnitude star. With averted vision the galaxy appears to swell to about 3′ in diameter and is oriented slightly northeast to southwest. At 72×, the galaxy appears double: the brightest component is the galaxy's core (which is a bright, irregularly round amorphous patch); the dimmer part is a strange extension to the south. This extension is most likely a bright arc of starlight lining one of its spiral arms.

Stop. Do not move the telescope.

4. NGC 4449 (H I-213)

Type	Con	RA	Dec	Mag	Dim	Rating
Barred irregular galaxy	Canes Venatici	12^h 28.2^m	+44° 06′	9.6	5.5′ × 4.1′	4

4'

General description

NGC 4449, the famous Box Galaxy, is a prominent irregular galaxy just 3° north–northwest of Beta Canum Venaticorum. The galaxy shines with the light of an unfocused 9.6-magnitude star and has a high surface brightness, making it a good target for small-telescope users. Under a dark sky, with patience, it is visible in 7×50 binoculars as a tiny diffuse glow. Telescopically, it is a very complex and intriguing sight.

Directions

Using Chart 37, return to Beta Canum Venaticorum, then, with your unaided eyes or binoculars, look 2¼° northwest for 6th-magnitude 4 Canum Venaticorum. Center 4 Canum Venaticorum in your telescope at low power, then switch to Chart 37b. From 4 Canum Venaticorum, move about 20′ northeast to 7th-magnitude Star a. Now make another 20′ hop in the same direction to 8.5-magnitude Star b. NGC 4449 is a little more than 1° further to the northeast.

The quick view

At 23×, NGC 4449 is a 5′-long oval adorned with a brightly dappled center. A diffuse glow of patchy light surrounds this oval core. At 72×, clumps of nebulous matter pop into view, branding the entire galaxy with a distinct X-shaped pattern. Overall, the galaxy is box-shaped, with a Y-shaped asterism of bright condensations crossing the galaxy's major axis.

Star charts for fifth night

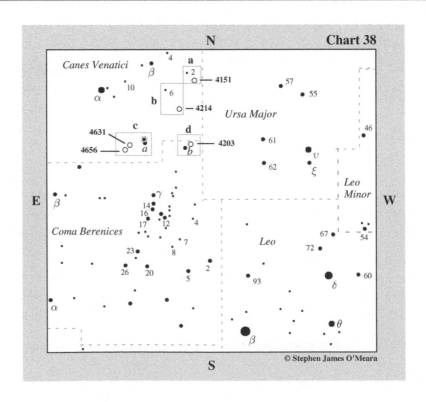

© Stephen James O'Meara

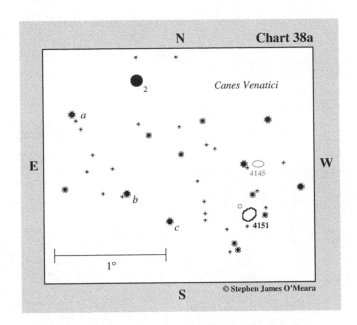

Chart 38a

N

E

W

S

Canes Venatici

2

a

b

c

4145

4151

1°

© Stephen James O'Meara

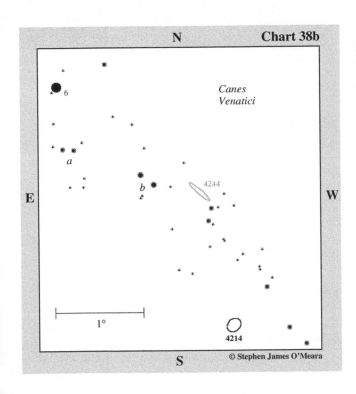

Chart 38b

N

E

W

S

Canes Venatici

6

a

b

4244

4214

1°

© Stephen James O'Meara

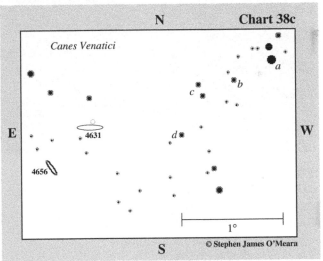

Chart 38c

N

E

W

S

Canes Venatici

a

b

c

d

4631

4656

1°

© Stephen James O'Meara

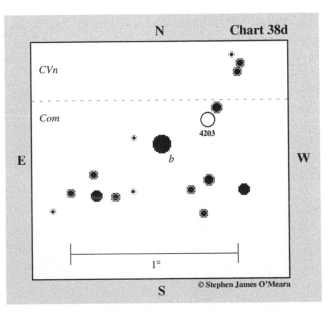

Chart 38d

N

E

W

S

CVn

Com

4203

b

1°

© Stephen James O'Meara

FIFTH NIGHT

1. NGC 4151 (H I-165)

Type	Con	RA	Dec	Mag	Dim	Rating
Mixed spiral galaxy	Canes Venatici	12^h 10.5^m	+39° 24′	10.8	6.4′ × 5.5′	3

General description
NGC 4151 is a small but reasonably obvious Seyfert galaxy about 4$\frac{1}{2}$° west–southwest of 4th-magnitude Beta (β) Canum Venaticorum. Its light is very condensed, making it a good object for suburban locations.

Directions
Use Chart 38 to locate 4th-magnitude Beta Canum Venaticorum. Now use your unaided eyes or binoculars to locate 6th-magnitude 2 Canum Venaticorum about 3$\frac{1}{2}$° to the west–southwest. Center 2 Canum Venaticorum in your telescope at low power, then switch to Chart 38a. From 2 Canum Venaticorum, move 40′ southeast to 7.5-magnitude Star a. Next, move a little more than 50′ south–southwest to similarly bright Star b. Now hop 30′ southwest to 7th-magnitude Star c. NGC 4151 is a little less than 50′ west, and slightly north, of Star c.

The quick view
At 23× in the 4-inch, NGC 4151 is an extremely condensed, almost starlike object. It swells a bit with averted vision. At 72×, the galaxy displays a bright (roughly 12th-magnitude) core surrounded by a small (3′) circular halo, which is best seen with averted vision. Note that the core can vary in intensity by about a magnitude, which might affect the object's detectability.

2. NGC 4214 (H I-95)

Type	Con	RA	Dec	Mag	Dim	Rating
Mixed irregular galaxy	Canes Venatici	12^h 15.6^m	+36° 20′	9.1	9.6′ × 8.1′	4

General description
NGC 4214 is a considerably bright and large irregular galaxy about 6° southwest of 4th-magnitude Beta (β) Canum Venaticorum, and 3° southwest of 5th-magnitude 6 Canum Venaticorum. It is easily spied in small telescopes and is a great target for suburban observers.

Directions
Use Chart 38 to locate 6 Canum Venaticorum, which is nearly 3° southwest of Beta Canum Venaticorum, then switch to Chart 38b. From 6 Canum Venaticorum, move about 40′ south and slightly west to a pair of 8th-magnitude suns (a), oriented east to west. Now swing your telescope about 1° southwest to a pair of roughly 7th-magnitude suns (b); the 10th-magnitude, edge-on galaxy NGC 4244 lies just 30′ west of Pair b. Center that galaxy, then make a slow and careful 1$\frac{1}{2}$° sweep south and slightly west to NGC 4214.

The quick view

At 23×, NGC 4214 appears as a bright oval disk, about 5′ northwest of a roughly 12th-magnitude star. It measures about 5′ across and is ever-so-slightly oriented northwest to southeast. At first, the disk may appear uniform in brightness, but with averted vision it gets progressively brighter toward the middle, though not to a sharp stellar point. The inner core is simply a bright oval of light within a fainter one. At 72×, the inner oval seems mottled. With time and concentration a dual core – two fuzzy pearls of light – may be seen.

3. NGC 4631 (H V-42)

Type	Con	RA	Dec	Mag	Dim	Rating
Barred spiral galaxy	Canes Venatici	12ʰ 42.1ᵐ	+32° 32′	9.2	14.7′× 3.5′	4

5′

General description

NGC 4631 is a delicately bright, elongated galaxy about $5\frac{1}{4}°$ northeast of 4th-magnitude Gamma (γ) Comae Berenices. It is easily seen in a small telescope under a dark sky. Light pollution will affect its view. It is a stunning sight in larger telescopes, being reminiscent of M82 in Ursa Major.

Directions

Use Chart 38 to locate Gamma Comae Berenices – the northernmost star in the famous inverted Y asterism that forms the core of the Coma Berenices Star Cluster. Now use your unaided eyes or binoculars to look about 5° north–northeast

of Gamma Comae Berenices for a pair of 6th-magnitude stars (a), which are separated by 10′ and oriented north–northeast to south–southwest. Center Pair a in your telescope at low power, then switch to Chart 38c. From Pair a, move 25′ southeast to 9th-magnitude Star b. Now move 20′ east–southeast to a pair of 9th-magnitude stars (c), oriented north–northeast to south–southwest. From Pair c, drop 30′ southeast to 9th-magnitude Star d. NGC 4631 is about 55′ due east of Star d.

The quick view

At 23×, NGC 4631 is a 10′-long cigar-shaped glow, oriented east to west. With imagination, it looks like a glowing meteor train. Its central region is a bright oval that weakens and tapers away from the core. At 72×, the galaxy is a beautifully detailed cloud of light and dark vapors entwined in a delicate embrace. Star clumps pepper the disk like snowballs tossed onto the side of a house. The 12.4-magnitude companion galaxy, NGC 4627, is visible in the 4-inch under a dark sky 3′ northwest of NGC 4631's center.

Stop. Do not move the telescope. Your next target is nearby!

4. NGC 4656 (H I-176)

Type	Con	RA	Dec	Mag	Dim	Rating
Barred spiral galaxy	Canes Venatici	12ʰ 44.0ᵐ	+32° 10′	10.5	18.8′× 3.2′	3

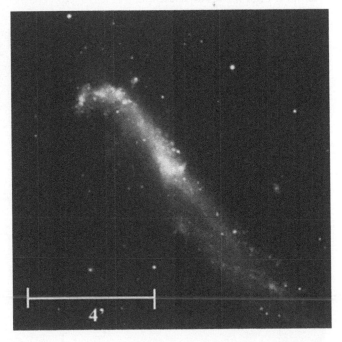

4′

General description

NGC 4656 is a somewhat large and elongated galaxy near NGC 4631. It has a low surface brightness, so it may be a challenge to small-telescope users if they are not under a dark sky. The view is also sensitive to slight changes in atmospheric clarity. Also, it might appear less than impressive if it is poorly positioned in a dark sky. The galaxy is best seen with moderate magnification and averted vision. The galaxy has a warped appendage that juts out at right angles to the main shaft like the hooked blade of a hockey stick. Thus, NGC 4656 is aptly called the Hockey Stick Galaxy.

Directions

Use Chart 38c. NGC 4656 lies only $\frac{1}{2}°$ southeast of NGC 4631.

The quick view

On clear and and transparent evenings, NGC 4656 is visible in the 4-inch at 23× as a thin shaft of light. It is a dim glow, but if you concentrate with averted vision, the galaxy does not appear uniformly lit. Like M82 in Ursa Major, it appears segmented, like three ghosts walking toward you hand in hand. At 72× and averted vision, the galaxy's segments are more enhanced. The brightest segment appears just northeast of where it seems the nucleus should be. There is a slightly fainter "oval" of light adjacent to what may be considered its nuclear region; the galaxy is very faint on the southwestern side of the "nuclear region" and looks like a dim tentacle of light.

Stop. Do not move the telescope.

4'

General description

NGC 4203 is a small and round lenticular galaxy $5\frac{1}{2}°$ northwest of Gamma Comae Berenices, near a 5.5-magnitude star. It is best seen at moderate magnification in small telescopes and looks like a tiny little comet.

Directions

Using Chart 38, from NGC 4656, return to 6th-magnitude Pair *a*. Now use your unaided eyes or binoculars to find 5.5-magnitude Star *b*, which is located $3\frac{1}{2}°$ to the west. Center Star *b* in your telescope at low power, then switch to Chart 38d. NGC 4203 is only 20′ northwest of Star *b*, immediately south of an 8th-magnitude star.

The quick view

At 23× in the 4-inch, NGC 4203 is a small (2′), round, comet-like glow – a dim comet with a bright inner coma, which becomes more apparent with averted vision. At 72×, the galaxy is a small, round, diffuse glow with a starlike nucleus.

5.   NGC 4203 (H I-175)

Type	Con	RA	Dec	Mag	Dim	Rating
Mixed lenticular galaxy	Coma Berenices	12^h 15.1^m	+33° 12′	10.9	3.5′ × 3.4′	2

Star charts for sixth night

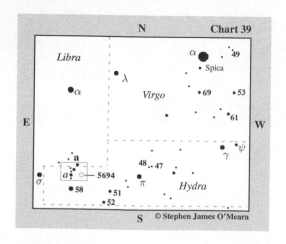

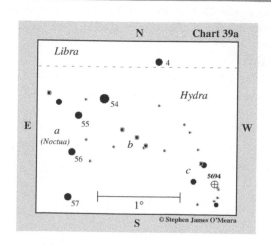

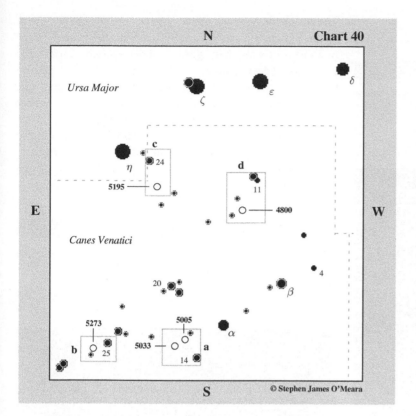

Chart 40

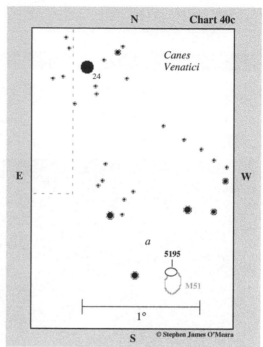

Chart 40c

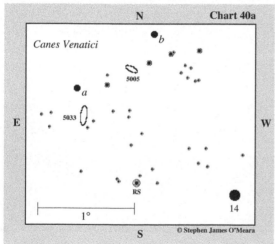

Chart 40a

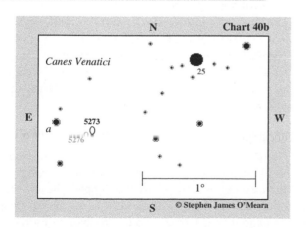

Chart 40b

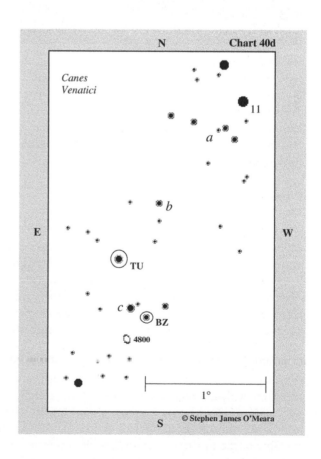

Chart 40d

SIXTH NIGHT

1. NGC 5694 (H II-196)

Type	Con	RA	Dec	Mag	Diam	Rating
Globular cluster	Hydra	14^h 39.6^m	−26° 32′	10.2	4.3	3

General description

NGC 5694 is a small and relatively dim globular cluster buried in the extreme northeastern corner of Hydra, near the now-obsolete constellation Noctua, the Night Owl. The cluster appears starlike at low powers in small telescopes. Fortunately, it lies in a rich field of stars, which makes finding the object a relatively simple task.

Directions

Use Chart 39 to locate 3rd-magnitude Alpha (α) Librae, which is a little more than 20° east–southeast of 1st-magnitude Alpha Virginis (Spica). Now use your unaided eyes or binoculars to look 10° south of Alpha Librae, for Asterism a – a lovely 3½°-long chain of 4th- to 6th-magnitude stars. Center Asterism a in your telescope at low power, then switch to Chart 39a. Asterism a is the now obsolete constellation Noctua, the Night Owl: the Owl's body is composed of the stars 4 Librae and 54, 55, 56, and 57 Hydrae. You want to move about 45′ westward to the 20′-long asterism of four 9th-magnitude stars (b), which looks like a miniature version of the constellation Sagitta, the Arrow. Next, carefully move about 50′ southwest to a pair of 7.5-magnitude stars (c), which are oriented northwest to southeast. The northern member of the pair has a 9th-magnitude companion to the east. NGC 5694 is a little less than 20′ southwest of Pair c.

The quick view

In the 4-inch at 23×, NGC 5694 is the northernmost "star" in a chain of suns that "glows" under averted vision. This glow should draw your eye to the cluster. At 72×, NGC 5694 is a small, compact glow, which looks like a swollen star. With averted vision, a condensed core with a starlike center can be seen. Some concentrations of light can also be detected, with effort, across the globular's face, but these must be superimposed stars, because the globular's brightest members shine at magnitude 15.5, so they cannot be resolved in small telescopes.

2. NGC 5033 (H I-97)

Type	Con	RA	Dec	Mag	Dim	Rating
Spiral galaxy	Canes Venatici	13^h 13.4^m	+36° 36′	10.2	10.5′× 5.1′	3.5

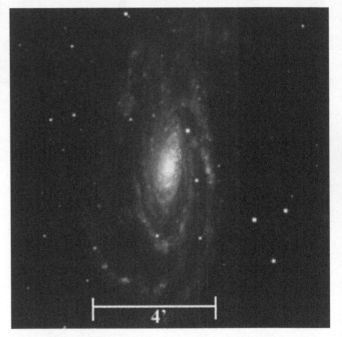

General description

NGC 5033 is a moderately large and bright galaxy about 3° east–southeast of the fine 3rd-magnitude double star Alpha (α) Canum Venaticorum. The galaxy has a bright core, which stands out well in small telescopes. Under a dark sky the core can be seen in 7×50 binoculars with averted vision.

Directions

Use Chart 40 to locate Alpha Canum Venaticorum, which is a little less than 15° southwest of Eta (η) Ursae Majoris, the easternmost star in the Big Dipper's handle. Now use your unaided eyes or binoculars to locate 5th-magnitude 14 Canum Venaticorum about 3° to the southeast. Center 14 Canum Venaticorum in your telescope at low power, then switch to Chart 40a. From 14 Canum Venaticorum, make a slow and careful sweep 1° east–northeast to the 8th-magnitude eclipsing binary star RS Canum Venaticorum. (The star varies between 8th- and 9th-magnitude every 4.8 days.) NGC 5033 is about 50′ north–northeast of that RS Canum Venaticorum, and a little less than 20′ south of 6.5-magnitude Star *a*.

Field note

Take care when identifying NGC 5033, because just $\frac{1}{2}°$ northwest of it is its telescopic "twin," NGC 5005 (see the next Herschel 400 object).

The quick view

At 23×, NGC 5033 is a moderately large (5′) elliptical haze with a bright round core that has an outer diffuseness when seen with averted vision. At 72×, the galaxy is a fine lens-shaped object that becomes increasingly more condensed and round toward the center. The lens is enveloped by a diffuse elliptical halo.

Stop. Do not move the telescope. Your next target is nearby.

3. NGC 5005 (H I-96)

Type	Con	RA	Dec	Mag	Dim	Rating
Barred spiral galaxy	Canes Venatici	13^h 10.9^m	+37° 03′	9.8	5.8′× 2.8′	4

General description

NGC 5005 is a moderately large and bright barred spiral galaxy near NGC 5033. Under a dark sky, it is visible in 7×50 binoculars with averted vision. Telescopically it is a nice, soft, oval haze.

Directions

Use Chart 40a. NGC 5005 is 40′ northwest of NGC 5033. It is also about 35′ west–northwest of Star *a*, and a little less than 30′ southeast of 6.5-magnitude Star *b*.

The quick view

At 23×, NGC 5055 is a soft, oval haze with a distinct brightening at the center. At times some fleeting details appear in the galaxy's outer envelope. At 72×, the galaxy has some ill-defined arcs of light, indicative of spiral structure; these can only be seen with effort.

Stop. Do not move the telescope.

4. NGC 5273 (H I-98)

Type	Con	RA	Dec	Mag	Dim	Rating
Lenticular galaxy	Canes Venatici	13^h 42.1^m	+35° 39′	11.6	2.8′× 2.4′	2

General description

NGC 5273 is a very small and faint lenticular galaxy near 5th-magnitude 25 Canum Venaticorum. A dark sky is required to see it in a small telescope. It appears ill-defined and small even in larger instruments. Be patient, work with moderate magnification, and use averted vision.

Directions

Use Chart 40 (and binoculars, if necessary) to locate 5th-magnitude 25 Canum Venaticorum which is about 8°

east–southeast of Alpha Canum Venaticorum. It is the southernmost of two 5th-magnitude stars in that area separated by 1° and oriented northwest to southeast. Center 25 Canum Venaticorum in your telescope at low power, then switch to Chart 40b. NGC 5273 is about $1\frac{1}{4}^{\circ}$ southeast of 25 Canum Venaticorum, and a little less than 20' west–southwest of 8th-magnitude Star *a*.

The quick view

At 23×, NGC 5273 is a very small (1') and diffuse glow that appears slightly irregular with averted vision. At 72×, the galaxy is just a dim round glow with a hint of a brightening toward the center. Larger telescopes will show the galaxy more concentrated toward the center.

5. NGC 5195 (H I-186)

Type	Con	RA	Dec	Mag	Dim	Rating
Peculiar irregular galaxy	Canes Venatici	13^h 30.0^m	+47° 16'	9.6	6.4'× 4.6'	4

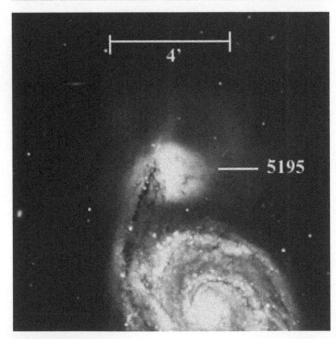

General description

NGC 5195 is the bright companion galaxy to Messier 51. Although it is widely recognized today as a Herschel object, it was Pierre Mechain who first discovered it in 1773 and Messier observed it several times afterwards.

Directions

Use Chart 40 to locate 2nd-magnitude Eta (η) Ursae Majoris, the easternmost star in the Big Dipper's handle. Now use your unaided eyes or binoculars to find 5th-magnitude 24

Ursae Majoris, which is $2\frac{1}{4}^{\circ}$ east–southeast of Eta Ursae Majoris. Center 24 Ursae Majoris in your telescope at low power, then switch to Chart 40c. From 24 Ursae Majoris, make a slow and careful $1\frac{1}{2}^{\circ}$ sweep southwest to a 40'-wide equilateral triangle (*a*) of 7th-magnitude suns. NGC 5195 is 20' east of the southernmost star in Triangle *a*, a little less then 5' north of M51's nucleus.

The quick view

At 23× in the 4-inch, NGC 5195 is a 3'-wide circular glow, like a 9th-magnitude comet without a tail. At 72×, the galaxy becomes a tad more oval shaped, oriented east to west. It has a distinct nebulous core surrounded by a circular inner "coma" nestled in an slight ellipse of fainter light. Larger telescopes will show all manner of irregularities in its halo, including what appears to be bright nebulous arcs on the northern and southern edges, and an array of wispy tendrils that jut mostly to the north.

6. NGC 4800 (H I-211)

Type	Con	RA	Dec	Mag	Dim	Rating
Spiral galaxy	Canes Venatici	12^h 54.6^m	+46° 32'	11.5	1.6'× 1.1'	2.5

General description

NGC 4800 is a very small and dim galaxy with a bright central condensation. It requires a dark sky to see at low powers in small telescopes and light pollution will definitely affect its visibility. Be prepared to use moderate magnification and to look for a "fuzzy" star with averted vision.

Directions

Use Chart 40 to locate 2nd-magnitude Epsilon (ε) Ursae Majoris in the handle of the Big Dipper. Now use your unaided eyes or binoculars to look about 7° south of Epsilon Ursae Majoris for 6.5-magnitude 11 Canum Venaticorum, which is the southwestern component of a pair of 6.5-magnitude stars (see Box **d**) separated by 20′ and oriented northeast to southwest. Center 11 Canum Venaticorum in your telescope at low power, then switch to Chart 40d. From 11 Canum Venaticorum, move 20′ southeast to a crooked 35′-long line of four 9th-magnitude stars (a.) Now move 40′ southeast to 9th-magnitude Star b.

Next, hop 35′ southeast to the semi-regular variable star TU Canum Venaticorum, which fluctuates between 5.5-magnitude and 6.6-magnitude every 50 days. Now hop 25′ south–southeast to 7.5-magnitude Star c. NGC 4800 is just 15′ south of Star c.

The quick view

At 23×, NGC 4800 is a 1′-wide, round, and diffuse glow that is visible only with averted vision. At 72×, the galaxy is an easy sight, appearing as a 1.5′-wide condensed, circular glow with an intense, though fuzzy, core.

Star charts for seventh night

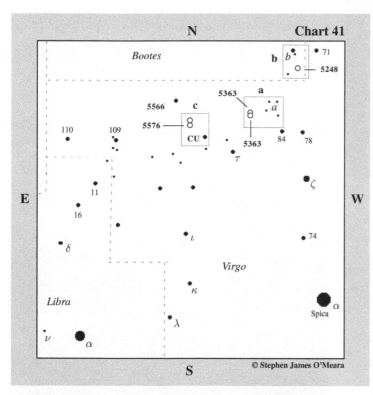

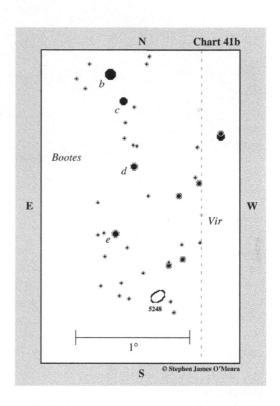

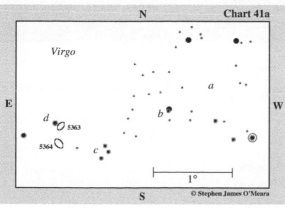

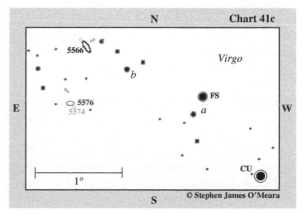

SEVENTH NIGHT

1. NGC 5363 (H I-6)

Type	Con	RA	Dec	Mag	Dim	Rating
Peculiar galaxy	Virgo	13^h 56.1^m	+05° 15′	10.1	4.7′× 3.2′	4

General description

NGC 5363 is a moderately small but bright galaxy that hides in the midsection of the bleak and lonely corridor of space between 1st-magnitude Alpha (α) Virginis (Spica), and −0-magnitude Alpha (α) Bootes (Arcturus–not shown). It is a compact glow, making it a fine target for small-telescopes users. From a dark-sky site, it can be picked up in a sweep. To find it, though, you will need to be patient and take the time to confirm its guidestars with binoculars.

Directions

Use Chart 41 to locate Alpha Virginis (Spica). A little more than 10° (a fist width) north–northwest is 3rd-magnitude Zeta (ζ) Virginis. Next, use your unaided eyes or binoculars to find 5th-magnitude 78 Virginis 4° due north of Zeta Virginis, then look for 6th-magnitude 84 Virginis 2° to the east. Now use your binoculars to find a 1½°-long kite-shaped asterism (a) 2½° to the northeast; the Kite is oriented northeast to southwest and comprises four 7th-magnitude suns. It's time to switch switch to Chart 41a. You want to center the Kite's easternmost star (b) in your telescope at low power; note that Star b is a fine telescopic double. From Star b, make a slow and careful 1° sweep southeast to a 10′-wide right triangle of 9th- and 10th-magnitude stars (c). Now swing

your scope 45′ northeast to 8.5-magnitude Star d. NGC 5363 is about 4′ southwest of Star d.

The quick view

At 23×, NGC 5363 is immediately obvious as a 3′-wide glow with a sharp core and a halo that brightens gradually toward the middle. With little effort, I can see that it is elongated slightly northwest–southeast. At 72×, the core of NGC 5363 is extremely sharp – a star surrounded by a circular inner iris, which is surrounded by a more diffuse and slightly elliptical lens of light.

Field note

A 12th-magnitude star is superimposed about 5″ southwest of NGC 5363's true (and equally bright) nucleus.

Stop. Do not move the telescope. Your next target is nearby.

2. NGC 5364 (H II-534)

Type	Con	RA	Dec	Mag	Dim	Rating
Spiral galaxy	Virgo	13^h 56.2^m	+05° 01′	10.5	6.6′× 5.1′	3

General description

NGC 5364 is a moderately bright spiral galaxy near NGC 5363. NGC 5364 is slightly larger and slightly fainter than NGC 5363. In small telescopes it is best seen with averted vision and moderate magnification.

Directions

Use Chart 41a. NGC 5364 is a little less than 15′ due south of NGC 5363.

The quick view

At 23×, NGC 5364 looms into view like a flaring comet. Its amorphous glow, and its large size makes it less obvious in the 4-inch than neighboring NGC 5363, though, with averted vision, NGC 5364 draws attention to itself. At 72×, the galaxy has a slightly swollen and brighter center with an irregularly bright surrounding halo.

Stop. Do not move the telescope.

3. NGC 5248 (H I-34)

Type	Con	RA	Dec	Mag	Dim	Rating
Mixed spiral galaxy	Bootes	13^h 37.5^m	+08° 53′	10.3	6.2′× 4.6′	3.5

4′

General description

NGC 5248 is a moderately bright and large spiral galaxy almost 10° north–northeast of 3rd-magnitude Zeta (ζ) Virginis. Its core is condensed, making it a reasonable target for small-telescope users, but a dark sky is required. Again, patience (and binoculars at first) are needed to locate the galaxy's field, which is in another stellar void.

Directions

Use Chart 41 to return to 78 Virginis. Note that 78 Virginis marks the southern apex of a 7°-long isosceles triangle with 5.6-magnitude 71 Virginis and 6th-magnitude Star *b*. Center Star *b* in your telescope at low power, then switch to Chart 41b. From Star *b*, move about 15′ southwest to 7th-magnitude Star *c*. Next, hop about 30′ south–southwest to 7.5-magnitude Star *d*. Now move 35′ south–southeast to 8th-magnitude Star *e*. NGC 5248 is about 40′ southwest of Star *e*.

The quick view

In the 4-inch at 23×, NGC 5248 is a small, somewhat weak glow with a sharply condensed core surrounded by a pale oval disk. It is only 3′ in extent and oriented roughly east to west. At 72×, the galaxy looks more-or-less the same way it did at low power, but a bit more intense. No definite structural details can be seen. With averted vision, however, it does appear mottled. A 13th-magnitude star borders the galaxy to the south–southwest.

4. NGC 5566 (H I-144)

Type	Con	RA	Dec	Mag	Dim	Rating
Barred spiral galaxy	Virgo	14^h 20.3^m	+03° 56′	10.6	5.7′× 2.1′	3.5

5566

4′

General description

NGC 5566 is a moderately small and bright galaxy about 5° northeast of 4th-magnitude Tau (τ) Virginis. Its core is nicely condensed, making it a reasonable target for small-telescope users.

Directions

Use Chart 41 to locate Tau Virginis, which is about 7° northeast of 3rd-magnitude Zeta (ζ) Virginis. From Tau Virginis, use your unaided eyes or binoculars to locate 5th-magnitude CU Virginis. Center CU Virginis in your telescope at low power, then switch to Chart 41c. From CU Virginis, make a careful 1° sweep northeast to a pair of 6.5- and 7.5-magnitude stars (*a*); the northernmost star in the pair is the irregular variable FS Virginis. Now make another careful 1° sweep further to the northeast, to

7th-magnitude Star b. NGC 5566 is 30′ northeast of Star b, and 10′ southeast of an 8th-magnitude star with a 10th-magnitude companion to the northwest.

The quick view

At 23×, NGC 5566 is a 2′-long and narrow glow, oriented northeast to southwest, with a bright core and starlike nucleus. At 72×, the galaxy has a bright round core with a sharp, starlike center. Otherwise, the disk is an amorphous glow, which, with averted vision, looks sharper and more tapered toward the northeast. Larger telescopes should reveal its two companion galaxies, NGC 5560 to the northwest, and NGC 5569 to the northeast.

Stop. Do not move the telescope. Your next target is nearby!

5. NGC 5576 (H I-146)

Type	Con	RA	Dec	Mag	Dim	Rating
Elliptical galaxy	Virgo	14ʰ 21.1ᵐ	+03° 16′	11.0	3.0′× 2.4′	2.5

General description

NGC 5576 is a small and somewhat dim elliptical galaxy near NGC 5566. It is best seen under a dark sky with moderate magnification and averted vision.

Directions

Chart 41c. NGC 5576 is about 40′ south–southeast of NGC 5566.

The quick view

At 23× in the 4-inch under a dark sky, NGC 5576 appears a little larger (3′) and more diffuse than neighboring NGC 5566. With averted vision, It swells into an oval glow, oriented west to east. At 72×, the galaxy's core seems round. It also contains a dim starlike core. The surrounding halo looks fuzzy and out of round. A 12th-magnitude companion galaxy, NGC 5574, can be seen at this magnification 3′ to the southwest (see photo).

Summer

7 · July

Star charts for first night

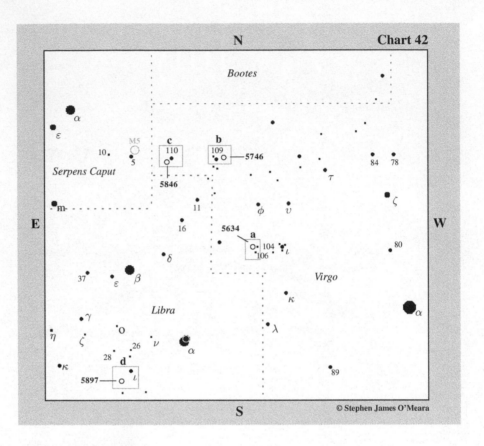

Chart 42

N

Bootes

Serpens Caput

α

ε

M5

10.

5

c
110
5846

b
109
5746

84   78

τ

ζ

φ   υ

E   W

m

11

16

5634

a
104
106   ι

80

δ

Virgo

37

ε   β

κ

Libra

γ

λ

η   ζ   o

ν

26   α

28   **d**

κ   ι

5897

α

89

S

© Stephen James O'Meara

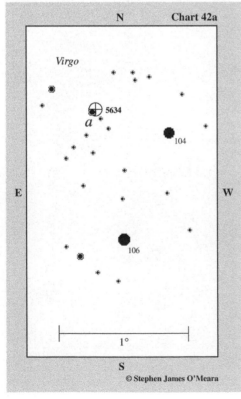

Chart 42a

N

Virgo

5634

a

104

E   W

106

1°

S

© Stephen James O'Meara

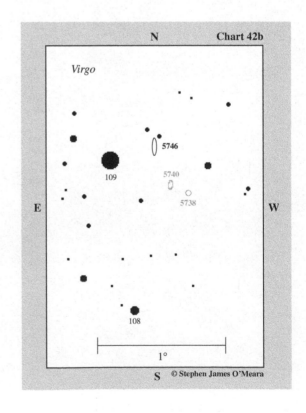

Chart 42b

N

Virgo

5746

109   5740

5738

E   W

108

1°

S   © Stephen James O'Meara

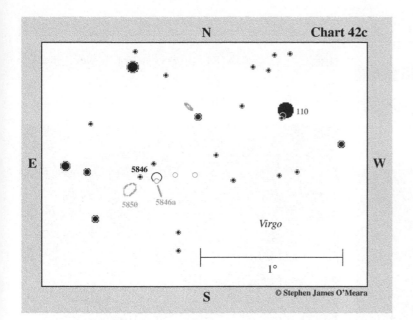

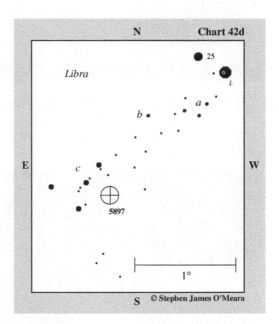

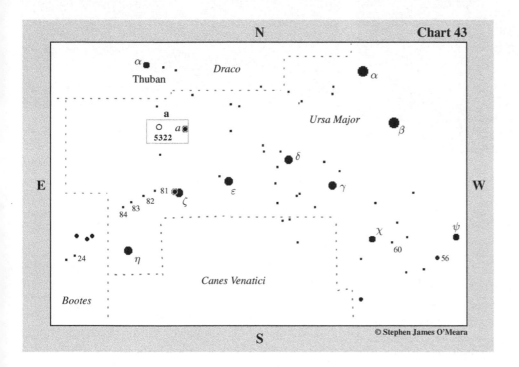

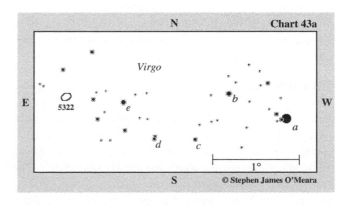

FIRST NIGHT

1. NGC 5634 (H I-70)

Type	Con	RA	Dec	Mag	Diam	Rating
Globular cluster	Virgo	14^h 29.6^m	−05° 59′	9.5	5.5′	3.5

General description

NGC 5634 is a moderately small and condensed globular cluster near 6th-magnitude 104 Virginis. It is also very close to an 8.5-magnitude star, making it relatively easy to find in a small telescope.

Directions

Use Chart 42 to locate 1st-magnitude Alpha (α) Virginis (Spica). Now look about 13° northeast for 4th-magnitude Iota (ι) Virginis, which sits at the center of an upright Y-shaped asterism of binocular stars. Now use your unaided eyes or binoculars to find 5th-magnitude 106 Virginis 3° to the southeast. Center this star in your telescope at low power, then switch to Chart 42a. From 106 Virginis, move 50′ north–northwest to 6th-magnitude 104 Virginis. NGC 5634 is a little less than 35′ east–northeast of 104 Virginis, just about 1.5′ west–northwest of 8.5-magnitude Star a.

The quick view

At 23×, NGC 5634 is part of a roughly 4′-wide isosceles triangle with Star a and a 10th-magnitude star to the south-west. With averted vision, it is a small object that swells to about 3′ in extent. It is best to boost the power to at least 101×, when the cluster stands out clearly from Star a. The cluster is very condensed but unresolvable, though it does gradually brighten towards the center. It looks more like a small comet than a globular cluster.

2. NGC 5746 (H I-126)

Type	Con	RA	Dec	Mag	Dim	Rating
Mixed spiral galaxy	Virgo	14^h 44.9^m	+01° 57′	9.3	8.1′ × 1.4′	3.5

General description

NGC 5746 is a beautiful edge-on system near 4th-magnitude 109 Virginis. It is one of the most neglected galaxies in Virgo and a real visual gem that's easy to find. A dark sky is necessary to appreciate its beauty in a small telescope, because the galaxy has a low surface brightness.

Directions

Use Chart 42 to locate 109 Virginis, which is about 13° northwest of 2.6-magnitude Beta (β) Librae. You can also follow the stream of stars that flow to it from Beta Librae, namely Delta (δ), 16, and 11 Virginis. Once you find 109 Virginis, center the star in your telescope at low power, then switch to Chart 42b. NGC 5746 is only about 20′ west and slightly north of 109 Virginis.

The quick view

At 23× in the 4-inch under a dark sky, NGC 5746 all but fades from view with a direct glance, but just like other edge-on galaxies, with averted vision it suddenly swells into view as a fine sliver of light; NGC 5746 is 6′ in length and oriented north–northwest to south–southeast. In fact, NGC 5746 is more apparent than NGC 891, appearing like a sharp needle with a bright bead at its core. With 72× and averted vision, the galaxy's nucleus is very apparent – a star in a thin wafer of light. The disk is not uniform, especially to the south, where there is a definite star or enhancement. With time I can also make out a waviness to the galaxy's bulge on the larger western side.

Stop. Do not move the telescope.

3. NGC 5846 (H I-128)

Type	Con	RA	Dec	Mag	Dim	Rating
Elliptical galaxy	Virgo	15ʰ 06.4ᵐ	+01° 36′	10.0	3.0′ × 3.0′	3

General description

NGC 5846 is a small but relatively conspicuous elliptical galaxy near 4th-magnitude 110 Virginis. It is like a condensed fuzzy star that is best seen at moderate magnifications in small telescopes.

Directions

Use Chart 42 to locate 110 Virginis, which is 4° east of 109 Virginis. Center 110 Virginis in your telescope at low power, then switch to Chart 42c. NGC 5846 is 1° southeast of 110 Virginis.

The quick view

Under a dark sky in the 4-inch at 23×, NGC 5846 is a 2′-wide and relatively obvious fuzzy patch that brightens to a core. At 72×, the galaxy is a round condensation with a starlike nucleus. Larger telescopes should clearly show the highly condensed 13.5-magnitude starlike galaxy NGC 5846 40″ south of NGC 5846's core. The 11th-magnitude galaxy NGC 5850 can also be seen 10′ to the southeast.

4. NGC 5897 (H VI-19)

Type	Con	RA	Dec	Mag	Diam	Rating
Globular cluster	Libra	15ʰ 17.4ᵐ	−21° 01′	8.4	11.0′	3.5

General description

NGC 5897 is an unusually loose globular cluster among the dim and unassuming stars of Libra, near 4.5-magnitude Iota (ι) Librae. The cluster can be seen on clear, moonless nights – away from city lights – as a large amorphous glow nearly one-third the apparent size of Omega (ω) Centauri. From a dark-sky site, it is visible in 7 × 50 binoculars. Even the smallest telescope will shows it as a beautiful but dim cometary glow near an inverted-Y-shaped asterism comprising four 8th-magnitude stars. I call NGC 5897 the Ghost Globular because it resembles a ghost image of the globular M55 in Sagittarius.

Directions

Use Chart 42 to find Iota Librae, which is about 6° southeast of the wide binocular double star Alpha¹, ² (α ¹, ²) Librae. If you cannot see Iota Librae with your unaided eye, use binoculars to find it; Iota Librae is the only star of that magnitude in the region, and 6th-magnitude 25 Librae lies only about 17′ to the northeast (see Chart 42d), so you will see a fine pair of bright stars in your binoculars or telescope. Once you identify Iota Librae, place it in your telescope at low power, then switch to Chart 42d. From Iota Librae, move about 30′ south–southeast to a 15′-wide triangle of three roughly 9.5-magnitude stars (a). Now hop about 30′ east and slightly south to solitary 9th-magnitude Star b. The 25′-long, inverted-Y-shaped Asterism c is less than 1° southeast of Star b. NGC 5897 is 15′ west Asterism c.

The quick view

At 23×, NGC 5897 is a perfect comet without a tail, reminiscent of M56 in Lyra. With averted vision, a central pip swims in and out of view. Also, a multitude of phantom stars float across the cluster's periphery. At 72×, NGC 5897's oblate disk appears buffed and featureless, as if its stars had been scrubbed away with bleach. But with a little concentration the glow resolves into a patchwork of glittering hazes that lacks a strong central condensation. The cluster's brightest stars shine at a magnitude of 13.3, and they appear to form slight needles of light that border minute "holes" at the core.

5. NGC 5322 (H I-256)

Type	Con	RA	Dec	Mag	Dim	Rating
Elliptical galaxy	Ursa Major	13^h 49.3^m	+60° 12'	10.2	6.1'× 4.1'	3.5

4'

Star charts for second night

General description
NGC 5322 is a moderately small but very condensed elliptical galaxy almost midway between 2nd-magnitude Zeta (ζ) Ursae Majoris in the Big Dipper's handle and 3.6-magnitude Alpha (α) Draconis (Thuban). Because it is so condensed, it is a good target for small-telescope users.

Directions
Use Chart 43 to locate 2nd-magnitude Epsilon (ε) Ursae Majoris and Alpha Draconis. Now look midway between them for solitary 5th-magnitude Star a. Center Star a in your telescope at low power, then switch to Chart 43a. Star a is easy to identify in a telescope because it is the brightest star in a 5' group of four stars; two of the other three stars shine at 8.5-magnitude, while the fourth shines at 10th magnitude. From Star a you could move about 45' east–northeast to 7th-magnitude Star b, then make a generous sweep about 1$\frac{1}{4}$° east–southeast to 7.5-magnitude Star e. Otherwise you could make shorter hops. From Star b, move 40' southeast to 8.5-magnitude Star c. Next, hop 30' east to a close pair of 9th-magnitude stars (d). A 35' swing northeast will take you to 7.5-magnitude Star e. NGC 5322 is 40' east of Star e.

The quick view
At 23× in the 4-inch, NGC 5322 is a 3'-wide oval glow, oriented east to west, that gradually, then suddenly, becomes much brighter toward the center to a very starlike nucleus. At 72×, the galaxy has an extremely bright and sharp stellar core in a 1.5'-wide, condensed inner lens that's surrounded by an elliptical halo, which, with averted vision, appears somewhat irregular – an illusion due to superimposed faint stars.

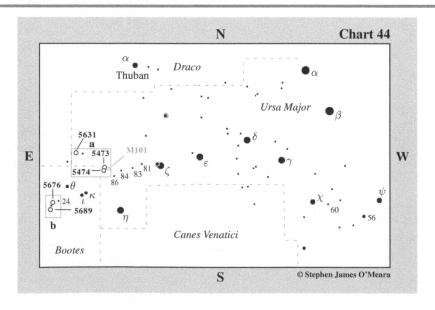

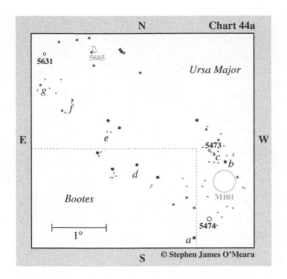

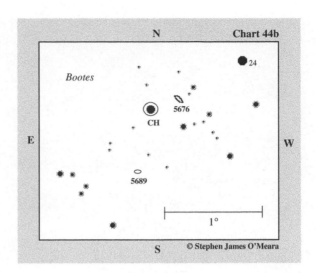

SECOND NIGHT

1. NGC 5474 (H I-214)

Type	Con	RA	Dec	Mag	Dim	Rating
Peculiar spiral galaxy	Ursa Major	14ʰ 05.0ᵐ	+53° 40′	10.8	6.0′× 4.9′	2

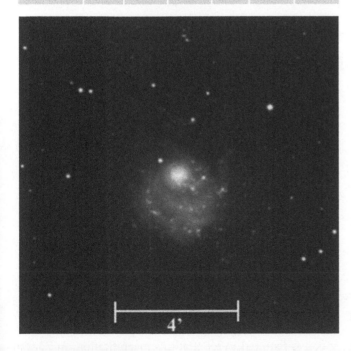

4′

General description

NGC 5474 is a moderately small and dim galaxy near M101, a magnificent open-face spiral galaxy about 5½° east and slightly south of 2nd-magnitude Zeta (ζ) Ursae Majoris in

the handle of the Big Dipper. A dark sky and low power is required to see it in a small telescope.

Directions

Use Chart 44 to locate Zeta Ursae Majoris. Now, using your unaided eyes or binoculars, locate 5th-magnitude 86 Ursae Majoris; note that it lies at the southeast end of a 3½°-long chain of four 5th-magnitude stars that starts about 1½° east–northeast of Zeta Ursae Majoris, at 81 Ursae Majoris. Center 86 Ursae Majoris in your telescope at low power, then sweep 1½° east–northeast to 7.9-magnitude M101. Center that bright galaxy in your telescope, then switch to Chart 44a. NGC 5474 is about 45′ south–southeast of M101's core and about 25′ north–northwest of 7th-magnitude Star *a*.

The quick view

At 23×, NGC 5474 is a faint, diffuse round patch of uniform light 3′ in diameter. In small telescopes, higher powers only diffuse the light to near invisibility. It is best seen under a dark sky at low power. Larger telescopes make it appear more obvious, slightly more concentrated but no more defined.

Stop. Do not move the telescope. Your next target is nearby!

2. NGC 5473 (H I-231)

Type	Con	RA	Dec	Mag	Dim	Rating
Mixed spiral galaxy	Ursa Major	14ʰ 04.7ᵐ	+54° 54′	11.4	2.2′× 1.7′	2

General description

NGC 5473 is an extremely small and dim galaxy near M101. Although it is smaller and fainter than NGC 5474 (see above), it is more condensed, so it looks like a slightly fuzzy star.

Directions

Using Chart 44a, from NGC 5474, return to M101. Now move about 20′ north–northwest of M101's center to 8th-magnitude Star *b*. A little less than 20′ northeast is a roughly 9th-magnitude Star *c*. NGC 5473 is about 5′ further to the northeast, about 1′ northeast of a 10.5-magnitude star.

The quick view

At 23× in the 4-inch under a dark sky, NGC 5473 is a very small (~1′) diffuse and featureless glow. Higher powers do little to enhance the view in a small telescope. Larger telescopes will show the object as a tiny elliptical glow, oriented roughly north to south, with a very condensed core (which is about all you see in small telescopes). A roughly 14th-magnitude star lies east–northeast of the bright nuclear condensation.

Stop. Do not move the telescope. Your next target is nearby!

3.   NGC 5631 (H I-236)

Type	Con	RA	Dec	Mag	Dim	Rating
Spiral galaxy	Ursa Major	14ʰ 26.6ᵐ	+56° 35′	11.5	1.8′ × 1.8′	2

General description

NGC 5631 is a small and dim galaxy about 4° northeast of M101. You will need to make a careful star hop to it, then look for a small fuzzy "star" at its location. A dark sky and moderate magnification is required to see it at all in a small telescope.

Directions

Using Chart 44a, from NGC 5473, return to M101. Now make a slow and careful 1½° sweep east and slightly north to a 50′-long arc of three roughly 7.5-magnitude stars (*d*), which is oriented east to west. About 45′ northeast of Arc *d*, you'll find a roughly 35′-long acute Triangle *e*, which comprises three 7th-magnitude stars and is oriented north–northwest to south–southeast. Now make another very slow and careful sweep about 50′ northeast to a pair of 8th-magnitude stars (*f*), oriented north–northwest to south–southeast and separated by 20′. Pair *f* forms the southwestern side of a roughly 35′-long acute triangle, whose northeastern apex is 7.5-magnitude Star *g*. NGC 5631 is a little more than 30′ north–northwest of Star *g*.

The quick view

Under a dark sky in the 4-inch at 23×, NGC 5631 is a very dim 2′-wide oval glow of low surface brightness. Averted vision is required to see it at this magnification. At 72×, it becomes clear that the 2′-wide oval is actually the combined glow of NGC 5631 and some nearby stars. At 72×, the galaxy is a 1′-wide circular glow that gradually, then suddenly, becomes brighter in the middle, though not to a stellar nucleus. Interestingly, larger telescopes show a starlike nucleus.

4. NGC 5676 (H I-189)

Type	Con	RA	Dec	Mag	Dim	Rating
Spiral galaxy	Bootes	14ʰ 32.8ᵐ	+49° 28′	11.2	3.7′ × 1.6′	2

General description

NGC 5676 is a small and dim galaxy near 6th-magnitude 24 Bootis. It is best seen at moderate magnification in a small telescope under a dark sky.

Directions

Use Chart 44 to find 2nd-magnitude Eta (η) Ursae Majoris, the tip of the Great Bear's tail. Now use your unaided eyes or binoculars to find the 2°-wide acute triangle, consisting of the 4th- and 5th-magnitude stars Kappa (κ), Theta (θ), and Iota (ι) Bootis. Then look 2° south–southeast for 24 Bootis. You want to center that star in your telescope at low power, then switch to Chart 44b. From 24 Bootis, make a careful sweep a little more than 1° southeast to the 6th-magnitude irregular variable star CH Bootis. NGC 5676 is a little less than 20′ northwest of CH Bootis.

The quick view

At 23×, NGC 5676 is barely visible even under a dark sky. It becomes a little more pronounced at 72×, appearing as a very dim oval glow, 2′ wide, oriented northeast to southwest. There's a very slight enhancement toward the galaxy's center.

Stop. Do not move the telescope. Your next target is nearby!

5. NGC 5689 (H I-188)

Type	Con	RA	Dec	Mag	Dim	Rating
Barred spiral galaxy	Bootes	14ʰ 35.5ᵐ	+48° 45′	11.9	3.7′ × 1.0′	1.5

General description

NGC 5689 is a very small and dim galaxy about $1\frac{1}{2}°$ southeast of 6th-magnitude 24 Bootis. A dark sky and moderate magnification is required to see it in a small telescope. Think, "small, fuzzy star."

Directions

Using Chart 44b, from NGC 5676, return to CH Bootis. NGC 5689 is a little less than 40′ to the south–southeast.

The quick view

At 23× in the 4-inch, NGC 5689 is merely a 1′-wide puff of dim light. At 72×, the galaxy displays a bright, fuzzy starlike core with dim extensions to the east and west. It is best seen with averted vision.

Star charts for third night

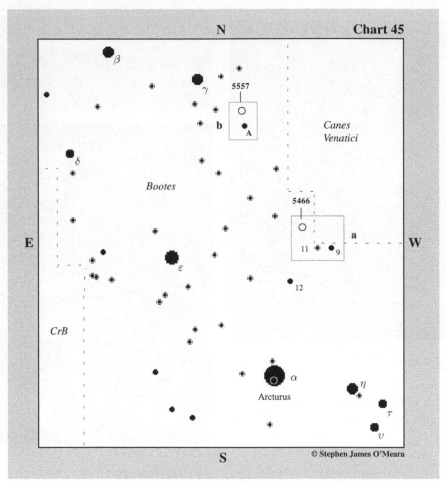

Chart 45

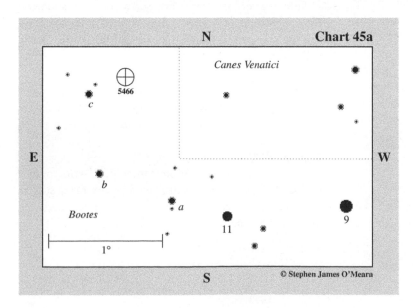

Chart 45a

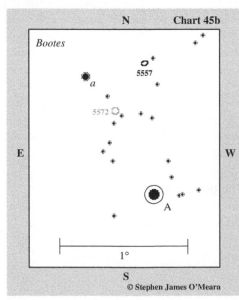

Chart 45b

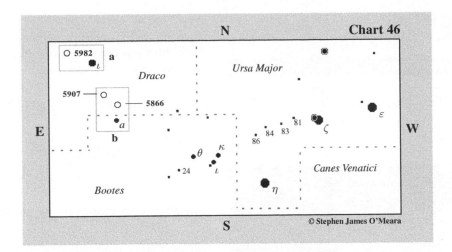

Chart 46

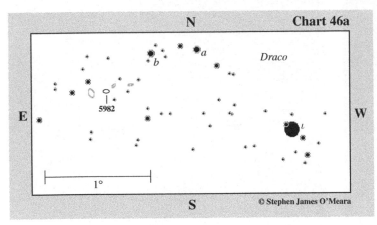

Chart 46a

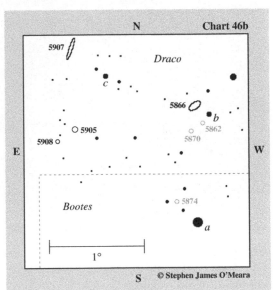

Chart 46b

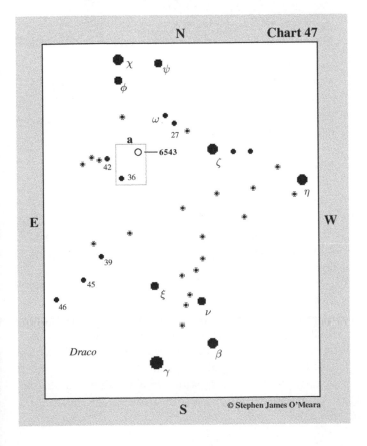

Chart 47

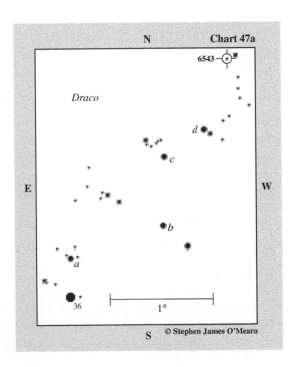

Chart 47a

Third night

1. NGC 5466 (H VI-9)

Type	Con	RA	Dec	Mag	Diam	Rating
Globular cluster	Bootes	14ʰ 05.5ᵐ	+28° 32′	9.2	9.0′	3.5

4′

General description

NGC 5466 is a beautiful globular cluster almost 10° north–northwest of −0-magnitude Alpha (α) Bootis (Arcturus). In a small telescope under a dark sky it will appear as a tailless 9th-magnitude comet.

Directions

Use Chart 45 to locate Alpha Bootis. Next, use your unaided eyes or binoculars to find 5th-magnitude 12 Bootis (about 6° to the north–northwest). Now look 4° northwest for 5th-magnitude 9 Bootis. Center 9 Bootis in your telescope at low power, then switch to Chart 45a. From 9 Bootis, make a gentle 1° sweep east–southeast to 6.5-magnitude 11 Bootis. Next, move 30′ east–northeast to 7th-magnitude Star a. A 40′ sweep to the east–northeast will bring you to 8th-magnitude Star b. And a roughly 45′ swing to the north–northeast takes you to 7th-magnitude Star c. NGC 5466 is only 20′ west–northwest of Star c.

The quick view

At 23×, NGC 5466 is an 8′-wide round and uniform, comet-like glow with no central condensation. With averted vision, the globular's shape becomes irregular and somewhat frazzled along the edges. At 72×, the globular cluster breaks up into dim mottled patches with some irregularities. Larger scopes will begin to resolve some of its brightest members, which shine at a magnitude of 13.8, though the horizontal branch magnitude is a dim 16.6.

2. NGC 5557 (H I-99)

Type	Con	RA	Dec	Mag	Dim	Rating
Elliptical galaxy	Bootes	14ʰ 18.4ᵐ	+36° 30′	11.0	2.2′ × 2.0′	2

4′

General description

NGC 5557 is a very small and faint galaxy near 3rd-magnitude Gamma (γ) Bootis. A dark sky, averted vision, and moderate magnification is required to see it in a small telescope. Think, "fuzzy star."

Directions

Use Chart 45 to locate Gamma Bootis. Next, look about 4° southwest for 5th-magnitude A Bootis. Center A Bootis in your telescope at low power, then switch to Chart 45b. From A Bootis, make a slow and careful 1° sweep northeast to 7th-magnitude Star a. NGC 5557 is 30′ west–northwest of Star a.

The quick view

At 23× in the 4-inch under a dark sky, NGC 5557 is a very small (~1′) and slightly condensed circular glow with averted vision. At 72×, the galaxy has a very small, swollen starlike core in a slightly oval halo (oriented east to west) that swells to about 2′ with averted vision. Larger telescopes may show a dim star immediately southeast of the nucleus.

3. NGC 5982 (H II-764)

Type	Con	RA	Dec	Mag	Dim	Rating
Elliptical galaxy	Draco	15ʰ 38.7ᵐ	+59° 21′	11.1	3.0′ × 2.2′	1.5

General description

NGC 5982 is a very small and very faint galaxy near 3rd-magnitude Iota (ι) Draconis. A dark sky is a must to see this exceedingly low-surface-brightness object in a small telescope. You need to know exactly where to look to see it with averted vision.

Directions

Use Chart 46 to find 2nd-magnitude Eta (η) Ursae Majoris, the star at the end of the Big Dipper's handle. Now use your unaided eyes or binoculars to find the 2°-wide acute triangle, consisting of the 4th- and 5th-magnitude stars Kappa (κ), Theta (θ), and Iota (ι) Bootis. Iota Draconis is about 10° (one fist held at arm's length) northeast of Theta Bootis. Center Iota Draconis in your telescope at low power, then switch to Chart 46a. From Iota Draconis, make a slow and careful 1¼° sweep northeast to 8th-magnitude Star a. Now move a little less than 30′ east–southeast to 7.5-magnitude Star b. NGC 5982 is 30′ southeast of Star b.

Field note

Although other galaxies populate the field, NGC 5982 is the brightest.

The quick view

Under a dark sky in the 4-inch at 23×, NGC 5982 is visible only as a speck of fuzzy lint (about 30″ in diameter) as seen with averted vision. At 72×, the galaxy is a difficult object of low surface brightness.

4. NGC 5866 (H I-215)

Type	Con	RA	Dec	Mag	Dim	Rating
Lenticular galaxy	Draco	15ʰ 06.5ᵐ	+55° 46′	9.8	7.3′ × 3.5′	4

General description

NGC 5866 is a moderately sized and highly condensed lenticular galaxy about 4° southwest of 3rd-magnitude Iota (ι) Draconis. Under a dark sky, it is just visible with 7 × 50 binoculars with effort. You have to know exactly where to look. The galaxy is easy to see in the smallest of telescopes, but it is not something that calls attention to itself in a sweep. Again, you have to know exactly where to look to see it. It is a wonder in all larger telescopes.

Directions

Use Chart 46 to find 3rd-magnitude Iota Draconis. Now look for 5th-magnitude Star a just about 5° southwest. Use binoculars if you have to, but Star a should be easy to identify, since it is the brightest star in the region. Center Star a in your telescope, then switch to Chart 46b. NGC 5866 is just a little more than 1° due north of Star a, and 10′ northeast of 7.5-magnitude Star b.

The quick view

At 23×, NGC 5866 is immediately obvious as a condensed elliptical glow, oriented northwest to southeast, with white winglike extensions. The galaxy's light is almost combined with that of an 11.5-magnitude star about 1.5′ to the northwest. The galaxy, however, stands out much more brightly.

Its core is very small (~2') and very compact. At 72×, the galaxy's spindle shape is nicely revealed. The lens looks extremely uniform and smooth. At 303×, a dust lane can be seen (a whisper of darkness) that cleanly splits the central bulge in two.

Stop. Do not move the telescope. Your next target is nearby!

5. NGC 5907 (H II-759)

Type	Con	RA	Dec	Mag	Dim	Rating
Spiral galaxy	Draco	15^h 15.9^m	+56° 20'	10.3	11.5'× 1.7'	3

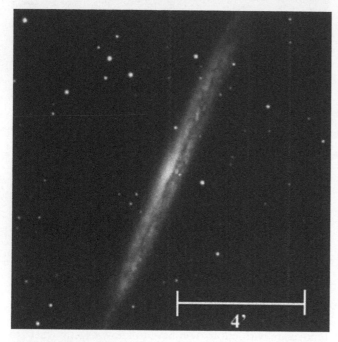

General description

NGC 5907 is a moderately large and extremely thin edge-on galaxy near NGC 5866. Commonly called the Splinter Galaxy (I call it the Cat Scratch Galaxy), it will be a challenge for small-telescope users who are not under a dark sky. But the galaxy is a glorious sight in all larger instruments.

Directions

Using Chart 46b, from NGC 5866, move 55' east–northeast to 7.5-magnitude Star c. NGC 5907 is 30' further to the east–northeast.

The quick view

At 23× in the 4-inch under a dark sky, NGC 5907 is a 10'-long sliver of light (oriented north–northwest to south–southeast). With averted vision, it is a needle, like a tiny tear in the fabric of space. At 72×, the galaxy is not uniformly bright. At the center is a slight bulge that gradually gets brighter toward the center to a starlike nucleus. The disk stretches out on either side and appears broken in places. At

101×, the fragmented appearance makes sense, as a very thin dust lane can be seen, with averted vision, slicing through the galaxy's major axis.

6. NGC 6543 (H IV-37)

Type	Con	RA	Dec	Mag	Dim	Rating
Planetary nebula	Draco	17^h 58.6^m	+66° 38'	8.1	23'× 17'	4

General description

NGC 6543 is a bright planetary nebula 5° east of 3rd-magnitude Zeta (ζ) Draconis. It is highly condensed and appears stellar at low power, but moderate to high magnifications will show its greenish disk to most who seek it.

Directions

Use Chart 47 to locate 5.5-magnitude 36 Draconis, which is about 6° east-southeast of Zeta Draconis. Center 36 Draconis in your telescope at low power, then switch to Chart 47a. From 36 Draconis, move 20' north to 7.5-magnitude Star a. Next, make a careful 1° sweep west–northwest to 7.5-magnitude Star b. Now move about 40' due north to 7th-magnitude Star c. Next, hop about 30' northwest to 7.5-magnitude Star d. NGC 6543 is a little more than 40' north–northwest of Star d, just 3' east–southeast of an 8th-magnitude star.

The quick view

At 23× in the 4-inch, NGC 6543 looks stellar with direct vision and swells with averted vision. The nebula shines with a pale green color. At higher powers, the planetary nebula shows an 11th-magnitude central star inside a bright luminous ring that is surrounded by a diffuse shell of pale light.

Star charts for fourth night

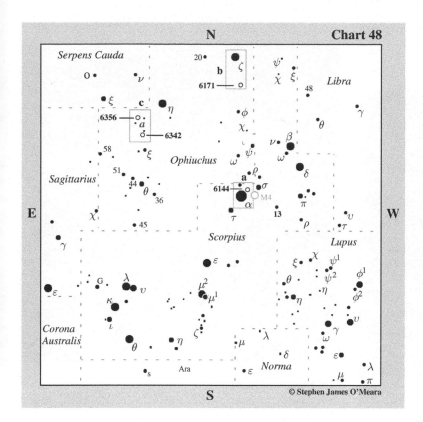

Chart 48

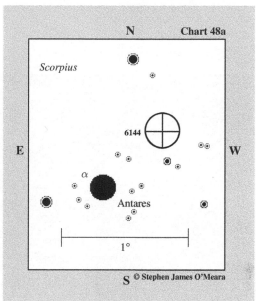

Chart 48a

© Stephen James O'Meara

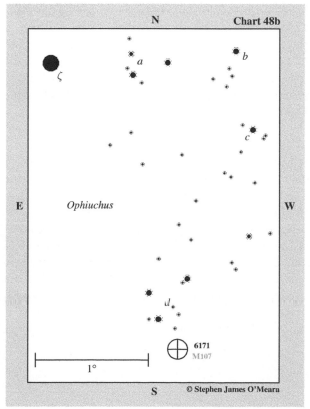

Chart 48b

© Stephen James O'Meara

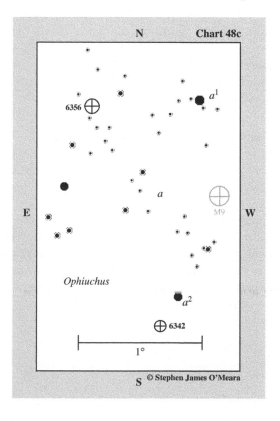

Chart 48c

© Stephen James O'Meara

FOURTH NIGHT

1. NGC 6144 (H VI-10)

Type	Con	RA	Dec	Mag	Diam	Rating
Globular cluster	Scorpius	16^h 27.2^m	−26° 01′	9.0	7.4′	3.5

General description
NGC 6144 is a nice globular cluster near 1st-magnitude Alpha (α) Scorpii (Antares) and open cluster M4. It is equal in brightness to some of the fainter Messier globulars in Sagittarius. It may be washed out if light pollution is present. It's also so close to Alpha Scorpii that moderate powers show it best.

Directions
Use Chart 48 to locate Alpha Scorpii. Center that star in your telescope at low power, then switch to Chart 48a. NGC 6144 is only about 40′ northwest of Alpha Scorpii.

The quick view
At 23× in the 4-inch under a dark sky NGC 6144 is a beautiful, 5′-wide circular glow, somewhat concentrated. It's easy to see with averted vision, if you can ignore the glare from Alpha Scorpii. At 101×, the cluster starts to resolve, especially around the edges. It shows a slightly brighter core that has no real central condensation.

2. NGC 6171 (H VI-40) = M107

Type	Con	RA	Dec	Mag	Diam	Rating
Globular cluster	Ophiu-chus	16^h 32.5^m	−13° 03′	7.8	13′	4

General description
NGC 6171, now known as M107, is a reasonably bright globular cluster between Zeta (ζ) and Phi (φ) Ophiuchi. Although William Herschel was the first to catalog the object in 1784, it was actually discovered by Pierre Mechain in April 1782 – though that fact did not surface until 1947 when Canadian astronomer Helen Sawyer Hogg found the discovery reference in a letter written by Mechain to Bernoulli in 1783. It is a nice object for small telescopes and a wonder in large instruments.

Directions
Use Chart 48 to locate 2.6-magnitude Zeta Ophiuchi, which is 12° northeast of 2.6-magnitude Beta (β) Scorpii. Center Zeta Ophiuchi in your telescope at low power, then switch to Chart 48b. From Zeta Ophiuchi, move 50′ west to 20′-wide Triangle a, which comprises three 7.0- to 8th-magnitude suns. Next move a little less than 40′ west–northwest to 7.5-magnitude Star b. Now drop about 45′ south–southwest, to 7th-magnitude Star c. You must now make a slow and careful 1$\frac{1}{2}$° sweep south–southeast to 20′-wide Triangle d, which comprises three roughly 7th-magnitude suns. NGC 6171 is about 25′ due south of Triangle d.

The quick view

At 23×, NGC 6171 appears as a 5′-wide sparkling glow with a weakly condensed core. With averted vision the cluster is elongated east to west and has at least four 11th-magnitude stars around it in the form of a crucifix. At 72× and 101×, the cluster's outlying stars stand out boldly against hazy wisps of light that look like ruffled hair. The cluster's core is boxy and patchy, marred with dark lanes.

3. NGC 6356 (H I-48)

Type	Con	RA	Dec	Mag	Diam	Rating
Globular cluster	Ophiu-chus	17^h 23.6^m	−17° 49′	8.2	10.0′	4

General description

NGC 6356 is a relatively bright globular cluster about 4° southeast of 2nd-magnitude Eta (η) Ophiuchi, near the 8th-magnitude globular cluster M9. It is nicely condensed making it a good target for small telescopes. Under a dark sky, it rivals M9 for attention.

Directions

Use Chart 48 to find Eta Ophiuchi. Now use your unaided eyes or binoculars to find 1½°-wide Triangle a, about 4° to the southeast. It is composed of three roughly 6.5-magnitude suns. Now switch to Chart 48c. You want to center the northernmost star (a^1) in the Triangle a, in your telescope at low power. NGC 6356 is a little more than 50′ east and slightly south of Star. Note that similarly bright M9 lies about the same distance to the south-southwest of Star a^1.

The quick view

At 23×, NGC 6356 is bright and condensed (about 3′ across), appearing almost as prominent as M9. It has a tight core, in a concentrated halo that with averted vision

seems mottled. At 72×, the cluster's core is not so condensed as it is soft, like a comet that has lost its nucleus and is beginning to fade. With averted vision it is mottled. The brightest stars in the cluster shine at 15th-magnitude. But clumps of them condense to form a mottled haze.

Stop. Do not move the telescope. Your next target is nearby!

4. NGC 6342 (H I-149)

Type	Con	RA	Dec	Mag	Diam	Rating
Globular cluster	Ophiu-chus	17^h 21.2^m	−19° 35′	9.5	4.4′	2

General description

NGC 6342 is a small and dim globular cluster about 4½° south–southeast of 2nd-magnitude Eta (η) Ophiuchi, near the 8th-magnitude globular cluster M9. It is an amorphous, feeble glow that is difficult to see even under a dark sky. Light pollution will definitely wash this target out.

Directions

Use Chart 48 to find Eta Ophiuchi. Now use your unaided eyes or binoculars to find 1½°-wide Triangle a, about 4° to the southeast. It is composed of three roughly 6.5-magnitude suns. Now switch to Chart 48c. You want to center the southernmost star (a^3) in Triangle a in your telescope at low power. NGC 6342 is a little less than 20′ southeast of Star a^2.

The quick view

At 23× in the 4-inch under a dark sky, NGC 6342 is a faint glow, about 1′ across, that requires averted vision to see well. At 72×, the cluster appears slightly elongated north–northeast to south–southwest with a roughly 12.5-magnitude star about 1′ to the south–southwest.

Star charts for fifth night

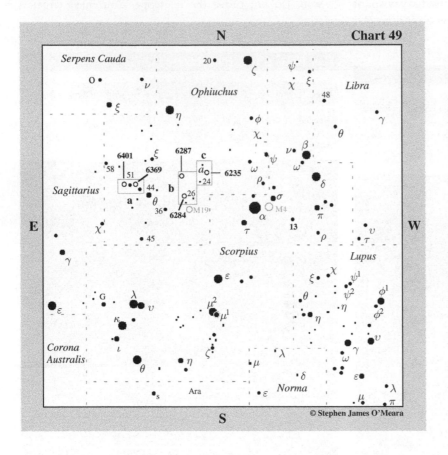

Chart 49

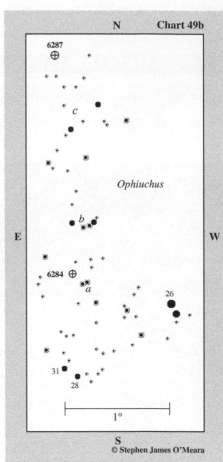

Chart 49b

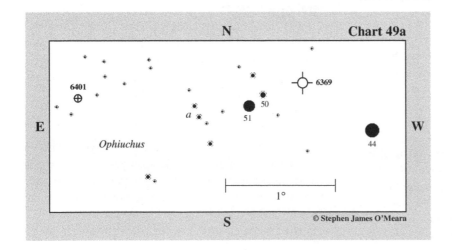

Chart 49a

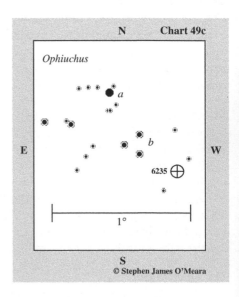

Chart 49c

FIFTH NIGHT

1. NGC 6369 (H IV-11)

Type	Con	RA	Dec	Mag	Dim	Rating
Planetary nebula	Ophiu-chus	17^h 29.3^m	−23° 46′	11.4	58″ ×34″	4

4′

General description
NGC 6369, the Little Ghost Nebula, is a beautiful ringed planetary nebula in the rich Ophiuchus Milky Way, about 2° northeast of 3rd-magnitude Theta (θ) Ophiuchi, near the 5th-magnitude star 51 Ophiuchi. Its high-surface-brightness ring is best seen with high magnification.

Directions
Use Chart 49 to locate Theta Ophiuchi. A little more than 1° to its northeast is 4th-magnitude 44 Ophiuchi. And a little more than 1° east and a little north of it you'll find 51 Ophiuchi. (Use your binoculars, if you must, to confirm this roughly 2½° arc of obvious stars.) You want to center 51 Ophiuchi in your telescope at low power, then switch to Chart 49a. Note that 51 Ophiuchi has a 7th-magnitude companion (50 Ophiuchi) 10′ to the northwest. NGC 6369 is simply 30′ northwest of 51 Ophiuchi, or 20′ west–northwest of 6th-magnitude 50 Ophiuchi.

The quick view
At 23× in the 4-inch under a dark sky, NGC 6369 is projected against the blackness of a dark nebula, Barnard 77, so, as NGC 6369's nickname implies, it appears like a translucent spirit materializing in the darkness. It appears very

stellar at low power but swells with averted vision. With any moderate magnification, the nebula is a perfect opal – smooth, round, and, well, opalescent. At 303× (under good seeing), the ring has a sharp interior that gradually dims outward. The central star, which shines at 16th magnitude, cannot be seen.

Stop. Do not move the telescope. Your next target is nearby!

2. NGC 6401 (H I-44)

Type	Con	RA	Dec	Mag	Diam	Rating
Globular cluster	Ophiu-chus	17^h 38.6^m	−23° 54′	7.4	4.8′	3.5

4′

General description
NGC 6401 is a very interesting globular cluster in a rich swath of Milky Way about 2° east of NGC 6369. At low power, it looks like an enhancement in the Milky Way but magnification will make the cluster stand out as a small round glow.

Directions
Using Chart 49a, from NGC 6369, return to 5th-magnitude 51 Ophiuchi. Now move 30′ east and slightly south to a pair of roughly 9th-magnitude stars (a). NGC 6401 lies about 1¼° east and slightly north of Pair a.

The quick view
At 23×, NGC 6401 looks like a 5′-wide knot in the Milky Way. At 72×, the cluster stands out as a slightly elongated, mottled haze. Its brightest stars shine at 15.5 magnitude but clumps of them make the cluster appear partially resolved at 101×.

Stop. Do not move the telescope.

3.   NGC 6284 (H VI-11)

Type	Con	RA	Dec	Mag	Diam	Rating
Globular cluster	Ophiu-chus	17^h 04.5^m	−24° 46′	8.9	6.2′	3

General description

NGC 6284 is a small and dim globular cluster a little less than 4° west and slightly north of 3rd-magnitude Theta (θ) Ophiuchi, near the 6.5-magnitude binocular double 26 Ophiuchi. Moderate magnification and averted vision will show this small round glow the best.

Directions

Use Chart 49 to find 26 Ophiuchi, which is about 5° west of Theta Ophiuchi, then switch to Chart 49b. NGC 6284 is 1° northeast of 26 Oph and just 10′ northeast of a pair of 9.5-magnitude stars (a) oriented west–northwest to east–southeast.

The quick view

At 23× in a 4-inch, NGC 6284 is simply a dim, 2′-wide diffuse glow. At 72×, the cluster appears round and has a slightly brighter core. The unresolved cluster swells a bit with averted vision. Its horizontal branch stars shine at 16.6 magnitude.

Stop. Do not move the telescope. Your next target is nearby!

4.   NGC 6287 (H II-195)

Type	Con	RA	Dec	Mag	Diam	Rating
Globular cluster	Ophiu-chus	17^h 05.1^m	−22° 42′	9.3	4.8′	2.5

General description

NGC 6287 is a moderately small and dim globular cluster a little more than 2° north and slightly east of NGC 6284. In a small telescope, it is a round, cometlike glow that could be washed out from suburban locations.

Directions

Using Chart 49b, from NGC 6284, move 30′ north–northwest to a 12′-wide arc (b) of four 7.5- to 9.5-magnitude suns oriented east to west. Now make a generous 1° sweep due north to a pair of 7.5-magnitude stars (c), separated by 20′ and oriented northwest to southeast. NGC 6287 lies 40′ north–northeast of Pair c.

The quick view

At 23× in the 4-inch under a dark sky, NGC 6287 is a round, cometlike glow, about 3′ in diameter. Averted vision is needed to see it well. It is much better at 72×, appearing as a uniformly bright circular glow with some mottling suspected on averted vision. With concentration, a diffuse, unresolved core can be seen.

Stop. Do not move the telescope.

5. NGC 6235 (H II-584)

Type	Con	RA	Dec	Mag	Diam	Rating
Globular cluster	Ophiu-chus	16ʰ 53.4ᵐ	−22° 11′	8.9	5.0′	2.5

4′

General description

NGC 6235 is a moderately small and dim globular cluster about $1\frac{1}{4}°$ northwest of 6th-magnitude 24 Ophiuchi. Light pollution will affect this target, which is a low-contrast glow at low power.

Directions

Using Chart 49b, from NGC 6287, return to 26 Ophiuchi. Now use Chart 49 and binoculars, if necessary, to find 24 Ophiuchi, which is 2° to the north–northwest. Next look a little more than $1\frac{1}{2}°$ further to the north–northwest for 7th-magnitude Star a. Center Star a in your telescope and switch to Chart 49c. From Star a, move 25′ southwest to an 8′-wide triangle of three 8th- to 9th-magnitude suns (b). NGC 6235 is about 20′ west–southwest of Triangle b.

The quick view

At 23× in the 4-inch under a dark sky, NGC 6235 is a moderately diffuse, low-contrast glow, about 2′ round. With averted vision it appears speckled. At 72×, the cluster has a soft gray pallor. With concentration, the soft uniform glow has a slightly brighter soft uniform core. This cluster does not take power well. Its brightest stars shine at 14.5 magnitude.

Star charts for sixth night

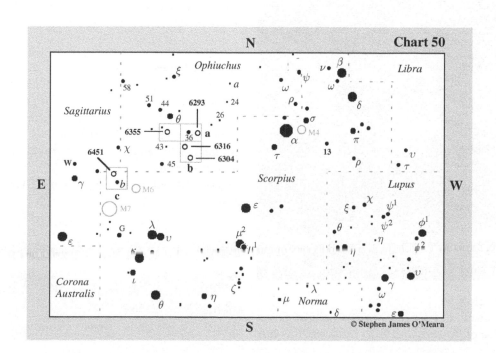

Chart 50

© Stephen James O'Meara

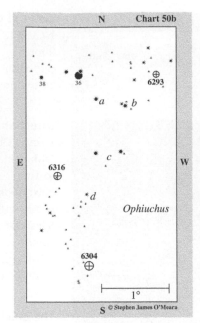

Chart 50b

© Stephen James O'Meara

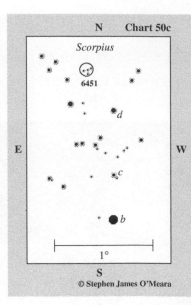

Chart 50c

© Stephen James O'Meara

Chart 50a

© Stephen James O'Meara

SIXTH NIGHT

1. NGC 6293 (H VI-12)

Type	Con	RA	Dec	Mag	Diam	Rating
Globular cluster	Ophiu-chus	17^h 10.2^m	−26° 35′	8.3	8.2′	3

General description

NGC 6293 is a small but relatively obvious globular cluster less than 1$\frac{1}{4}$° west of 4th-magnitude 36 Ophiuchi. It is very condensed and can be seen with direct vision at low power under a dark sky.

Directions

Use Chart 50 to locate 36 Ophiuchi, which is about 2° southwest of 3rd-magnitude Theta (θ) Ophiuchi. Center 36 Ophiuchui in your telescope, then switch to Chart 50a. You can either make a direct sweep 1$\frac{1}{4}$° west to nab the cluster, or you can make a series of short hops. From 36 Ophiuchi, move a little more than 25′ southwest to 7th-magnitude Star a. Now make a roughly 30′ hop west–southwest to similarly bright Star b. NGC 6293 lies a little less than 40′ northwest of Star b.

The quick view

At 23×, NGC 6293 is a small and tight ball of light measuring about 3′ in diameter. At 72×, the cluster's core is bright and is surrounded by a mottled and irregular halo that seems to sizzle with dim starlight. The cluster's brightest stars shine around 14th magnitude, so they are within reach of modest-sized telescopes.

Stop. Do not move the telescope. Your next target is nearby!

2. NGC 6355 (H I-46)

Type	Con	RA	Dec	Mag	Diam	Rating
Globular cluster	Ophiu-chus	17^h 24.0^m	−26° 21′	8.6	4.2′	3.5

General description

NGC 6355 is a modestly bright but obvious globular cluster $1\frac{1}{2}°$ south–southeast of 3rd-magnitude Theta (θ) Ophiuchi, or about 2° east–northeast of 4th-magnitude 36 Ophiuchi. It is best seen at moderate magnifications.

Directions

Using Chart 50a, from NGC 6293, return to 36 Ophiuchi. Now move about 30′ east to 7th-magnitude 38 Ophiuchi. A gentle sweep about 40′ east–northeast will bring you to 7th-magnitude Star c. A 10′-wide arc (d) of three roughly 8th-magnitude stars (oriented north to south) lies less then 15′ further to the northeast. NGC 6355 is a little more than 30′ east–northeast of Arc d.

The quick view

At 23×, NGC 6355 looks like a moderately small (2′) and compact glow with a bright, though diffuse, center. At 72×, the globular cluster is a pale cometlike glow that gradually gets brighter toward the center but not to a nucleus. The inner region remains a compact uniform glow. No stars can be resolved. Its horizontal branch magnitude is 17.2.

Stop. Do not move the telescope.

General description

NGC 6316 is a small and reasonably obvious globular cluster about $1\frac{1}{2}°$ south–southeast of 4th-magnitude 36 Ophiuchi. It is slightly condensed and is best seen with averted vision.

Directions

Using Chart 50a, from NGC 6355, return to 36 Ophiuchi, then switch to Chart 50b. From 36 Ophiuchi move a little more than 25′ southwest to 7th-magnitude Star a. Now look about 50′ south–southwest for a pair of 6th-magnitude stars (c) separated by about 20′ and oriented roughly east to west. NGC 6316 is only 40′ southeast of the easternmost star in Pair c.

The quick view

At 23× in a 4-inch, NGC 6316 is a small and somewhat condensed glow, about 2′ in diameter. It lies just northeast of an 11th-magnitude sun, which, when seen together, makes the cluster appear elongated at a glance. At 72×, the cluster is simply a round glow with a slightly brighter middle. It is not resolved. The cluster's brightest stars shine at 15th magnitude.

Stop. Do not move the telescope. Your next target is nearby!

3.	NGC 6316 (H I-45)						
Type		Con	RA	Dec	Mag	Diam	Rating
Globular cluster		Ophiu-chus	17^h 16.6^m	−28° 08′	8.1	5.4′	3

4.	NGC 6304 (H I-147)						
Type		Con	RA	Dec	Mag	Diam	Rating
Globular cluster		Ophiu-chus	17^h 14.5^m	−29° 28′	8.3	8.0′	2

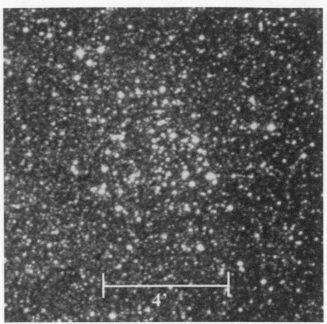

General description

NGC 6304 is a moderately small and faint globular cluster a little less than $1\frac{1}{2}°$ south–southeast of NGC 6316. In a small telescope, it is best seen under a dark sky at moderate magnification.

Directions

Using Chart 50b, from NGC 6316, move about 30′ west–southwest to the solitary 8.5-magnitude Star d. Now make a very slow sweep 1° due south, where you should see NGC 6304 as a dim puff of light with averted vision.

The quick view

At 23× in the 4-inch under a dark sky, NGC 6304 is a moderately faint 2′ glow with some central condensation visible with averted vision and time. The cluster is better seen at 72×. The core is dramatically elongated in a roughly east–west direction, and it is surrounded by an irregular halo of dim light. The core also sparkles a bit with averted vision. The cluster's brightest stars shine around 14.5 magnitude, so they are within range of moderate-sized telescopes.

5. NGC 6451 (H VI-13)

Type	Con	RA	Dec	Mag	Diam	Rating
Open cluster	Scorpius	17^h 50.7^m	−30° 13′	8.2	8.0′	3.5

General description

NGC 6451 is a small but reasonably bright open star cluster in Scorpius. It lies about 3° northeast of the magnificent open cluster M6 and is a nice compact blaze of stars in a rich Milky Way field.

Directions

Use Chart 50. To find NGC 6451, start by locating M6 about 5° north–northeast of the stinger stars at the end of the Scorpion's tail. Now look 2° east–northeast for solitary 5th-magnitude Star b. Center Star b in your telescope at low power, then switch to Chart 50c. From Star b, move a little less than 30′ due north to 8th-magnitude Star c, which has a 9.5-magnitude companion immediately to its southwest. Now move 40′ due north again, to 7th-magnitude Star d. NGC 6451 is 30′ northeast of Star d.

The quick view

At 23× in the 4-inch, NGC 6451 is a very nice sight, appearing as a 5′-wide wedge of well-resolved starlight. At 72×, the cluster is an irregular assortment of some 80 stars of 12th magnitude and fainter. The core is elongated, and random arms can be seen sprawling into the Milky Way background.

Star charts for seventh night

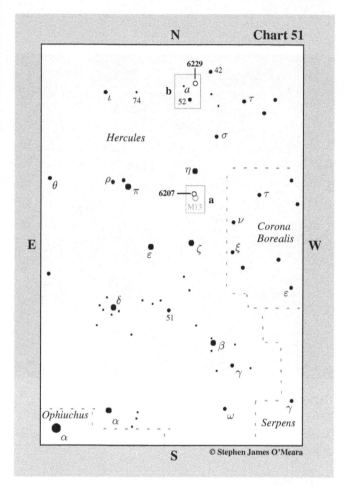

Chart 51

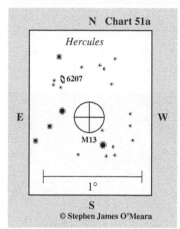

N   Chart 51a

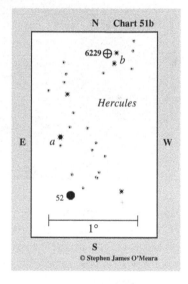

N   Chart 51b

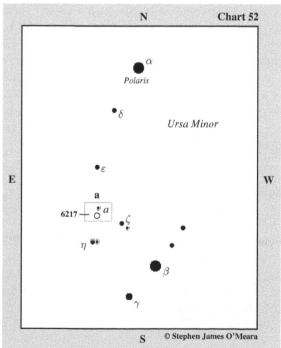

N   Chart 52

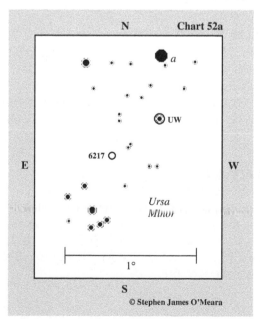

N   Chart 52a

SEVENTH NIGHT

1. NGC 6207 (H II-701)

Type	Con	RA	Dec	Mag	Dim	Rating
Spiral galaxy	Hercules	16ʰ 43.1ᵐ	+36° 50′	11.6	3.0′× 1.1	3

General description

NGC 6207 is a small ellipse of light near M13, the great globular cluster in the Keystone of Hercules. In a small telescope it is best seen with moderate magnification and averted vision.

Directions

Use Chart 51 to locate M13, which is about $2\frac{1}{2}$° south of 3.5-magnitude Eta (η) Herculis; under a dark sky, M13 can be seen as a fuzzy, 5th-magnitude star with the unaided eye. Center M13 in your telescope at low power, then switch to Chart 51a. NGC 6207 is about 30′ north–northeast of M13's core.

The quick view

At 23× in the 4-inch, NGC 6207 is just visible under a dark sky, though you must know exactly where to look. At 72×, the galaxy is a weak 2′-long ellipse that pops into view with averted vision. The ellipse is oriented northeast to southwest and larger telescopes will show it to have a starlike core.

2. NGC 6229 (H IV-50)

Type	Con	RA	Dec	Mag	Diam	Rating
Globular cluster	Hercules	16ʰ 46.9ᵐ	+47° 32′	9.4	4.5′	3.5

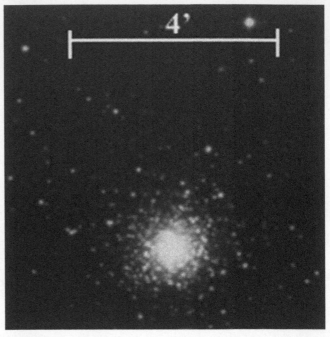

General description

NGC 6229 is a small but bright globular cluster about $1\frac{1}{2}$° north–northwest of 5.5-magnitude 52 Herculis. It is a good object for city observers.

Directions

Use Chart 51 to find 52 Herculis, which is about 7° north of 3.5-magnitude Eta (η) Herculis. It also forms the north-eastern apex of a near equilateral triangle with the 4th-magnitude stars Sigma (σ) and Tau (τ) Herculis. Center 52 Herculis in your telescope at low power, then switch to Chart 51b. From 52 Herculis, move 40′ north–northeast to 7th-magnitude Star a. Now make a slow sweep 1° northwest to a pair of 8th-magnitude stars (b), oriented north–northwest to south–southeast and separated by 5″. NGC 6229 is only 5′ east–northeast of Pair b.

The quick view

At 23×, NGC 6229 is a very small (1′) glow. But it is so condensed that it is relatively bright. At 72×, the cluster is a nice and tight concentration of light. This is the cluster's core, which, with some concentration, can be seen surrounded by a larger, though much fainter, halo. With averted vision, the halo's edges are irregular and form a starfish-shaped pattern.

3. NGC 6217 (H I-280)

Type	Con	RA	Dec	Mag	Dim	Rating
Spiral galaxy	Ursa Minor	16^h 32.6^m	+78° 12′	11.2	3.3′ ×3.3′	2.5

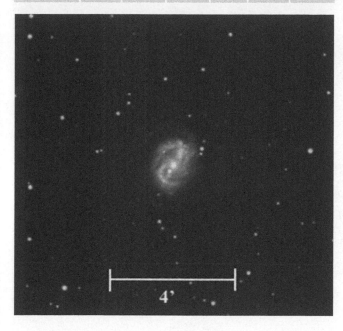

4′

General description

NGC 6217 is a small and dim galaxy about $2\frac{1}{2}°$ east–northeast of 5th-magnitude Zeta (ζ) Ursae Minoris. A dark sky, magnification, and averted vision is needed to see it in a small telescope.

Directions

Use Chart 52 to find Zeta Ursae Majoris, the northernmost star the Little Dipper's bowl. Now use binoculars or your unaided eyes to spot solitary, 5.5-magnitude Star a, which is a little less than $2\frac{1}{2}°$ to the northeast. Center Star a in your telescope at low power, then switch to Chart 52a. NGC 6217 is 50′ southeast of Star a.

The Quick View

At 23× in a 4-inch under a dark sky, NGC 6217 is just visible as a tiny speck of light once the object's position is known. At 72×, the galaxy is very dim and requires averted vision to see. Its core also appears slightly elongated north–northwest to south–southeast, like a bar.

8 · August

Star charts for first night

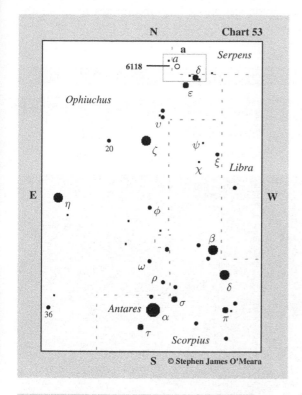

Chart 53

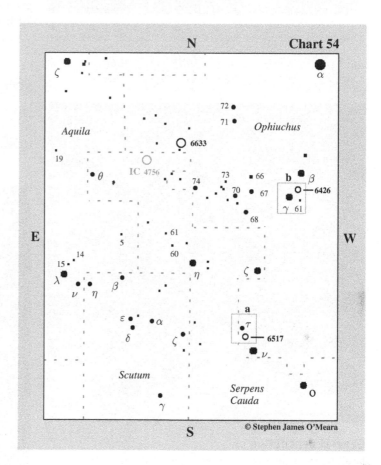

Chart 54

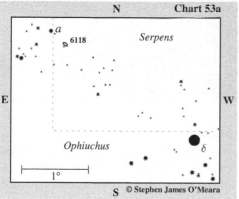

Chart 53a

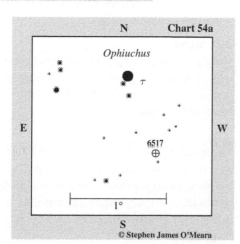

Chart 54a

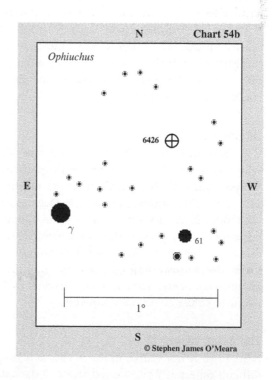

Chart 54b

FIRST NIGHT

1. NGC 6118 (H II-402)

Type	Con	RA	Dec	Mag	Dim	Rating
Spiral galaxy	Serpens (Cauda)	16^h 21.8^m	−02° 17′	11.7	4.6′ × 1.9′	1

4′

General description

NGC 6118 is an extremely dim and difficult galaxy for a small telescope, even under a dark sky. It is about $2\frac{1}{4}°$ northeast of 3rd-magnitude Delta (δ) Ophiuchi, near a 6.5-magnitude star. Patience and time is required to see this galaxy in a small telescope; it is one of the most difficult objects in the Herschel 400 list. Plan on spending some time confirming your sighting of it.

Directions

Use Chart 53 to locate 2.6-magnitude Zeta (ζ) Ophiuchi, which is about 15° north of 1st-magnitude Alpha (α) Scorpii (Antares). Now look about 8° northwest for the pair of roughly 3rd-magnitude stars Delta (δ) and Epsilon (ε) Ophiuchi. You want to center Delta Ophiuchi in your telescope at low power, then switch to Chart 53a. Now, using your binoculars, look for 6.5-magnitude Star a, which marks the northwestern end of a 40′-long line of slightly fainter stars. Center Star a in your telescope. NGC 6118 lies a little less than 20′ to the southwest.

The quick view

NGC 6118 is not visible in the 4-inch at 23×. It is a most difficult object at 72×. Averted vision, a dark sky, and lots

of time breathing rhythmically and lightly tapping the telescope tube (to set the object in motion) will help to bring it out. Look for a slightly round core with dim extensions to the southwest and northeast. A roughly 12th-magnitude star lies a few arcminutes to its south–southwest.

2. NGC 6517 (H II-199)

Type	Con	RA	Dec	Mag	Diam	Rating
Globular cluster	Ophiuchus	18^h 01.8^m	−08° 57′	10.1	4.0′	1.5

4′

General description

NGC 6517 is a very dim and tiny globular cluster between 3.5-magnitude Nu (ν) and 4th-magnitude Tau (τ) Ophiuchi. The cluster is a difficult object for small telescopes even under dark skies. Use moderate magnification and averted vision.

Directions

Use Chart 54 to locate Tau Ophiuchi, which is about $7\frac{1}{2}°$ west of Alpha (α) Scuti. Center Tau in your telescope at low power, then switch to Chart 54a. NGC 6517 is about 50′ southwest of Tau.

The quick view

At 23× in the 4-inch under a dark sky, NGC 6517 is barely visible with averted vision as a faint 2′ haze about 6′ north–northeast of a 10th-magnitude star. At 72×, the cluster does not appear much better, remaining as a dim, uniformly bright glow.

3. NGC 6426 (H II-587)

Type	Con	RA	Dec	Mag	Diam	Rating
Globular cluster	Ophiu-chus	17ʰ 44.9ᵐ	+03° 10′	10.9	4.2′	1.5

General description

NGC 6426 is another small and dim globular cluster in Ophiuchus, located between 3rd-magnitude Beta (β) and 4th-magnitude Gamma (γ) Ophiuchi. It is very difficult to see in a small telescope even under a dark sky.

Directions

Use Chart 54 to locate Gamma Ophiuchi. Center that star in your telescope at low power, then switch to Chart 54b. NGC 6426 is a little more than 50′ northwest of Gamma Ophiuchi. You can also look 35′ north–northeast of the beautiful 6th-magnitude double star 61 Ophiuchi, which is about 50′ west–southwest of Gamma Ophiuchi.

The quick view

Even under a dark sky NGC 6426 is not visible at 23× in the 4-inch. At 72×, the cluster is small (about 2′) and very difficult to see, requiring a lot of breathing and the use of averted vision. It is nothing more than an ill-defined puff of faint light.

4. NGC 6633 (H VIII-72)

Type	Con	RA	Dec	Mag	Diam	Rating
Open cluster	Ophiu-chus	18ʰ 27.2ᵐ	+06° 30′	4.3	20.0′	4

General description

NGC 6633 is one of two magnificent naked-eye open clusters – the other being IC 4756 in Serpens – standing side by side on the banks of the river Milky Way. When seen upside down from the Northern Hemisphere, the two clusters stand only 3° apart "under" a treelike asterism of stars, whose broad, 10°-wide canopy is formed by the stars Beta (β), Gamma (γ), 67, 68, 70, and 74 Ophiuchi, and whose trunk is marked by the stars 71 and 72 Ophiuchi. These same stars comprise the now defunct constellation Taurus Poniatovii, the Polish Bull.

Directions

Use Chart 54 to find NGC 6633. If you are under a dark sky, just look for two hazy 4th-magnitude globes halfway between 4.6 magnitude Theta (θ) Serpens (Cauda) – a glorious lemon-yellow double star resolvable in binoculars – and the similarly bright pair of stars, 71 and 72 Ophiuchi, which mark the tip of Taurus Poniatovii's western horn. From suburban locations, just raise your binoculars to this spot; the clusters will be unmistakable. NGC 6633 is the northwesternmost of the two clusters.

The quick view

In the 4-inch at 23×, NGC 6633 displays a fine, 30′-long neck of stars oriented northeast to southwest. Look for a perpendicular row of some two dozen suns (between 8th- and 11th-magnitude) at the northeastern end of that neck, and for a bright and tight ellipse of starlight at the southwestern end.

Star charts for second night

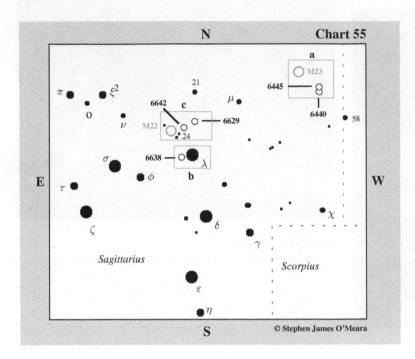

Chart 55

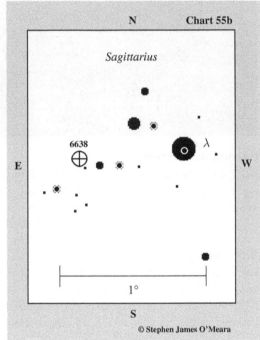

Chart 55b

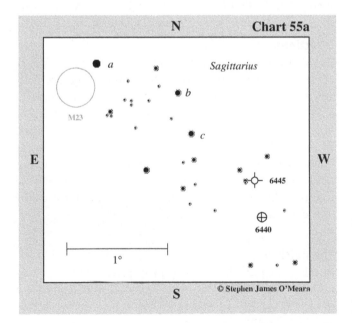

Chart 55a

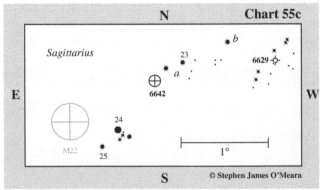

Chart 55c

SECOND NIGHT

1. NGC 6445 (H II-586)

Type	Con	RA	Dec	Mag	Dim	Rating
Planetary nebula	Sagittarius	17ʰ 49.2ᵐ	−20° 01′	11.2	3.0′ × 1.0′	4

General description

NGC 6445 is a fine planetary nebula only 2° southwest of the 5th-magnitude open cluster M23. The nebula lies in a region dappled with dimly glowing star clouds and veins of dark nebulosity. The smallest of telescopes will show the nebula as a swollen star. And high magnification will reveal its ring structure.

Directions

Use Chart 55 to locate M23; use binoculars if necessary. Center M23 in your telescope at low power, then switch to Chart 55a. A 6.5-magnitude star (a) lies a little less than 20′ northwest of M23. Center Star a, then move 55′ west–southwest to 7th-magnitude Star b. Next, move a little more than 25′ southwest to equally bright Star c. NGC 6445 is about 45′ southwest of Star c, just 5′ west of an 8th-magnitude star.

The quick view

The nebula is immediately obvious in the 4-inch at 23× as a non-stellar glow, which appears all the more obvious because it is so close to a bright star. At 72×, the planetary nebula looks annular, actually, it's more like two offset arcs of light. The northwest side is more difficult to observe

because of the proximity of the 8th-magnitude star. Increasing magnification to 101× enhances the arcs and brings out some fainter stars. Higher power works really well with this nebula. Do not hope to see the nebula's illuminating central star, which shines at a dim 19th magnitude.

Stop. Do not move the telescope. Your next object is nearby!

2. NGC 6440 (H I-150)

Type	Con	RA	Dec	Mag	Diam	Rating
Globular cluster	Sagittarius	17ʰ 48.9ᵐ	20° 21′	9.3	4.4′	4

General description

NGC 6440 is a nicely condensed globular cluster near NGC 6445. It can be seen in the smallest of telescopes as a well concentrated glow and, though it is not difficult to detect, it does not resolve into stars in a small telescope.

Directions

Use Chart 55a. NGC 6440 is just 20′ south–southwest of NGC 6445.

The quick view

At 23× in the 4-inch under a dark sky, NGC 6440 is a concentrated 2′-wide glow that is quite obvious under dark skies. At 72×, the globular cluster has a roughly 1′ core in a diffuse halo twice as large. No stars are resolved. The cluster does not get any easier to see with magnification.

3. NGC 6638 (H I-51)

Type	Con	RA	Dec	Mag	Diam	Rating
Globular cluster	Sagittarius	18^h 30.9^m	−25° 30′	9.2	7.3′	3.5

4. NGC 6642 (H II-205)

Type	Con	RA	Dec	Mag	Diam	Rating
Globular cluster	Sagittarius	18^h 31.9^m	−23° 28′	8.9	5.8′	3

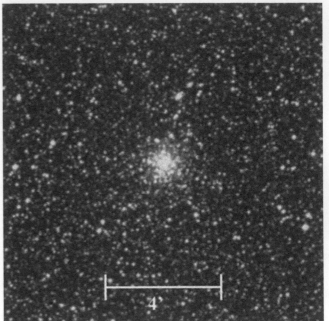

General description

NGC 6638 is a small but bright globular cluster near 3rd-magnitude Lambda (λ) Sagittarii. It is very condensed, almost starlike, but more like a very condensed 9th-magnitude comet. It is a nice object for a small telescope, and it is best seen at moderate to high magnifications.

Directions

Use Chart 55 to locate Lambda Sagittarii. Center that star in your telescope at low power, then switch to Chart 55b. NGC 6638 is only about 40′ east and slightly south of Lambda Sagittarii.

The quick view

In the 4-inch at 23×, the globular cluster is a tiny (2′) speck of light, shining like a slightly swollen 9th-magnitude star. It is very small but very condensed, so it is easy to see. At 72×, the cluster is a very nice sight, appearing as a round and highly condensed glow with a bright starlike pip at the center. At 101×, the cluster looks like a brighter globular cluster seen at low power, having a small core surrounded by a speckled halo of light. The cluster's brightest stars shine at 14th magnitude, and these can be seen swimming around the halo at high magnification and averted vision.

Stop. Do not move the telescope.

General description

NGC 6642 is a small and somewhat faint globular cluster a little more than 1° northwest of the magnificent globular cluster M22. Seeing these two globulars in the same field of view presents an awesome spectacle, showing the dynamics of the depths of space. It may be difficult to see in a small telescope at low power, especially from suburban skies, so it's best to get the object's location, then use moderate magnification to spy your target, which looks more like a planetary nebula than a globular cluster.

Directions

From NGC 6638, return to Lambda Sagittarii. Then use Chart 55 to locate 5.5-magnitude 24 Sagittarii and the stunning globular cluster M22 about 2° to the northeast. Center 24 Sagittarii in your telescope at low power, then switch to Chart 55c. NGC 6642 is about 45′ northwest of 24 Sagittarii, and about 15′ southeast of a pair of 7th- and 8th-magnitude stars, the brightest of which is 23 Sagittarii.

The quick view

In the 4-inch at 23×, NGC 6642 is a small (2′) glow about 10′ southeast of an 8th-magnitude star (a). The combined light of this star and NGC 6642 should attract attention. At 72×, the cluster looks like a uniformly round planetary nebula with a bright core.

Stop. Do not move the telescope. Your next target is nearby!

5. NGC 6629 (H II-204)

Type	Con	RA	Dec	Mag	Diam	Rating
Planetary nebula	Sagittarius	18^h 25.7^m	−23° 12′	11.3	15″	3

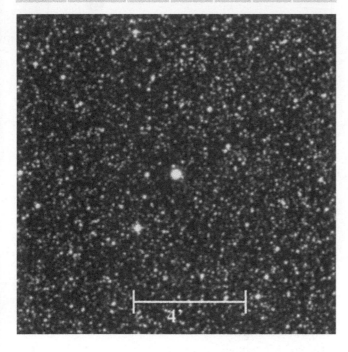

4′

General description
NGC 6629 is a very small but highly condensed planetary nebula nearly 3° north–northwest of 3rd-magnitude Lambda (λ) Sagittarii and nearly 1½° west–northwest of globular cluster NGC 6642. It is very stellar and you need to use high power to see its tiny disk. So be prepared to get the object's field and play with magnification.

Directions
Using Chart 55c, from NGC 6642, move 20′ northwest to 7th-magnitude 23 Sagittarii. Now move about 35′ further to the northwest to 7.5-magnitude Star b. NGC 6629 is 35′ west–southwest of Star b, immediately northwest of a 10th-magnitude star.

The quick view
Because of its proximity to the 10th-magnitude star, NGC 6629 is best noticed at 72× in my 4-inch. Still, at this power, the nebula appears very stellar and shines with the light of an 11th-magnitude star. You need to use the highest power you can muster to really notice the planetary's tiny (15″) disk.

Star charts for third night

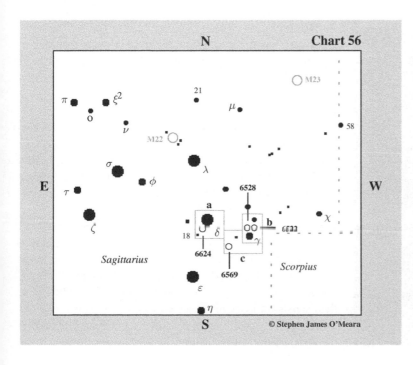

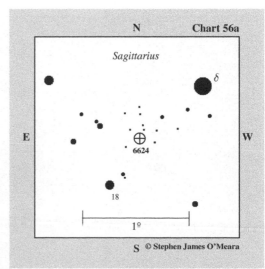

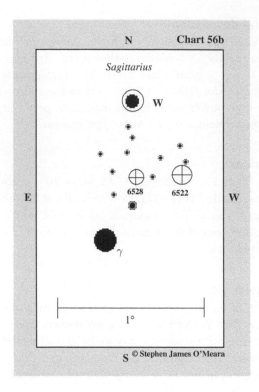

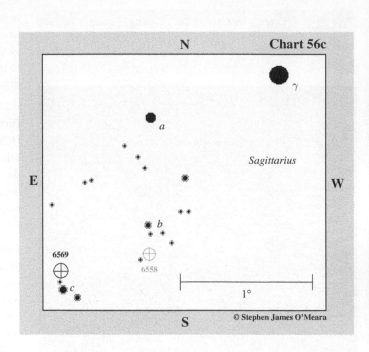

THIRD NIGHT

1. NGC 6624 (H I-50)

Type	Con	RA	Dec	Mag	Diam	Rating
Globular cluster	Sagittarius	18^h 23.7^m	−30° 22′	7.6	8.8′	4

General description

NGC 6624 is a small but bright globular cluster basking in the glow of 3rd-magnitude Delta (δ) Sagittarii. It is easily picked up as a fuzzy spot with a 3-inch telescope, and, under a dark sky, it is visible as a starlike object in 7×50 binoculars. In fact, NGC 6624 is just as bright as the much-sought-after Sagittarius globular clusters M54, M69, and M70, and it is a full magnitude brighter than M75, also in Sagittarius.

Directions

Use Chart 56 to locate Delta Sagittarii. Center that star in your telescope at low power, then switch to Chart 56a. NGC 6624 is a little more than 45′ southeast of Delta Sagittarii.

The quick view

In the 4-inch at 23×, NGC 6624 is tiny and bright; it is also easy to sweep over if you move the telescope too fast, so take your time scanning. The globular cluster's core is highly condensed, but it becomes more globelike with averted vision. If you still have trouble spotting it in a sweep, just move to the globular cluster's location, then use averted vision to scan the field for a starlike object surrounded by a tenuous haze. You'll find NGC 6624 at the southwest end of a 7′-wide chain of three "stars"; the other two stars shine at around 10th magnitude, so they are reasonably dimmer. Then, immediately put in a higher power. At 72×, some field stars scan be seen projected against the cluster's outer halo, especially to the west, northwest, and east. The brightest stars in the cluster shine at 14th magnitude, so the combined light of clumped members should pop in and out of view, especially in the outer halo. The cluster is so condensed that it takes power well.

2. NGC 6528 (H II-200)

Type	Con	RA	Dec	Mag	Diam	Rating
Globular cluster	Sagittarius	18ʰ 04.8ᵐ	−30° 03′	9.6	5.0′	3.5

General description

NGC 6528 is a somewhat faint globular cluster near 3rd-magnitude Gamma (γ) Sagittarii in a magnificent star-rich Milky Way field. NGC 6528 is paired with the slightly more obvious globular cluster NGC 6522 (see above right). Together these globulars look like lint on a snow-covered sweater. In a small telescope, NGC 6528 is best seen at moderate magnification.

Directions

Use Chart 56 to locate Gamma Sagittarii. Center that star in your telescope at low power, then switch to Chart 56b. NGC 6528 is about 25′ north–northwest of Gamma Sagittarii.

The quick view

In the 4-inch at 23×, NGC 6528 appears very small (~1′), like a star with a tiny halo projected against the rich Milky Way. Averted vision brings it out. At 72×, NGC 6528 remains small and dim. It has a bright core and the halo appears mottled with averted vision. The cluster's brightest stars shine at magnitude of 15.5

Stop. Do not move the telescope. Your next target is nearby!

3. NGC 6522 (H I-49)

Type	Con	RA	Dec	Mag	Diam	Rating
Globular cluster	Sagittarius	18ʰ 03.6ᵐ	−30° 02′	9.9	9.4′	3.5

General description

NGC 6522 is a moderately large and surprisingly obvious globular cluster near Gamma Sagittarii and globular cluster NGC 6528. Although it is dimmer than NGC 6528, it is the more obvious of the cluster pair, which I liken to the M81 and M82 of globular clusters. NGC 6522 is a fine object in small telescopes and is very nice at magnification.

Directions

Use Chart 56b. NGC 6622 is a little less than 20′ west, and a little north, of NGC 6528.

The quick view

In the 4-inch at 23×, NGC 6522 is much easier to see than NGC 6528; it also appears about twice as large (~2′). The cluster has a bright core surrounded by a swarm of dim stars. At 72×, the cluster is slightly elongated, probably owing to a roughly 13th-magnitude star 40″ to its southwest. The cluster's halo sparkles with dim suns that mix with the Milky Way background. The cluster's brightest stars shine at around 14th magnitude.

Stop. Do not move the telescope.

4. NGC 6569 (H II-201)

Type	Con	RA	Dec	Mag	Diam	Rating
Globular cluster	Sagittarius	18^h 13.6^m	$-31°$ $49'$	8.4	6.4'	3.5

General description

NGC 6569 is a moderately small and somewhat diffuse globular cluster about $2\frac{1}{4}°$ southeast of 3rd-magnitude Gamma (γ) Sagittarii. It is somewhat paradoxical, appearing small and condensed at low power but large and diffuse at high power. In a small telescope, it may be washed out under suburban skies.

Directions

Using Chart 56b, from NGC 6522, return to Gamma Sagittarii and center it in your telescope. Now switch to Chart 56c. From Gamma Sagittarii, make a gentle sweep about 1° southeast, where you'll find 5.5-magnitude Star *a*. Now dip about 50' south to 8th-magnitude Star *b*. Note that an 8.6-magnitude globular cluster, NGC 6558 (a non-Herschel 400 object) lies only about 12' south of Star *b*. Your target lies about 45' southeast of Star *b*, nearly 10' north of 7th-magnitude Star *c*.

The quick view

In the 4-inch at 23×, NGC 6569 is a small (2') and condensed circular glow. At 72×, the cluster swells to about 4' and appears as a circular patch of uniform light. With averted vision a dim core can be seen, like a comet with a dim inner coma. The cluster's brightest stars shine at 15th magnitude, so any stars you might suspect in a small telescope are most likely field stars.

Star charts for fourth night

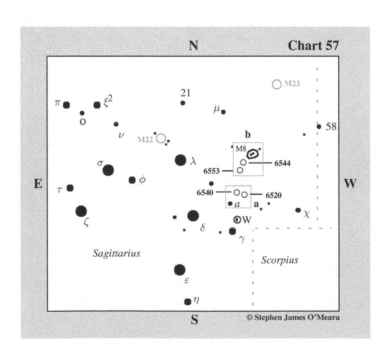

© Stephen James O'Meara

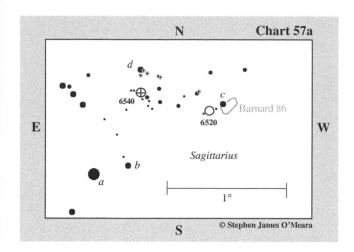

Chart 57a

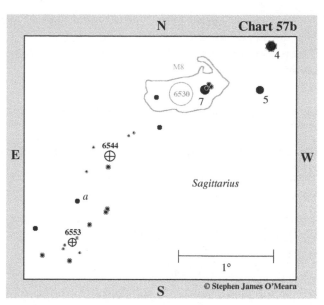

Chart 57b

FOURTH NIGHT

1. NGC 6520 (H VII-7)

Type	Con	RA	Dec	Mag	Diam	Rating
Open cluster	Sagittarius	18ʰ 03.4ᵐ	−27° 53′	7.6	5.0′	4

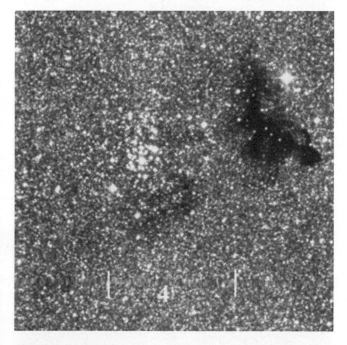

General description
NGC 6520 is a small but magnificent open cluster in one of the most glorious fields of the Milky Way, about $2\frac{1}{2}°$

north–northwest of 3rd-magnitude Gamma (γ) Sagittarii at the the tip of the Teapot's spout. It can be spied as a slightly fuzzy star in binoculars, if you know where to look.

Directions
Use Chart 57 to locate Gamma Sagittarii. Nearly 1° to the north–northwest is the bright Cepheid variable star W Sagittarii; it ranges in brightness from a magnitude of 4.3 to 5.1 every 7.6 days. Now look $1\frac{1}{4}°$ northeast of W Sagittarii for 4.5-magnitude Star a. Center Star a in your telescope at low power, then switch to Chart 57a. From Star a, move a little less than 20′ west–northwest to 7th-magnitude Star b. Now make a gentle 1° sweep northwest to 7th-magnitude Star c, with a 7.6-magnitude companion about 7′ to the southeast; the 7.6-magnitude companion is the compressed glow of NGC 6520. There are no stars as bright as Star c in the region out to nearly 2°, so there is no mistaking it.

The quick view
In the smallest of telescopes, NGC 6520 will look elongated, like a little spit of starlight. In the 4-inch at 23×, the cluster's six brightest stars lie along a northwest–southeast-trending line. The dark nebula Barnard 86 can be seen to the north-west like an ink stain on the bright skirt of the Milky Way. The star in middle of the line is a ruby among diamonds. With averted vision a circular halo of fainter suns surround the line like a crowd during a parade. At 72×, about 16 stars are obvious with others popping in and out of view. It's a very tight cluster so high magnification is a must. I found the view best with a 5-mm Nagler and a 3× Barlow (303×).

At this power, a stunning circlet of stars surrounds the red ruby heart of the cluster.

Stop. Do not move the telescope. Your next target is nearby!

2. NGC 6540 (H II-198)

Type	Con	RA	Dec	Mag	Diam	Rating
Globular cluster	Sagittarius	18^h 06.1^m	−27° 46′	~10	~5′	3

General description

NGC 6540 is a moderately small and diffuse globular cluster embedded in a rich swath of Milky Way near NGC 6520 – just 5° from the galactic center! William Herschel discovered the object on May 24, 1784, and cataloged it as a faint nebula. Per Collinder included it in his 1930 catalog as an open star cluster (Cr 364), and S. Djorgovski suspected it to be a globular cluster in 1986. Eduardo Bica, Sergio Ortolani, and Beatriz Barbuy confirmed NGC 6540 as a globular cluster in 1994 (*Astronomy and Astrophysics*, vol. 283, no. 1, pp. 67–75). According to Brian Skiff (Lowell Observatory), this is the first NGC/IC object to be identified as a globular cluster since the 1930s. Skiff also notes that there have been no good measurements of the visual magnitude of NGC 6540 to date. Most catalogs list the object's photographic magnitude as 14.6. But this measurement is extremely misleading! These listings, however, do not reflect the object's visual appearance (so do not be put off by them). Instead, wash the slate clean and imagine that you are trying to seek out a 5′-wide, 10th-magnitude comet with a 1.5′-wide central concentration – that is what NGC 6540 looks like

in a 4-inch telescope. The only difficulty in seeing it is that this dim glow is projected against a bright Milky Way starfield, so it requires a dark sky and averted vision to isolate the cluster's glow in a small telescope at low power. Amateurs under dark skies have detected NGC 6540 in telescopes as small as 2-inches!

Directions

Use Chart 57a. NGC 6540 is a little less than 40′ east, and slightly north, of NGC 6520 and about 15′ due south of 7.5-magnitude Star d.

The quick view

In the 4-inch at 23×, NGC 6540 is a 5′ diffuse glow shining against a bright Milky Way background. If the Milky Way were not so bright here, the cluster would be easier to see. As it is, though, a dark sky and averted vision is needed to see the cluster's apparent diffuse glow, which appears as an enhancement in the Milky Way. At 72×, the cluster is a 1.5′-wide concentration of light. When seen with with averted vision, this concentration becomes what appears to be a strong east to west oriented lane of stars surrounded by a north to south oriented wedge of unresolved starlight. Bica *et. al.* do not rule out the possibility that NGC 6540 may be a merger of two old globular clusters.

3. NGC 6544 (H II-197)

Type	Con	RA	Dec	Mag	Diam	Rating
Globular cluster	Sagittarius	18^h 07.3^m	−25° 00′	7.5	9.2′	4

General description

Globular cluster NGC 6544 is one of those easy-to-see objects that's all but lost in the crowd of bright clusters and nebulosities that populate the Milky Way's hub around M8, the renowned Lagoon Nebula – the Orion Nebula of the South. NGC 6544 is a fine binocular object, a tiny raft of light about 1° southeast of the visually weathered shores of M8. In 7×50 binoculars under a dark sky, the globular cluster is a fairly round but compact glow with a bright, starlike nucleus. It is a modest marvel in all telescopes.

Directions

Use Chart 57 to locate 3rd-magnitude M8 – the brilliant Lagoon Nebula, about 5° west and slightly north of 3rd-magnitude Lambda (λ) Sagittarii at the top of the Teapot. Center the nebula in your telescope at low power, then switch to Chart 57b. NGC 6544 is 1° southeast of NGC 6530 – the bright 14′-wide cluster of stars embedded in the Lagoon Nebula.

The quick view

In the 4-inch at 23×, NGC 6544 is a large, faint glow with a bright and highly condensed core, like the nucleus of an old comet surrounded by the faint and tenuous glow of its halo. An 8th-magnitude star lies 6.5′ south–southeast. At 101×, the cluster is very mottled. Its brightest stars (of 12.8 magnitude) form a northwest–southeast-trending line across the core. A dimmer arm extends from the southeastern end of that line and connects to the solitary 12th-magnitude star. So with averted vision, the brightest portions of the globular cluster has a sideways V-shaped pattern. The cluster takes power well; I find the best views between 182× and 303×. With a horizontal branch magnitude of 15.0, observers with even modest-sized telescopes should be able to resolve many of the cluster's fainter stars – more so than in some of the dimmer Messier globulars.

Stop. Do not move the telescope. Your next target is nearby!

4′

4. NGC 6553 (H IV-12)

Type	Con	RA	Dec	Mag	Diam	Rating
Globular cluster	Sagittarius	18ʰ 09.3ᵐ	−25° 54′	8.3	9.2′	3.5

General description

NGC 6553 is a small and moderately bright globular cluster 2° southeast of M8, near globular cluster NGC 6544. It has a cometlike quality. It is less conspicuous than NGC 6544 and, in a small telescope, may be slightly challenging to see from a suburban location.

Directions

Using Chart 57b, from NGC 6544, move about 35′ southeast to 6.5-magnitude Star a. NGC 6553 is a little less than 30′ south and slightly east of Star a.

The quick view

In the 4-inch at 23×, NGC 6553 is a 3′-wide glow, like a small ball of gray haze that gets gradually brighter towards the middle, though it has no obvious concentration at this magnification. At 72×, a 12th-magnitude star flanks the cluster to the northwest. The core is dim but appears slightly elongated. With averted vision, the cluster starts to break apart into hazy clumps. The cluster's brightest stars shine at a magnitude of 15.3.

Star charts for fifth night

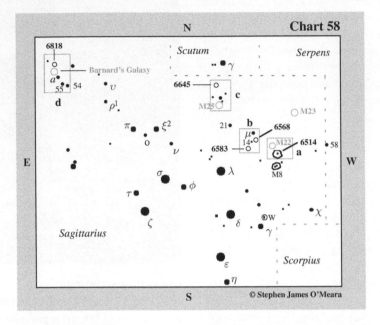

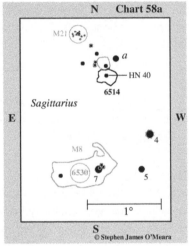

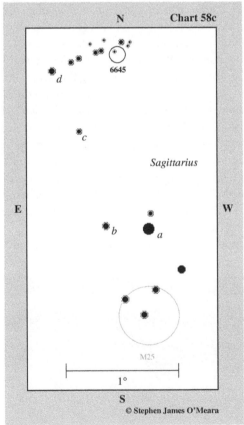

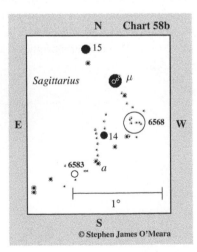

FIFTH NIGHT

1. NGC 6514 (H IV-41 = H V-10, 11, 12) = NEBULA ASSOCIATED WITH OPEN STAR CLUSTER M20

Type	Con	RA	Dec	Mag	Dim	Rating
Diffuse nebula	Sagittarius	18^h 02.5^m	−23° 02′	6.3	20.0′ × 20.0′	5

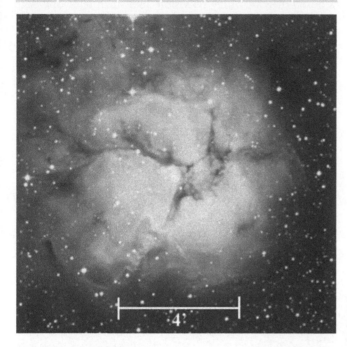

General description

M20 is widely known as the Trifid Nebula, but this is not true. When Charles Messier discovered M20 on June 5, 1764, he described it only as a cluster of stars, not a nebula. William Herschel was the first to notice the true nebulosity here. On July 12, 1784, he described it as "three nebulae faintly joined ... around a double star." He gave these three sections separate catalog numbers (H V-10,11,12) but provided only one position. He rediscovered the nebula on May 26, 1785, and cataloged it again – but this time as a planetary nebula (H IV-41). Unfortunately, the original NGC catalog lists Messier's discovery of the cluster and Herschel's discovery of the nebulosity under the same listing: NGC 6514. Some star charts confuse the matter further by erroneously assigning the label NGC 6514 to the cluster and M20 to the nebula. Since you are after Herschel's contribution, you need to spy the three brightest patches of nebulosity surrounding the cluster. William's son, John, first used the name "Trifid" to describe the trifold aspect of the nebula.

Directions

Use Chart 58 to locate 3rd-magnitude M8 – the brilliant Lagoon Nebula, about 5° west and slightly north of 3rd-magnitude Lambda (λ) Sagittarii at the top of the Teapot, then switch to Chart 58a. The nebula NGC 6514 is only 1$\frac{1}{2}$° north–northwest of M8, 40′ southwest of open cluster M21, and 10′ southeast of 6th-magnitude Star a.

The quick view

In the 4-inch at 23×, NGC 6514 is a clover of gas surrounding the 7.6-magnitude primary of the discrete triple star HN 40 – the nebula's primary source of illumination; another ball of gas surrounds a 7.4-magnitude star just 10′ to the north. The trifold aspect of the nebula you seek surrounds HN 40. All the nebulosity associated with the Trifid spans two-thirds that of the full Moon.

Stop. Do not move the telescope.

2. NGC 6568 (H VII-30)

Type	Con	RA	Dec	Mag	Diam	Rating
Open cluster	Sagittarius	18^h 12.7^m	−21° 35′	8.6	12.0′	3.5

General description

NGC 6568 is a moderately large and dim cluster near 4th-magnitude Mu (μ) Sagittarii. It has a low surface brightness, so it is best seen under a dark sky.

Directions

Use Chart 58 to locate Mu Sagittarii, which is about $3\frac{1}{2}°$ northeast of NGC 6514. Center Mu Sagittarii in your telescope at low power, then switch to Chart 58b. NGC 6568 is just 35′ southwest of Mu and about 20′ west–northwest of 5.5-magnitude 14 Sagittarii.

The quick view

At 23× in the 4-inch under a dark sky, NGC 6568 is a coarsely scattered, though well resolved open star cluster with some 50 members of 11th magnitude and fainter. These stars are spread across 12′ of sky. At 72×, the brightest members form a crown (like Corona Borealis) with a crooked extension of stars to the west.

Stop. Do not move the telescope. Your next target is nearby!

3.   NGC 6583 (H VII-31)

Type	Con	RA	Dec	Mag	Diam	Rating
Open cluster	Sagittarius	18^h 15.8^m	−22° 08′	10.0	5.0′	2.5

4′

General description

NGC 6583 is a very small and dim open star cluster. Dark skies and moderate magnification are required to show it well.

Directions

Using Chart 58b, from NGC 6568, swing over to 14 Sagittarii and look for a pretty, 20′-long cascade of 8.5- to 10th-magnitude suns (oriented north to south), immediately

to the star's east. Follow that cascade to its southern tip, which ends at a pair of 8.5-magnitude stars (a). NGC 6583 is about 15′ southeast of Pair a.

The quick view

In the 4-inch at 23× under a dark sky, NGC 6583 is a small (5′) patch of ill-defined light. With averted vision the patch dissolves into a loose gathering of very dim suns. At 72×, the cluster is a small patch of fuzzy light against a rich Milky Way background. NGC 6583 contains some 35 members, a dozen of which may be visible in a small telescope with concentration.

Stop. Do not move the telescope.

4.   NGC 6645 (H VI-23)

Type	Con	RA	Dec	Mag	Diam	Rating
Open cluster	Sagittarius	18^h 32.6^m	−16° 53′	8.5	15.0′	3.5

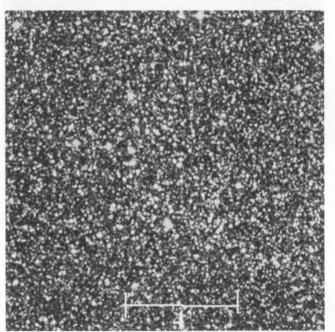

General description

NGC 6645 is a nice open cluster with a rich assortment of stars. It is rather diffuse, so it may be somewhat of a challenge to see under suburban skies; it is a cinch under dark skies. You have to squint to see it in the photo above.

Directions

Using Chart 58, return to Mu (μ) Sagittarii. Now look about $2\frac{3}{4}°$ east and slightly north for 5th-magnitude 21 Sagittarii. Just 2° further to the northeast is the beautiful open cluster M25; use binoculars to see it, if necessary. Center M25 in your telescope at low power, then switch to Chart 58c. From

M25 move 45′ north to 5th-magnitude Star *a*. Now hop 25′ east to 7th-magnitude Star *b*. Next make a careful 50′ sweep north–northeast to solitary 8.5-magnitude Star *c*. Now make a roughly 35′ sweep northeast to 7th-magnitude Star *d*, which is the southeasternmost star in a 40′-long chain of six roughly 8th-magnitude stars. NGC 6645 is at the northwestern end of that chain.

The quick view

In the 4-inch at 23×, NGC 6645 is a moderately large and diffuse congregation of dim suns surrounding a tight starlike core. At 72×, the cluster, which contains some 40 stars of 8th magnitude and fainter, appears somewhat rectangular, oriented north to south with irregularly bright members scattered haphazardly around its ragged halo of misty light.

Star charts for sixth night

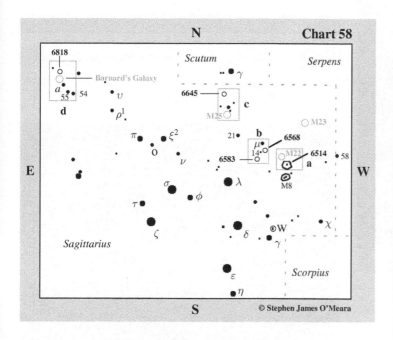

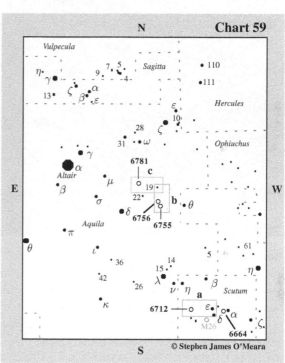

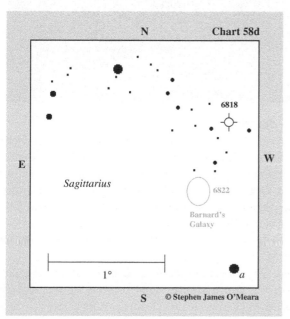

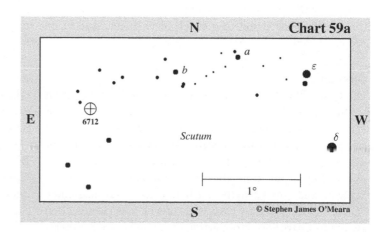

SIXTH NIGHT

1. NGC 6818 (H IV-51)

Type	Con	RA	Dec	Mag	Dim	Rating
Planetary nebula	Sagittarius	19^h 43.9^m	−14° 09′	9.3	22″× 15″	4

General description

Popularly known as the Little Gem, NGC 6818 is an attractive ring planetary nebula neatly tucked away in the north-eastern recesses of Sagittarius. It is about a Moon diameter north–northwest of one of the most celebrated dwarf galaxies visible from mid-northern latitudes – NGC 6822, Barnard's Galaxy. NGC 6818 is visible in 7 × 50 binoculars as a "star" among stars. The nebula is very condensed through a small telescope and is a good object for suburban skies. It takes magnification well. Finding it may take some time, though, because it lies in a lonely part of the sky, so be prepared to make a careful search for it.

Directions

Use Chart 58 to locate 3rd-magnitude Lambda (λ) Sagittarii at the top of the Teapot. About 10° to the east–northeast you'll find 3.5-magnitude Pi (π) Sagittarii, the easternmost star in the scoop of the Spoon. Now look 4° northeast for 4th-magnitude Rho[1] (ρ[1]) Sagittarii, then use your unaided eyes or binoculars to spy the 5.5-magnitude stars 54 and 55 Sagittarii and, just 40′ to the north–northeast, 6th-magnitude Star a. Center Star a in your telescope at low power, then switch to Chart 58d. From Star a, move about 45′ northeast to Barnard's Galaxy, which is quite large (15′) but can be swept up under a dark sky in a rich-field telescope; Edward Emerson Barnard discovered it visually with a 5-inch

refractor in 1884. NGC 6818 is 35′ to the north–northwest, about 12′ east–northeast of an 8th-magnitude star.

The quick view

In the 4-inch at 23× under a dark sky NGC 6818 is very stellar. But if you stare long enough with averted vision, you can see that the planetary nebula has some definition – a very slightly swollen disk. The nebula's round form is definite at 72×, appearing as a pale ashen globe, like the planet Neptune. Again, with some concentration and averted vision, the Little Gem's color changes to a pale ashen blue or steel gray. At 303×, the nebula's inner annulus looks beaded.

2. NGC 6664 (H VIII-12)

Type	Con	RA	Dec	Mag	Diam	Rating
Open cluster	Scutum	18^h 36.5^m	−08° 11′	7.8	12.0′	3.5

General description

NGC 6664 is a large, fuzzy, and scattered open cluster hidden in the glare of 4th-magnitude Alpha (α) Scuti. Under a dark sky it is easily spied in 7 × 50 binoculars, once you know where to look. But it most likely will be washed out from suburban locations because of its size and low surface brightness. Telescopically the cluster is sparse but well resolved.

Directions

Use Chart 59 to locate Alpha Scuti. Center that star in your telescope at low power. NGC 6664 is about 20′ due east of Alpha Scuti.

The quick view

In the 4-inch at 23×, the 60 or so stars in NGC 6664 are easily resolved. The brightest members are unevenly spread across

12′ and are aligned north–northeast to south–southwest – in the shape of a semicolon. At 72×, the cluster appears sparse without any central concentration. With imagination, the stars can be seen as Santa's sled or a swan boat.

Stop. Do not move the telescope. Your next target is nearby!

3. NGC 6712 (H I-47)

Type	Con	RA	Dec	Mag	Diam	Rating
Globular cluster	Scutum	18ʰ 53.1ᵐ	−08° 42′	8.3	9.8′	4

4′

General description

NGC 6712 is a moderately large and pretty bright globular cluster about $2\frac{1}{2}°$ east–southeast of 5th-magnitude Epsilon (ε) Scuti and $2\frac{1}{2}°$ northwest of open star cluster M26. Under a dark sky, NGC 6712 can be seen without difficulty in 7×50 binoculars as a pale, round glow. In a small telescope it looks like the head of comet. In larger telescopes many of its stars are nicely resolved.

Directions

Using Chart 59, from NGC 6664, return to Alpha Scuti. Now use your unaided eyes or binoculars to find Epsilon Scuti 2° to the east. Center Epsilon Scuti in your telescope at low power, then switch to Chart 59a; note that Epsilon Scuti has a 7th-magnitude companion about 5′ to the south–southeast. From Epsilon Scuti, make a careful sweep 50′ east–northeast to 7th-magnitude Star a. Next, hop a little more than 40′ east–southeast to equally bright Star b. NGC 6712 is 1° southeast of Star b. Sweep slowly.

The quick view

At 23× in the 4-inch, NGC 6712 is a very compact cometlike glow in a rich Milky Way field. With averted vision, the cluster's edges start to resolve – though it's hard to judge which stars belong to the cluster and which to the Milky Way background. Partial resolution is certainly possible in a 4-inch, for the brightest stars in the cluster shine at a magnitude of 13.3. The core screws in with a direct gaze and looks irregularly round with averted vision. At 72×, NGC 6712 looks like a misty globe bristling with fuzzy starlight. The cluster comes alive at a magnification of 101×. The outer halo is speckled like flecked metal. I couldn't see any central concentration – no bright stellar pip.

Star charts for seventh night

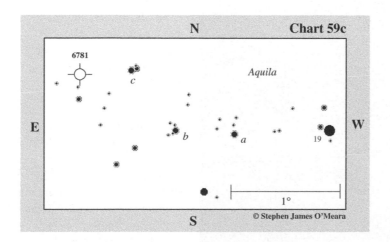

N   Chart 59c

6781

c

Aquila

E   W

b   a   19

1°

S

© Stephen James O'Meara

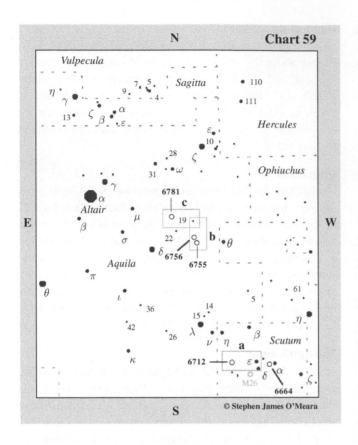

N   **Chart 59**

Vulpecula

Sagitta   • 110

η γ   7 5   • 111
   9   4
13 ζ β α   *Hercules*
        β ε

            ε
            10
       28   ζ   *Ophiuchus*
   31   ω

   γ
E α   W
Altair μ   6781
   β   c
           19
   σ   22   b
        δ 6756 θ
Aquila   6755

   π
θ   61
   ι   5
      36
      15 14
      42   λ   η
      26   ν η β   *Scutum*
   κ   a
        6712   ε
            M26   δ α
                     ζ
                  6664

S

© Stephen James O'Meara

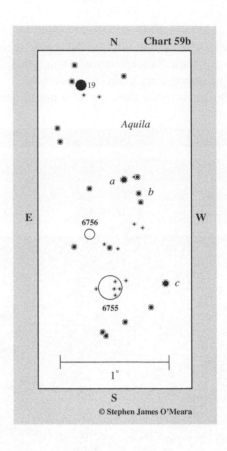

N   **Chart 59b**

        19

        Aquila

        a   b
E   6756   W

        6755

        c

        1°

S

© Stephen James O'Meara

SEVENTH NIGHT

1. NGC 6755 (H VII-19)

Type	Con	RA	Dec	Mag	Diam	Rating
Open cluster	Aquila	19ʰ 07.8ᵐ	+04° 16′	7.5	15.0′	3.5

4′

General description

NGC 6755 is a large and moderately bright open cluster near 5.5-magnitude 19 Aquilae. It is best seen in a small telescope under dark skies. It is a beautiful object in large telescopes at low power.

Directions

Use Chart 59 to find 3.4-magnitude Delta (δ) Aquilae. Now use your unaided eyes or binoculars to look $2\frac{1}{2}°$ northwest for 5.5-magnitude 22 Aquilae. About $2\frac{1}{4}°$ further to the northwest is similarly bright 19 Aquilae. Center 19 Aquilae in your telescope at low power, then switch to Chart 59b. From 19 Aquilae, move about 55′ south–southwest to 7th-magnitude Star *a*. About 10′ west–southwest is an 18′-long line of three 9th-magnitude stars (*b*). From Line *b* sweep 50′ south–southwest to 7.5-magnitude Star *c*. NGC 6755 is 30′ east of Star *c*.

The quick view

In the 4-inch at 23× under a dark sky, NGC 6755 is a very large and scattered cluster with an elongated core oriented roughly northeast to southwest. About three dozen stars are visible at 72×, and these are randomly scattered into bits and fragments. The cluster contains 150 members of 11th magnitude and fainter.

Stop. Do not move the telescope. Your next target is nearby!

2. NGC 6756 (H VII-62)

Type	Con	RA	Dec	Mag	Diam	Rating
Open cluster	Aquila	19^h 08.7^m	+04° 42′	10.6	4.0′	2.5

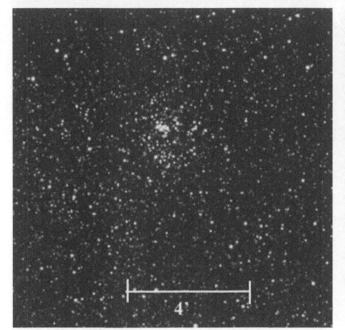

3. NGC 6781 (H III-743)

Type	Con	RA	Dec	Mag	Dim	Rating
Planetary nebula	Aquila	19^h 18.5^m	+06° 32′	11.4	1.9′ × 1.8′	4

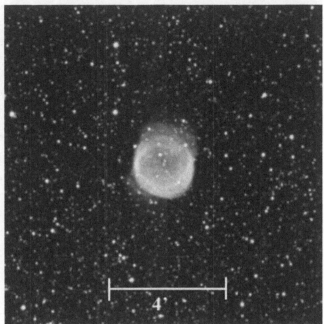

General description

NGC 6756 is a very small and somewhat dim open cluster about 1° south of 5th-magnitude 19 Aquilae near NGC 6755. Be prepared to use averted vision and look for a tiny condensed patch of light.

Directions

Use Chart 59b. NGC 6756 is 30′ north–northeast of NGC 6755.

The quick view

In the 4-inch at 23×, NGC 6756 is a small (4′), round, diffuse glow, like a 10th-magnitude comet without a tail. It lies almost midway between, and slightly north of, two 9th-magnitude stars separated by 20′ and oriented east to west. At 72×, the cluster remains a small uniform glow with no stars resolved at a glance. Its core, however, appears somewhat enhanced with averted vision. NGC 6756 contains some 40 stars of 13th magnitude and fainter.

Stop. Do not move the telescope.

General description

NGC 6781 is a fairly large and bright planetary nebula nearly $2\frac{1}{2}$° east–northeast of 19 Aquilae. From a dark sky, the nebula is apparent with direct vision in a small telescope, but it will probably be a challenge to see under suburban skies. I call it the Ghost of the Moon Nebula.

Directions

Using Chart 59b, from NGC 6756, return to 19 Aquilae, then switch to Chart 59c. From 19 Aquilae, move about 55′ east to 7th-magnitude Star a. Next, make a 30′ hop east to equally bright Star b. Now move a little more than 40′ northeast to 7th-magnitude Star c. NGC 6781 is a little less than 30′ east and slightly south of Star c.

The Quick View

At 23× in the 4-inch, NGC 6781 is a very round and somewhat concentrated 2′ glow that shines with a uniform luster. At 72×, the planetary is a beautiful ghostly glow that looks like the ghost of the naked-eye Moon.

9 · September

Star charts for first night

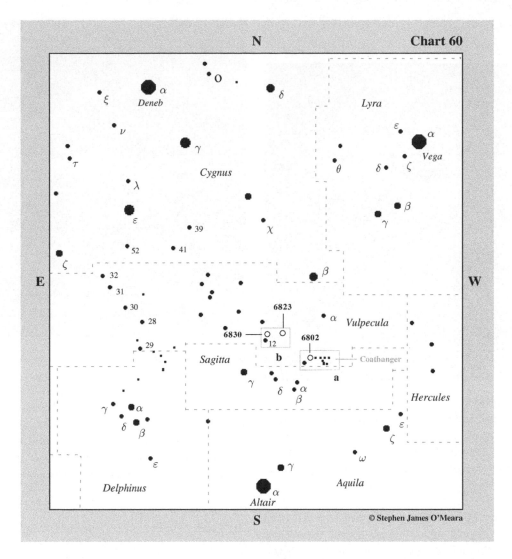

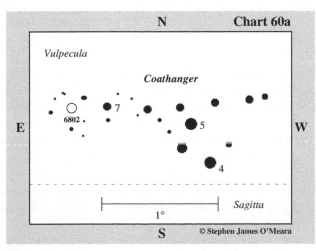

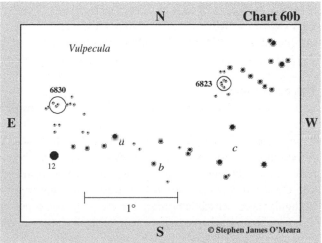

FIRST NIGHT

1. NGC 6802 (H VI-14)

Type	Con	RA	Dec	Mag	Diam	Rating
Open cluster	Vulpecula	19^h 30.6^m	+20° 16′	8.8	5.0′	3

General description

NGC 6802 is a small and somewhat dim open cluster immediately east of 6.5-magnitude 7 Vulpeculae, the easternmost star in the famous Coathanger asterism. Under a dark sky the cluster is a small puff of light that will vanish with any light pollution.

Directions

Use Chart 60 to find the Coathanger asterism. Start at the tail end of Sagitta's shaft – the 4th-magnitude stars Alpha (α) and Beta (β) Sagittae. You'll find the Coathanger just 4° northwest of Alpha Sagittae. Center the Coathanger in your telescope then switch to Chart 60a. NGC 6802 is less than 20′ east of 6th-magnitude 7 Vulpeculae, which lies at the eastern end of the 2°-long straight row of stars in the Coathanger asterism.

The quick view

In the 4-inch at 23× under a dark sky, NGC 6802 is simply a dim circular glow of uniformly faint light. At 72×, the 5′-wide cluster is actually extremely elongated north to south. Although the cluster's brightest stars shine at 14th magnitude, some members, or clumps of them, can be resolved. The cluster also breaks down into three segments: a little triangular core, which is bordered to the north and south by opposing triangles, making the overall appearance of the cluster similar to that of a little bow tie.

2. NGC 6830 (H VII-9)

Type	Con	RA	Dec	Mag	Diam	Rating
Open cluster	Vulpecula	19^h 51.0^m	+23° 06′	7.9	6.0′	3.5

General description

NGC 6830 is a small but very nice open cluster near 5th-magnitude 12 Vulpeculae, the western tip of the "Boot" of Vulpecula. Under a dark sky in a 3° field of view, the cluster shares the celestial stage with M27 (the Dumbbell Nebula). Look for a tiny tangle of suns immersed in a web of unresolved light. It is a small but rich cluster that is best appreciated with low power in a small telescope.

Directions

Use Chart 60 to locate 12 Vulpeculae, which forms the northern apex of a near-equilateral triangle with Gamma (γ) and Delta (δ) Sagittae in the Arrow's shaft. Center 12 Vulpeculae in your telescope at low power, then switch to Chart 60b. NGC 6830 is a little less than 30′ due north of that star.

The quick view

At 23× in the 4-inch, NGC 6830 is a pretty little cluster with a bright triangular core. The core is surrounded by a 5′-wide halo of dim suns; the cluster also appears wedged between two 9th-magnitude stars. At 72×, NGC 6830 looks like a cross with bent arms. A splash of bright stars lies at the cluster's core; with averted vision, these suns mingle with a multitude of well-resolved, fainter stars that seem to fade away in layers. The cluster has some 82 stars of 10th magnitude and fainter.

Stop. Do not move the telescope. Your next target is nearby!

3. NGC 6823 (H VII-18)

Type	Con	RA	Dec	Mag	Diam	Rating
Open cluster	Vulpecula	19^h 43.2^m	+23° 18'	7.1	7.0'	3.5

4'

General description

NGC 6823 is another moderately small but enticingly rich open cluster near 5th-magnitude 12 Vulpeculae and NGC 6830.

Directions

Using Chart 60b, from NGC 6830, return to 12 Vulpeculae, then move about 40' west–northwest to 7th-magnitude Star a. Next, turn 30' southwest to 7.5-magnitude Star b. Now swing about 50' west–northwest to a 30'-wide triangle (c) of three 6.5-magnitude suns. NGC 6823 is a little less than 30' north of the northernmost star in Triangle c.

The quick view

At 23× in the 4-inch, NGC 6823 is at the base of a 30'-wide L-shaped asterism of stars (oriented northeast to southwest) whose long axis comprises a cascade of a half dozen 9th-magnitude suns. The cluster itself has four stars that form a miniature L-shaped asterism – one that mirrors the larger L-shaped asterism to its northwest. The cluster's L-shaped core is surrounded by an irregularly shaped halo of dim starlight. At 72×, a nice double star illuminates the cluster's otherwise round core. Some 20 or so stars surround that core forming a series of broken arms against a faint background of unresolved or dim starlight. NGC 6823 has some 80 stars of 9th magnitude and fainter.

Star charts for second night

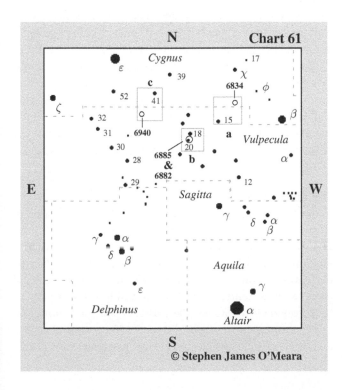

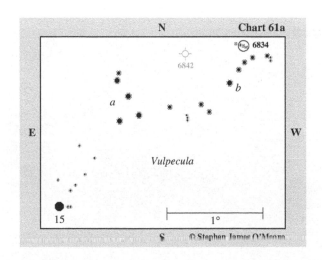

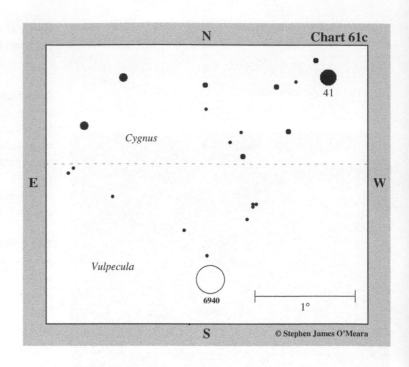

SECOND NIGHT

1. NGC 6834 (H VIII-16)

Type	Con	RA	Dec	Mag	Diam	Rating
Open cluster	Cygnus	19^h 52.2^m	+29° 24′	7.8	6.0′	2.5

General description

NGC 6834 is a small and relatively dim open star cluster midway between, and a little northeast of, a line between 5th-magnitude 15 Vulpeculae and 5th-magnitude Phi (φ) Cygni. Look for a dim haze centered on a little line of 10th- and 11th-magnitude stars.

Directions

Use Chart 61 to find 15 Vulpeculae, which is the northwestern tip of Vulpecula's Boot. Center 15 Vulpeculae in your telescope at low power, then switch to Chart 61a. From 15 Vulpeculae make a slow and careful $1\frac{1}{4}°$ sweep northwest to a 20′-wide inverted Y-shaped asterism (a) comprising four roughly 7th-magnitude suns. Next make another slow and careful sweep 1° west–northwest to 7.5-magnitude Star b. NGC 6834 is a little more than 25′ north–northwest of Star b, which is the southeasternmost star in a 20′-wide arc of 8th and 9th-magnitude suns.

The quick view

In the 4-inch at 23× under a dark sky, NGC 6834 is a small cluster (5′) that appears elongated east to west. It soon becomes apparent that the elongation is mainly due to a little line of 10th and 11th-magnitude suns that slice across the cluster's core. The cluster itself, however, is a small elliptical glow centered on that line and it looks like a faint background glow. At 72×, the cluster is odd – a sinuous line of five prominent suns split the center. With averted vision, that line is surrounded by a rich assortment of finely resolved stars. The cluster also swells with averted vision and has north and south extensions, like wings that flow out from the center of the line. The cluster is very rich, having 138 members in a disk 6′ across; the stars shine at 11th magnitude and fainter.

2 & 3. NGC 6885 & 6882 (H VIII-20 & H VIII-22)

Type	Con	RA	Dec	Mag	Diam	Rating
Open cluster	Vulpecula	20^h 11.6^m	+26° 28′	8.1	20.0′	4

General description
NGC 6885 is a two-in-one cluster. William Herschel first discovered NGC 6885 on September 9, 1784. The next night he discovered NGC 6882. But Harold G. Corwin (Infrared Processing and Analysis Center) believes William Herschel saw the same cluster on both nights. Herschel made a 15′ positional error in his second observation, which led to the belief (and confusion) that two clusters exist. But only one does: NGC 6885. Therefore, by making an observation of NGC 6885, you also make an observation of NGC 6882. Under a dark sky, the cluster is large and scattered but its stars are quite apparent, being huddled around 6th-magnitude 20 Vulpeculae.

Directions
Use Chart 61 to locate 20 Vulpeculae, which is about 10° (a fist) east–southeast of Beta (β) Cygni (Albireo) – the famous telescopic double star marking the Swan's nose. 20 Vulpeculae is huddled together with two other stars of similar magnitude: 18 and 19 Vulpeculae. Center 20 Vulpeculae in your telescope at low power and switch to Chart 61b to confirm you have the correct star. The densest

part of NGC 6885 should be apparent as a bar of fuzzy starlight 6′ northwest of 20 Vulpeculae.

The quick view
At 23× in the 4-inch, NGC 6885 looks like massive cluster that's been shredded by tidal forces. The brightest and most obvious sections are an east to west oriented spit of starlight 6′ northwest of 20 Vulpeculae, and a trapezoid of starlight just southwest of 20 Vulpeculae. At 72×, a line of equally spaced 11th- to 12th-magnitude suns slice through the spit of starlight, which also scintillates with fainter suns. The cluster contains only 60 members to 13th magnitude, the brightest of which shine at around 6th magnitude.

4. NGC 6940 (H VII-8)

Type	Con	RA	Dec	Mag	Diam	Rating
Open cluster	Vulpecula	20^h 34.5^m	+28° 17′	6.3	25.0′	4

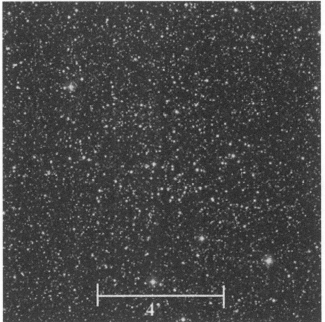

General description
NGC 6940 is a very bright, very large, and very rich open star cluster in the southern branch of the Milky Way bordering the Great Rift. You'll find it lost in the void between the southern wing of Cygnus and the Little Fox's hind quarters. It is a binocular treat and a telescopic wonder of stellar riches. In 7 × 30 binoculars it is an elliptical swarm of dim suns shining with a total light of a 6th-magnitude star. Some brighter field stars are superimposed on the cluster. In even the smallest of telescopes, the cluster looks like a dwarf elliptical galaxy, elongated northeast to southwest – as if it were being pulled apart by tidal forces from a close encounter with the Milky Way.

Directions

Use Chart 61 to locate 4th-magnitude 41 Cygni. The cluster marks the southern apex of a near-equilateral triangle with that star and equally bright 52 Cygni to the northeast. You can try to immediately locate this 6th-magnitude glow in binoculars, or you can use Chart 61c to star-hop to it.

The quick view

At 23× in the 4-inch (which is all the power you really need to scan this cluster), NGC 6940 is a highly elliptical body of shimmering lights with many star clumps that give off a rich, mottled glow. The cluster's stars are scatted across the field like highly reflective dust particles in a strong and erratic wind. The cluster contains some 170 members of 11th magnitude and fainter. These stars are, in addition, projected against a rich Milky Way background, so the view is visually exhausting.

Star charts for third night

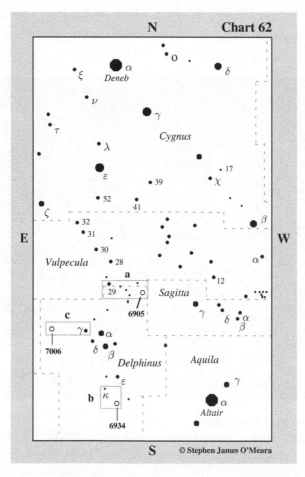

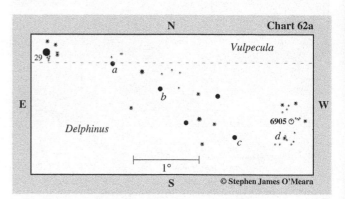

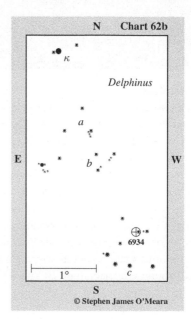

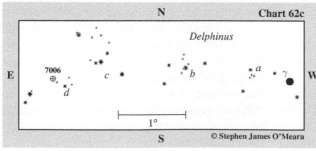

THIRD NIGHT

1. NGC 6905 (H IV-16)

Type	Con	RA	Dec	Mag	Dim	Rating
Planetary nebula	Delphinus	20^h 22.4^m	+20° 06'	11.1	42"× 35"	3

4'

General description

Commonly known as the Blue Flash Nebula, NGC 6905 is a beautiful planetary in the Delphinus Milky Way, near the borders of Vulpecula and Sagitta. Through a small telescope, it looks like a piece of lint trapped in a carpet of stars. Be prepared to make a very slow and careful search to find it.

Directions

Using Chart 62, start your search by looking midway between Epsilon (ε) Cygni, the southern wing tip of the Swan and the Diamond of Delphinus-Alpha (α), Beta (β), Gamma (γ), and Delta (δ) Delphini. There you should see a roughly 5°-long chain of four fifth-magnitude stars: 32, 31, 30, and 28 Vulpeculae (from northeast to southwest, respectively). Next look for the fifth-magnitude star 29 Vulpeculae, which lies 3° south of 28 Vulpeculae. Center 29 Vulpeculae in your telescope at low power, then switch

to Chart 62a. First move 1° west to 6th-magnitude Star a, right on the border between Vulpecula and Delphinus. A 50' hop southwest will bring you to another 6th-magnitude star (b), in Delphinus; it marks the northeast end of a 1¼°-long diamond of similarly bright suns. If you link all the stars just mentioned – from 32 Vulpeculae to this diamond – they form an asterism that I call the Diamond-headed Snake. Now center 6th-magnitude Star c at the southwest tip of the diamond (the Snake's nose) and move about 50' due west to 7th-magnitude Star d. NGC 6905 lies a little more than 15' to the north–northwest of Star d; it (and the trapezoid it resides in; see the photo to the left) marks the southeast facet of a 25'-wide diamond of 8th-magnitude stars.

Field note

NGC 6905 is small and somewhat dim. This is usually not a problem when such objects are far from the Milky Way. But NGC 6905 lies in a fabulously dense field of stars. My advice to beginners is to forget trying to sweep up a planetary nebula. Instead look for a clustering of 11th-magnitude and fainter stars that form a 4'-wide trapezoid in the object's position, then zoom in for the capture.

The quick view

In the 4-inch at 23× under a dark sky, NGC 6905 is incredibly dim; being the faintest member of a tiny trapezoid of suns. The nebula is also just east of a little L-shaped asterism of dim suns. If you really want to distinguish the planetary nebula from its surroundings, I suggest you immediately increase the magnification to 151×. At this power, the nebula is a pale puff of light caged in a trapezoid of three 11th- to 12th-magnitude suns and one dimmer one. It is hypnotic to try and pull the planetary nebula out from these stars. As the eye swims around, the planetary nebula wafts in and out of view like vapors from a moist mouth on a cold day. For this reason, I call NGC 6905 the Caged Spirit Nebula. In larger scopes, amateurs have had success seeing rich hues, as well as the 14th-magnitude central star, in 10-inch and larger telescopes.

2. NGC 6934 (H I-103)

Type	Con	RA	Dec	Mag	Diam	Rating
Globular cluster	Delphinus	20^h 34.2^m	+07° 24'	8.9	7.1'	3.5

General description

NGC 6934 is a moderately small but condensed globular cluster $3\frac{3}{4}^{\circ}$ south and slightly east of 4th-magnitude Epsilon (ε) Delphini. It is surprisingly obvious in 7×50 binoculars, lying just north of a sideways hook of stars. Through a telescope the globular cluster lies only 2′ west of a 9.5-magnitude star.

Directions

Use Chart 62. One way you can sweep up NGC 6934 is to put a low-power eyepiece in your telescope, center 4th-magnitude Epsilon Delphi (the tail of the Dolphin), and move $3\frac{3}{4}^{\circ}$ south. Otherwise you can star-hop. Start with finding 5th-magnitude Kappa (κ) Delphi, which is a little less than 2° southeast of Epsilon Delphi. Center Kappa Delphi in your telescope at low power, then switch to Chart 62b. From Kappa Delphi, make a slow and careful 1° sweep southwest to a roughly 20′-wide triangle (a) of three 7.5-magnitude suns. Now drop 30′ further to the southwest to a dimmer 18′-wide triangle (b) of 8th-magnitude suns. NGC 6934 is a generous $1\frac{1}{4}^{\circ}$ further to the southwest about 30′ north of the 50′-long sideways hook (c) of 7th- to 8th-magnitude stars.

The quick view

At 23× in the 4-inch, NGC 6934 looks like a 9th-magnitude comet kissing a 9th-magnitude star. A close pair of 10th-magnitude stars lies to the west. At 72×, the cluster has a dim core of dappled clumps of starlight surrounded by a 5′-wide halo of uniform light sprinkled with flecks of dim starlight. The brightest stars shine at a magnitude of 13.8, so they are within reach of small telescopes.

3. NGC 7006 (H I-52)

Type	Con	RA	Dec	Mag	Diam	Rating
Globular cluster	Delphinus	21ʰ 01.5ᵐ	+16° 11′	10.6	3.6′	3

General description

NGC 7006 is a dim – and remote (127,0000 light-years) – globular cluster $3\frac{1}{2}^{\circ}$ due east of 4.5-magnitude Gamma (γ) Delphini. Although NGC 7006 has been seen in telescopes as small as 3-inches, it's difficult to detect. Best to locate the field first, then use moderate magnification to seek it out.

Directions

Use Chart 62 to locate Gamma Delphini. Center Gamma Delphini in your telescope at low power, then switch to Chart 62c. From Gamma Delphini, move 35′ east–northeast to 7.5-magnitude Star a. Next make a careful 1° sweep due east to 7.5-magnitude Star b, which has an 11th-magnitude companion to the east. Now make another careful 1° sweep east to two more 7.5-magnitude stars (c) separated by about 20′ and oriented northeast to southwest. Now look about 45′ southeast for yet another 7.5-magnitude star (d). NGC 7006 is only about 12′ northeast of Star d.

The quick view

At 23× in the 4-inch under dark skies, NGC 7006 is a very faint glow measuring only about 3′ in diameter. It is better seen at 72×, when it looks like a round globe with hints of a bright core and faint outer extensions. There is no hint of resolution; the cluster's brightest stars shine at a magnitude of 15.6.

Star charts for fourth night

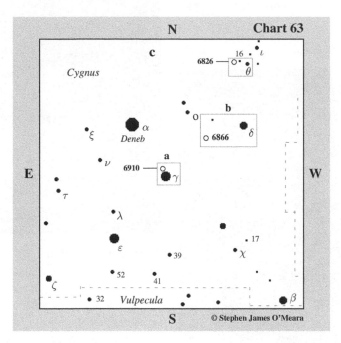

Chart 63

N

c

Cygnus

6826 — 16 · ι

θ

b

O 6866 · δ

ξ

α

Deneb

a

6910 — γ

ν

E W

τ

λ

ε

17

39 χ

52 41

ζ

32 Vulpecula β

S

© Stephen James O'Meara

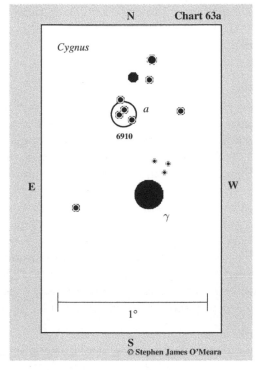

Chart 63a

N

Cygnus

a

6910

E W

γ

1°

S

© Stephen James O'Meara

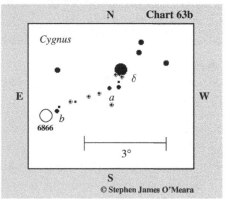

Chart 63b

N

Cygnus

δ

a

E W

b

6866

3°

S

© Stephen James O'Meara

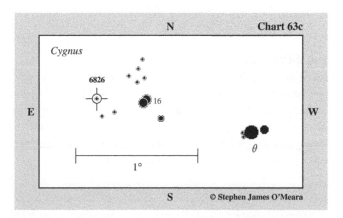

Chart 63c

N

Cygnus

6826

16

E W

θ

1°

S

© Stephen James O'Meara

FOURTH NIGHT

1. NGC 6910 (H VIII-56)

Type	Con	RA	Dec	Mag	Diam	Rating
Open cluster	Cygnus	20ʰ 23.2ᵐ	+40° 47′	7.4	10.0′	4

4'

General description
NGC 6910 is a moderately small but crisp little cluster just 30′ north–northeast of Gamma (γ) Cygni. A very nice aggregation of pretty suns.

Directions
Use Chart 63 to locate Gamma Cygni in the long axis of the Northern Cross. Center that star in your telescope at low power, then switch to Chart 63a. NGC 6910 is just 30′ north–northeast of Gamma Cygni and about 18′ southeast of a 6th-magnitude star.

The quick view
In the 4-inch at 23×, NGC 6910 is a very delicate cluster with two bright 7th-magnitude gems amidst an elongated spread of sharp stars. At 72×, the spread of stars comprises a sinuous stream of sparkling stars of similar magnitude that flow like water past the brighter stars superimposed on it. There are some 66 stars here of 9.6 magnitude and fainter.

2. NGC 6866 (H VII-59)

Type	Con	RA	Dec	Mag	Diam	Rating
Open cluster	Cygnus	20ʰ 03.9ᵐ	+44° 09′	7.6	15′	4

4'

General description
NGC 6866 is a marvellous open cluster about $3\frac{1}{2}°$ east–southeast of 3rd-magnitude Delta (δ) Cygni, in the north-western wing of the Swan. Caroline Herschel discovered it on the evening of July 23, 1783. In 7×50 binoculars it looks like an irregular amorphous glow that's somewhat mottled with averted vision. The smallest of telescopes will show the cluster as a partially resolved spine of stars oriented north–northwest to south-southeast against a dim mottled glow. It is a marvel in telescopes of all sizes.

Directions
Use Chart 63 to find Delta Cygni – the star that marks the northwestern arm of the Northern Cross – and the 4th-magnitude star Omicron (o) Cygni, about 5° east–northeast of Delta Cygni. NGC 6866 lies about 3.5° east–southeast of Delta Cygni and makes an isosceles triangle with Delta and Omicron Cygni. To star-hop to it, center Delta Cygni in your telescope at low power, then switch to Chart 63b. From Delta Cygni, move about 1° to the south–southeast, where you will find a pair of 7th-magnitude stars (a), which are oriented east to west and separated by about 25′. These stars are at the western end of a roughly $2\frac{1}{2}°$ chain of similarly bright suns, the eastern end of which is marked by 7.6-magnitude Star b. NGC 6866 is only about 20′ southeast of Star b.

The quick view
The view at 23× in the 4-inch mimics that of what Caroline Herschel saw. The cluster displays a small core that appears a bit bulbous, fuzzy, and bright on the northern end. This pouch of stars is surrounded by a dimmer crown of scattered suns that, with concentration, has sharp and long extensions to the east, west, and south. At 72×, the pouch transforms into a "pulsating" mass of starlight; if you sweep across this area with averted, then direct, vision the pouch puffs out, then contracts, in regular fits and starts. Aside from the obvious

affects of peripheral versus direct vision, there must be suns, or groups of suns, here at or near the limit of resolution.

Field note

When Caroline Herschel discovered this object, she thought it might be a nebula.

3. NGC 6826 (H IV-73)

Type	Con	RA	Dec	Mag	Dim	Rating
Planetary nebula	Cygnus	19^h 44.8^m	+50° 31′	8.5	27″× 24″	4

General description

NGC 6826, commonly known as the Blinking Planetary Nebula, is a bright and beautiful object immediately east of the beautiful 6th-magnitude double star 16 Cygni. The object swells and shrinks with averted and direct vision, respectively, thus its nickname. It is visible with binoculars under a dark sky and is a fine object in telescopes of all sizes.

Directions

Use Chart 63 to locate 4.5-magnitude Theta (θ) Cygni, which is about 5$\frac{1}{4}$° north–northwest of 3rd-magnitude Delta (δ) Cygni. Now locate 16 Cygni about 2$\frac{1}{4}$° to the northeast. There is no mistaking 16 Cygni because it is a fine double star whose components both shine at 6th-magnitude and are separated by 40″. Center 16 Cygni in your telescope at low power, then switch to Chart 63c. NGC 6826 is only about $\frac{1}{2}$° east of 16 Cygni.

The quick view

At 23× in the 4-inch, NGC 6826 is a starlike glow that should swell with averted vision and a few minutes of staring. At 72×, the planetary nebula is a small (25″) disk with a soft aqua hue, very much like that of the planet Neptune. At 189×, the planetary nebula's 10.6-magnitude central star shines forth from a diffuse shell of largely uniform light; the shell becomes a little bit brighter to a dim annulus toward the center with averted vision.

Star charts for fifth night

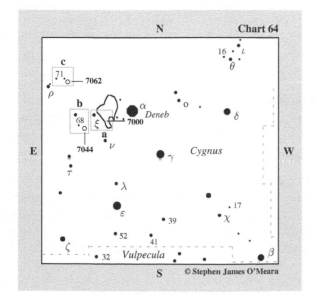

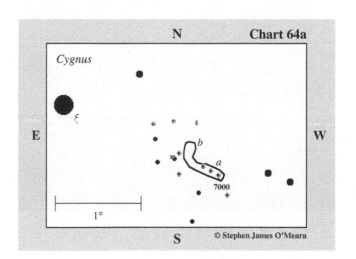

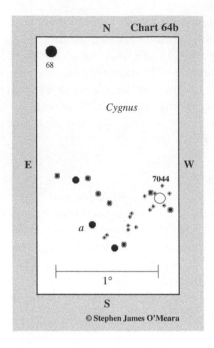

Chart 64b

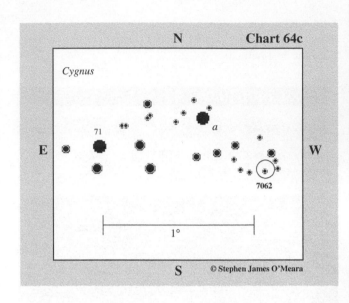

Chart 64c

FIFTH NIGHT

1. NGC 7000 (H V-37?)

Type	Con	RA	Dec	Mag	Diam	Rating
Emission nebula	Cygnus	$\sim21^h$ 00.0^m (H V-37?)	$\sim+43°$ $35'$ (H V-37?)	–	$\sim7'$	3.5

General description

NGC 7000, commonly called the North America Nebula, is a large but somewhat conspicuous emission nebula near Alpha (α) Cygni (Deneb), the tip of the Northern Cross. It is visible to the unaided eye under dark skies as an enhanced region in the Cygnus Milky Way. It is best seen in a wide field of view under a dark sky. But William Herschel did not discover the North America Nebula. His H V-37 is only a long streamer within that larger nebulosity; he described it as a "[v]ery large, diffused nebulosity, brighter in the middle, 7' or 8' long. 6' broad and losing itself very gradually and imperceptibly." His recorded position for this object is in question, though it is most likely somewhere along the bright strip of nebulosity comprising "Central America," just southwest of Xi (ξ) Cygni. So you should really be trying to see this "part" of the North America Nebula, which I've traced in Chart 64a; the section marked (b) appears in the photograph.

Directions

Use Chart 64 to locate 4th-magnitude Xi Cygni, then switch to Chart 64a. H V-37 is most likely the brightest 7'-long patch of light in the region outlined on this chart.

The quick view

In the 4-inch at 23× under a dark sky, The North America Nebula is a pale green glow (100' wide) covering a rich field of Milky Way, which includes some small and dim open star clusters. H V-37 is a long and bright strip of nebulosity that appears broken in the middle. The brightest section is at the southwestern end and measures about 10′ in length, oriented northeast to southwest; it is just southeast of a row of three 10th-magnitude suns (a) oriented the same way. The second, though slightly less conspicuous region is marked at position b.

Stop. Do not move the telescope.

2. NGC 7044 (H VI-24)

Type	Con	RA	Dec	Mag	Diam	Rating
Open cluster	Cygnus	21ʰ 13.1ᵐ	+42° 29′	12.0	7.0′	1.5

4′

General description

NGC 7044 is a very small, dim, and difficult open cluster about 2° southeast of 4th-magnitude Xi (ξ) Cygni. It will be extremely difficult to see in a small telescope under any light pollution. Dark skies are a must! Be prepared to make a slow and careful search for it.

Directions

Using Chart 64, from NGC 7000, return to Xi Cygni. Now look about 2½° due west for 5.5-magnitude 68 Cygni. Center 68 Cygni in your telescope at low power, then switch to Chart 64b. Now use binoculars to look for a roughly 50′ long arc of three 6.5-magnitude stars (a) oriented north–northeast to south–southwest. Center the two southwestern-most stars in Arc a in your telescope at low power. NGC 7004 is 40′ to the northwest of the mid point between those two stars.

The quick view

At 23× in the 4-inch, NGC 7044 is very small (~3′) and insignificantly faint round puff of gray light. The cluster is almost impossible to see at 72×, one roughly 9th-magnitude star flanking the cluster to the northeast is most apparent, as are a sprinkling of perhaps a half dozen suns. The cluster has some 60 members that shine at 15th magnitude and fainter, so the "stars" I saw were most likely clumps of starlight.

3. NGC 7062 (H VII-51)

Type	Con	RA	Dec	Mag	Diam	Rating
Open cluster	Cygnus	21ʰ 23.4ᵐ	+46° 23′	8.3	5.0	3.5

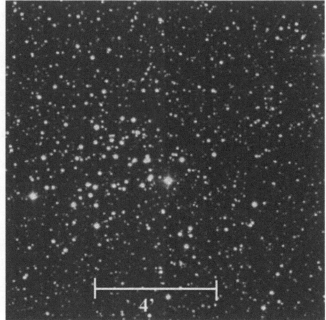

4′

General description

NGC 7062 is a small and somewhat dim open cluster about 2° northwest of 4th-magnitude Rho (ρ) Cygni, and a little more than 1° west–southwest of 5.5-magnitude 71 Cygni. Although it is dim, the cluster is condensed, so it is rather obvious.

Directions

Use Chart 64 to locate Rho Cygni, which is 9° (about a fist) east and slightly north of Alpha (α) Cygni. It is the brightest star in the immediate vicinity. Now use your unaided eyes or binoculars to locate 71 Cygni about 1¼° to the northwest. Center 71 Cygni in your telescope at low power, then switch to Chart 64c. From 71 Cygni, move about 45′ west–northwest to 6th-magnitude Star a. NGC 7062 is 30′ southwest of Star a.

The quick view

At 23× in the 4-inch, NGC 7062 is a small but condensed irregular glow centered on the west end of a 10'-long chain of 10th-magnitude stars. Although it is unresolved, the glow is fairly easy to see under a dark sky. At 72×, the cluster is a menagerie of irregularly bright suns scattered across an irregular disk measuring 5' across. The cluster has some 85 members shining at a magnitude of 10 and fainter. Averted vision is required to see the fainter members.

Star charts for sixth night

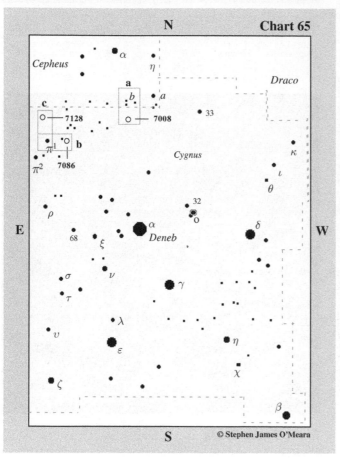

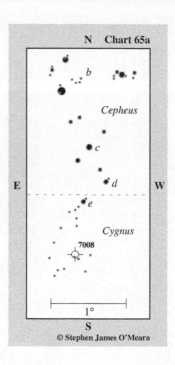

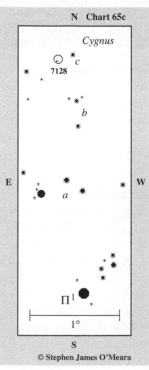

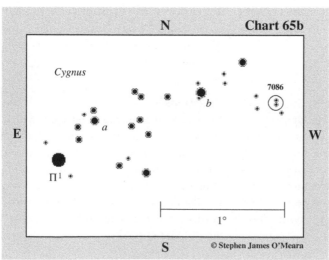

SIXTH NIGHT

1. NGC 7008 (H I-192)

Type	Con	RA	Dec	Mag	Dim	Rating
Planetary nebula	Cygnus	21ʰ 00.5ᵐ	+54° 33′	9.9	98″× 75″	4

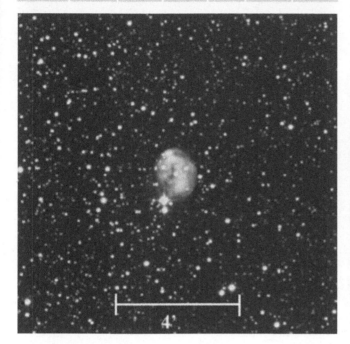

4′

General description
NGC 7008 is a bright planetary nebula in a 5°-wide dark nebula some 8° north–northeast of Alpha (α) Cygni (Deneb). The planetary nebula lies in the nondescript northern shores of this black lagoon, less than 1° from the Cepheus border. It is a fine sight in all telescopes. Be prepared to make a careful search for it, though.

Directions
Use Chart 65 to first locate 2nd-magnitude Alpha (α) Cepheii, the southwest cornerstone of the famous House asterism. Third-magnitude Eta (η) Cepheii lies about 3¾° to its west–southwest. From Eta Cepheii, look 4° south and slightly east for 4.5-magnitude Star a. Now use your binoculars to look about 2° east–southeast for a 50′-wide triangle of three 6th- and 7th-magnitude suns (b). Place Triangle b in your telescope at low power and confirm its identity using Chart 65a. You want to center the bright southeasternmost star in Triangle b, then sweep about 50′ southwest to 6.5-magnitude Star c. Next move 30′ southwest to 7th-magnitude Star d. Now swing about 25′ southeast to similarly bright Star e. NGC 7008 is 40′ south and a little east of Star e.

The quick view
In the 4-inch at 23×, the planetary is kissing a 9th-magnitude star to the south-southeast. This star also has a 10th-magnitude companion 18″ to the south-southwest.

Seen at a glance at low power, the nebula and these stars form a very small and fuzzy "double," like M40 in Ursa Major. So your target is easy to sweep over. At 72×, the planetary nebula comes to life, being a very beautiful object – a dainty skirt of light clinging to its 9th-magnitude neighbor.

2. NGC 7086 (H VI-32)

Type	Con	RA	Dec	Mag	Diam	Rating
Open cluster	Cygnus	21ʰ 30.5ᵐ	+51° 36′	8.4	12.0′	3.5

4′

General description
NGC 7086 is a moderately large and obvious open cluster a little less than 2° west–northwest of 5th-magnitude Pi¹ (π¹) Cygni. While the cluster is relatively dim, it is also very condensed. Indeed, it is visible in 7×50 binoculars under a dark sky, and it is a pretty object in telescopes of all sizes.

Directions
Use Chart 65 to locate Pi¹ Cygni, which is about 6° north of 4th-magnitude Rho (ρ) Cyg. Center Pi¹ Cygni in your telescope at low power, then switch to Chart 65b. From Pi¹ Cygni, move a little less than 30′ northwest to 7th-magnitude Star a. Next, move a little more than 50′ west–northwest to 6.5-magnitude Star b. NGC 7086 is a little less than 40′ west and a tad south of Star b.

The quick view
At 23× in the 4-inch, NGC 7086 is a fairly large (10′) and scattered open cluster with a wide range of well-resolved stars. The cluster's core is patchy, and it is surrounded by an irregularly round halo of dim suns. At 72×, the cluster has a definite wedge-shaped core with a north to south extension of stars. With averted vision, the cluster sports broad outer

arms in a maltese-cross fashion. Very pretty. The cluster contains some 80 stars of 10th magnitude and fainter.

Stop. Do not move the telescope.

3. NGC 7128 (H VII-40)

Type	Con	RA	Dec	Mag	Diam	Rating
Open cluster	Cygnus	21^h 43.9^m	+53° 43′	9.7	4.0′	2.5

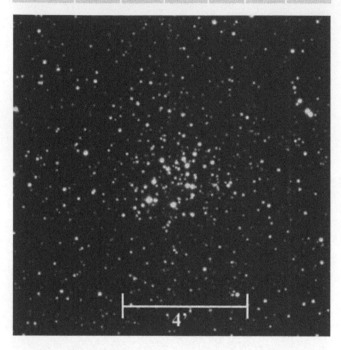

General description

NGC 7128 is a small and dim open cluster about $2\frac{1}{2}°$ north and slightly east of 5th-magnitude Pi^1 (π^1) Cygni. Although it is small, it is very condensed, so it can be seen under a dark sky with averted vision.

Directions

Using chart 65b, from NGC 7086, return to Pi^1 Cygni, then switch to Chart 65c. From Pi^1 Cygni, make a slow and careful sweep a little more than 1° north and slightly east to a 30′-wide arc of three 6.5- to 7th-magnitude suns (a), oriented east to west. Center the westernmost star in that arc, then move 50′ north to a pair of 8th-magnitude stars (b), oriented north to south and separated by about 15′. NGC 7128 is only 30′ north–northeast of the northernmost star in Pair b and 10′ southeast of 8th-magnitude Star c.

The quick view

At 23× in the 4-inch, NGC 7128 is a very small (4′) cluster but also bright and condensed when seen with averted vision. A tiny 10th-magnitude double star lies on the cluster's southeast flank. The cluster seems to swell from that pair like a bubble. At 72×, the cluster is a small, circular assemblage of suns with a hollow center (the bubble). The cluster has some 70 members that shine at 11.5 magnitude and fainter. About a dozen of these stars can be seen at 101× forming a warped, C-shaped pattern.

Star charts for seventh night

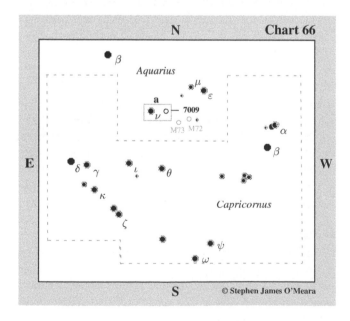

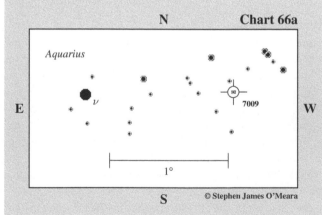

SEVENTH NIGHT

1.NGC 7009 (H IV-1)

Type	Con	RA	Dec	Mag	Dim	Rating
Planetary nebula	Aquarius	21^h 04.2^m	$-11°$ $22'$	8.0	$44'' \times$ $23''$	4

4'

General description

NGC 7009, the famous Saturn Nebula, is a small but very bright planetary nebula about $1\frac{1}{4}°$ west of Nu (ν) Aquarii, near M72 and M73. The nebula is easy to spot in 7×50 binoculars as an 8th-magnitude star, and it is a marvel in telescopes of all sizes.

Directions

Use Chart 66 to first locate 4.5-magnitude Nu (ν)-Aquarii, which is about 8° east–northeast of 3.5-magnitude Alpha (α) Capricorni. Center Nu Aquarii in your telescope at low power, then use Chart 66a to guide you on a gentle $1\frac{1}{4}°$ sweep due west of that star to the planetary nebula.

The quick view

In the 4-inch at 23×, the NGC 7009 appears as an 8th-magnitude star that swells slightly with averted vision. With concentration a fainter outer crown can be detected. At 72×, the nebula displays a tight inner annulus surrounded by a pale-green crown. At very high powers, the nebula's 11.5-magnitude central star rises from the depths of the nebulosity to shine like a beacon. Also use very high powers – as high as you can go – to see the nebula's appendages, which makes it look like Saturn seen with its rings edge on.

Fall

10 · October

Star charts for first night

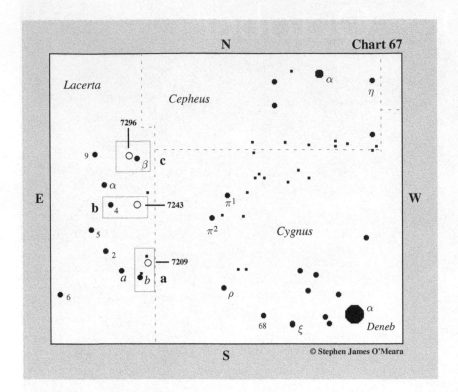

Chart 67

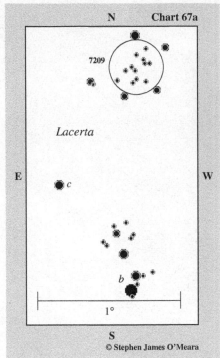

Chart 67a

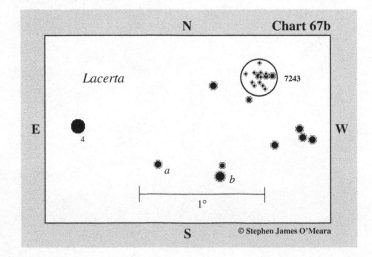

Chart 67b

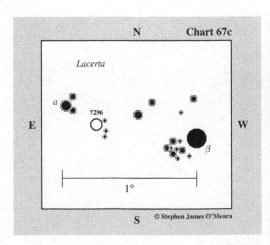

Chart 67c

FIRST NIGHT

1. NGC 7209 (H VII-53)

Type	Con	RA	Dec	Mag	Diam	Rating
Open cluster	Lacerta	22^h 05.1^m	+46° 29'	7.7	15.0'	4

General description

NGC 7209 is a large and reasonably bright open cluster about $5\frac{1}{2}$° southwest of 4th-magnitude Alpha (α) Lacertae. It also lies near a beautiful triple star in the Lizard's foot. The cluster is visible in binoculars under a dark sky, and it is a very attractive sight in telescopes of all sizes.

Directions

Use Chart 67 to find Alpha Lacertae, then 5th-magnitude Star b, which is about $6\frac{1}{2}$° to the south–southwest. Star b is easy to identify in binoculars because it is a beautiful triple star. Center this triple in your telescope at low power, then switch to Chart 67a. You could just sweep about $1\frac{1}{2}$° north of the triple to NGC 7209, or you could move about 45' northeast to 6th-magnitude Star c. NGC 7209 is a little less than 1° northwest of Star c.

The quick view

In the 4-inch at 23×, NGC 7209 is a beautiful hook-shaped cluster of well-resolved suns spread across 15' of sky. The field is extremely rich in starlight but not all the stars are cluster members, so the cluster actually looks almost twice as large as it really is. Still, there are nearly 100 cluster members here of 7.7 magnitude and fainter. At 72×, there's a lazy zigzagging line of stars running north to south through the cluster's center.

Stop. Do not move the telescope.

2. NGC 7243 (H VIII-75)

Type	Con	RA	Dec	Mag	Diam	Rating
Open cluster	Lacerta	22^h 15.0^m	+49° 54'	6.4	30'	4

General description

NGC 7243 is a very large and bright coarse scattering of suns $2\frac{1}{2}$° west–southwest of Alpha (α) Lacertae or about $1\frac{1}{2}$° west–northwest of 5th-magnitude 4 Lacertae. It can be seen with binoculars under a dark sky as a partially resolved arrowhead of fuzzy starlight that points to the northeast. Telescopically, the cluster lies in a rich starfield and is best seen at low power.

Directions

Using Chart 67, return your gaze to Alpha Lacertae. Now look $1\frac{1}{2}$° southwest for 5th-magnitude 4 Lacertae. Center 4 Lacertae in your telescope at low power, then switch to Chart 67b. From 4 Lacertae, you could make a slow sweep $1\frac{1}{2}$° west–northwest directly to NGC 7243, or you could move 40' southwest to 7th-magnitude Star a. Next hop 30' west–southwest to 6.5-magnitude Star b. NGC 7243 is about 50' north–northwest of Star b.

The quick view

At 23× in the 4-inch, NGC 7243 is a loose gathering of some 40 reasonably bright stars and twice as many dim ones spread across an area of sky equal to that of the full Moon. A slender diamond of stars cuts across the cluster's western flank from a triangular core of stars. Several other stars are arranged in long looping arms that look like the flapping wings of a seagull. At 72×, many fainter suns sparkle into view, giving the cluster added luster. The southeastern star in the central triangle is a fine double star for small apertures. It consists of a 9.3-magnitude primary with a 9.7-magnitude companion 9.2″ away. Stop. Do not move the telescope.

3. NGC 7296 (H VII-41)

Type	Con	RA	Dec	Mag	Diam	Rating
Open cluster	Lacerta	22ʰ 28.0ᵐ	+52° 19′	9.7	3.0′	3.5

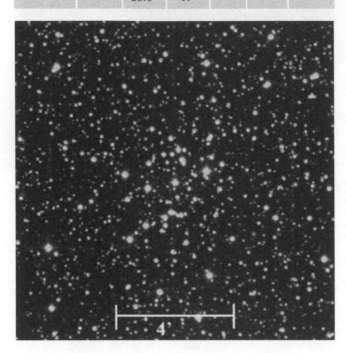

General description

NGC 7296 is a very small but rather obvious and pretty open cluster 40′ east and slightly north of 4.5-magnitude Beta (β) Lacertae. It may be difficult to see well under light-polluted skies at low power. The cluster looks best at moderate magnifications.

Directions

Using Chart 67, from NGC 7243, return to Alpha (α) Lacertae. Now look about $2\frac{1}{4}°$ to the northwest for Beta Lacertae. Center Beta Lacertae in your telescope at low power, then switch to Chart 67c. NGC 7296 is 40′ east of Beta Lacertae, and a little more than 15′ southwest of 6.5-magnitude Star a.

The quick view

At 23× in the 4-inch, NGC 7296 is a small but moderately condensed (2′) haze that swells to 3′ with averted vision. With direct vision, the cluster appears mottled. With averted vision it looks like a cape of faint stars blowing in a wind. The cluster is most beautiful at 72× – a splash of dim suns that flows southeastward from a 9.5-magnitude sun. At 101×, NGC 7296 becomes a series of arranged strings of stars. The cluster contains some 20 members of 10th magnitude and fainter.

Star charts for second night

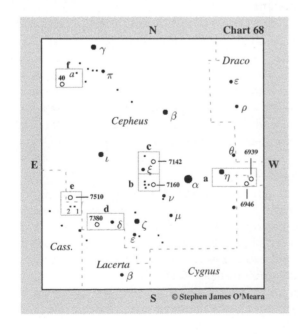

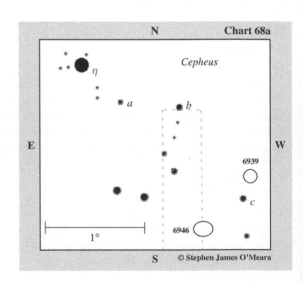

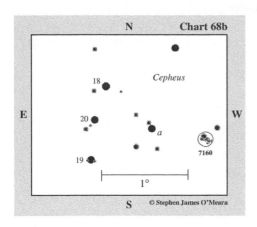

Chart 68b

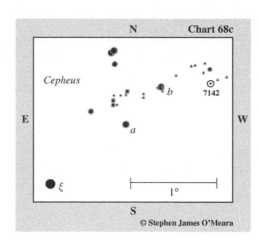

Chart 68c

SECOND NIGHT

1. NGC 6939 (H VI-42)

Type	Con	RA	Dec	Mag	Diam	Rating
Open cluster	Cepheus	20^h 31.5^m	+60° 40′	7.8	10.0′	4

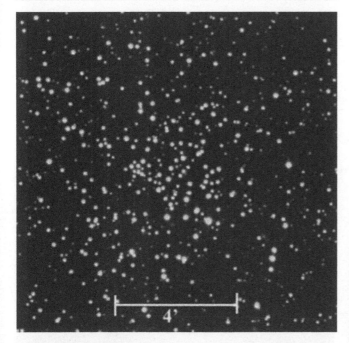

General description

NGC 6939 is a large and reasonably bright open cluster about 2° southwest of 4th-magnitude Eta (η) Cephei and just $\frac{2}{3}$° north-west of the slightly dimmer galaxy NGC 6946 (see right). These two objects make a stunning pair in all telescopes.

Directions

Use Chart 68 to find Eta Cephei. Center Eta Cephei in your telescope at low power, then switch to Chart 68a. From Eta Cephei, move 30′ southwest to 8.5-magnitude Star a. Next, move about 35′ west to 7.5-magnitude Star b. NGC 6939 is

1° southwest of Star b, and just about 12′ north–northwest of 7th-magnitude Star c.

The quick view

In the 4-inch at 23×, NGC 6939 is a 10′-wide round ball of scintillating points of light against an unresolved background of dimmer suns. At 72×, the cluster is cleanly resolved with a dramatic dark lane bordered to the west by a row of similarly bright suns. With some concentration and averted vision, the cluster's core is loosely concentrated with arms of starlight spiraling around it like a spinning pinwheel.

Stop. Do not move the telescope. Your next target is nearby!

2. NGC 6946 (H IV-76)

Type	Con	RA	Dec	Mag	Dim	Rating
Mixed spiral galaxy	Cepheus	20^h 34.8^m	+60° 09′	8.8	13.0′ × 13.0′	4

General description

NGC 6946 is a large and moderately bright galaxy that makes a beautiful companion to open cluster NGC 6939 (see above). Although moderately bright, the galaxy's low surface brightness makes it somewhat of a challenge to see, especially under light-polluted skies. Under dark skies in a small telescope, the galaxy looks like a ghost image of NGC 6939.

Directions

Using Chart 68a, From NGC 6939, move about 12′ south–southeast to 7th-magnitude Star c. NGC 6946 is only about 25′ southeast of Star c.

The quick view

At 23× in the 4-inch under a dark sky, NGC 6946 is a largely uniform glow with a bead-like core. With averted vision and time, the galaxy appears slightly elongated, oriented northeast to southwest. Some ripples of structure waft in and out of view, but the galaxy is largely featureless. At 72×, the galaxy all but vanishes. Thus, this galaxy is best seen in large telescopes or at low power in a small telescope.

3. NGC 7160 (H VIII-67)

Type	Con	RA	Dec	Mag	Diam	Rating
Open cluster	Cepheus	21ʰ 53.7ᵐ	+62° 36′	6.1	5.0′	4

General description

NGC 7160 is a bright but tiny little cluster of scattered gems a little less than 4° east of Alpha (α) Cephei and about $2\frac{1}{4}°$ south–southwest of 4.5-magnitude Xi (ξ) Cephei. It is visible as a tiny knot of starlight in binoculars and is very distinctive in telescopes of all sizes.

Directions

Use Chart 68 to locate Alpha Cephei, then Xi Cephei to the east–northeast. Now use your unaided eyes or binoculars to look for a roughly 45′-wide, sideways pyramid of starlight 2° to the south. Center that pyramid in your telescope at low power and confirm the field with Chart 68b; the pyramid's eastern base comprises three 5.5-magnitude suns: 18, 20, and 19 Cephei (from north to south) respectively. Now center the pyramid's western apex, Star a, and move 35′ west and slightly south to NGC 7160.

The quick view

At 23× in the 4-inch, NGC 7160 is a sideways Y-shaped cluster with two bright 7th-magnitude members on the eastern end, and a tight arc of three 9th-magnitude stars on the western end. A dusting of dimmer suns surround these stars. At 72×, NGC 7160 is an irregular scattering of about dozen suns, whose brightest members rest in a blanket of dimmer suns. The cluster contains some 60 members of 7th magnitude and fainter.

Stop. Do not move the telescope.

4. NGC 7142 (H VII-66)

Type	Con	RA	Dec	Mag	Diam	Rating
Open cluster	Cepheus	21ʰ 45.1ᵐ	+65° 46′	9.3	12.0′	3

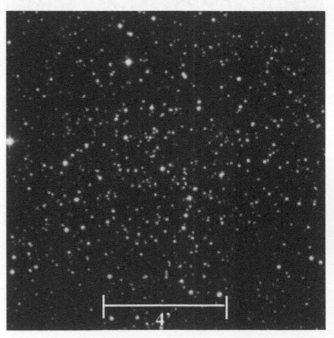

General description

NGC 7142 is a moderately large and dim open cluster about $2\frac{1}{4}°$ northwest of 4.5-magnitude Xi (ξ) Cephei. The cluster requires a dark sky to see well in a small telescope. It has a very low surface brightness, so any light pollution will greatly affect your chances of seeing it.

Directions

Using Chart 68, from NGC 7160, return to Xi Cephei, center it in your telescope at low power, then switch to Chart 68c. From Xi Cephei, move a little more than 1° northwest to 6th-magnitude Star *a*. Now move about 35′ further to the northwest, to 6.5-magnitude Star *b*. NGC 7142 is about 35′ west of Star *b*.

The quick view

At 23× in the 4-inch under a dark sky, NGC 7142 is an amorphous ghostly glow that's just visible above the sky background as seen with averted vision. Dim suns weave in and out of view across an area 10′ in extent. At 72×, the cluster is extremely difficult to observe. With time, about a dozen faint suns can be seen loosely scattered across the field. The brightest ones form a roughly 5′-wide oval oriented northeast to southwest.

Star charts for third night

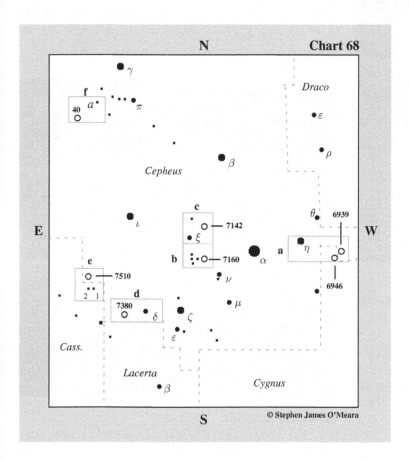

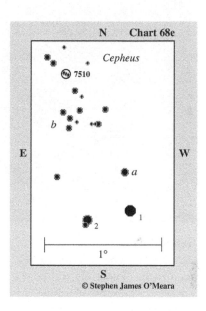

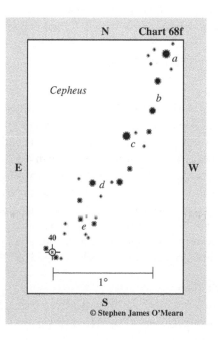

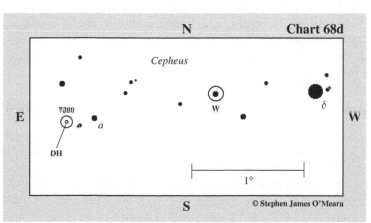

THIRD NIGHT

1. NGC 7380 (H VIII-77)

Type	Con	RA	Dec	Mag	Diam	Rating
Open cluster	Cepheus	22^h 47.3^m	+58° 08′	7.4	20.0′	4

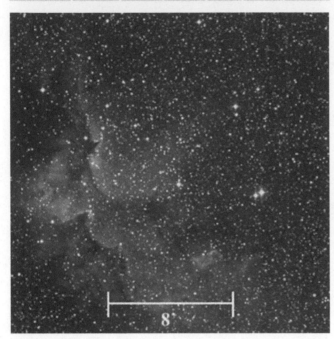

General description

Caroline Herschel discovered NGC 7380 in 1787. It is a fine assortment of coarsely scattered stars 2° east of the famous variable and double star Delta (δ) Cephei. It is easy to see in 7×50 binoculars as a small, condensed glow of uniform brightness. And through the smallest of telescopes, it appears as a delicate glow with a central condensation and faint asymmetries. In telescopes, the cluster is veiled in soft nebulosity – a very pretty sight.

Directions

Use Chart 68 to locate Delta Cephei, which is almost 10° (a fist) southeast of Alpha (α) Cephei. Center Delta Cephei in your telescope at low power, then switch to Chart 68d. Note that Delta Cephei is a beautiful double star with a 6.3-magnitude green companion 41″ from the yellow primary. NGC 7380 is 2° due east of Delta Cephei, so you can try making a great sweep for it. Otherwise, move 1° east to the 8th-magnitude irregular variable star W Cephei. Now carefully move a little more than 1° east–southeast to 6.5-magnitude Star a. NGC 7380 is about 15′ east and a little south of Star a, nearly centered on DH Cephei.

The quick view

In the 4-inch at 23×, NGC 7380 displays a triangular-shaped body of stars surrounded by nebulosity – both bright and dark. It is also in a rich Milky Way field. The glow and intensity of the nebula differs ever-so-slightly from that of the neighboring patches of Milky Way (something best appreciated under a dark sky). At a glance about 50 stars pop into view. At 72×, the main body of the cluster, again, has a triangular shape. There's also a main arc of three bright stars, the central one of which is a fine double with 7.7- and 8.6-magnitude components. The cluster contains about 125 stars of 10th magnitude and fainter, so it is a delight to see.

Stop. Do not move the telescope.

2. NGC 7510 (H VII-44)

Type	Con	RA	Dec	Mag	Diam	Rating
Open cluster	Cepheus	23^h 11.1^m	+60° 34′	7.9	7.0′	3.5

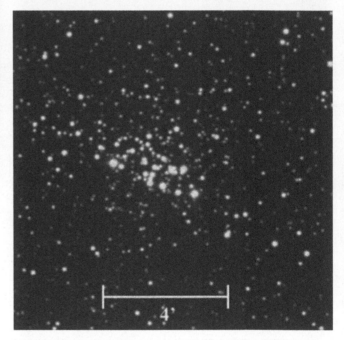

General description

NGC 7510 is a moderately small, moderately bright, open cluster about 5½° northeast of Delta (δ) Cephei, or a little more than 1° north–northeast of 5th-magnitude 1 Cephei. Its core consists of a very tight agglomeration of reasonably bright suns, so it's a good object to hunt down from suburban locations. Just remember that the bright core is tiny.

Directions

Using Chart 68, from NGC 7380 look 3° northeast for 5th-magnitude 1 Cephei, with 6th-magnitude 2 Cephei about 25′ to the west–southwest. Use binoculars, if necessary, to sight these stars. You want to center 1 Cephei in your telescope at low power, then switch to Chart 68e. From 1 Cephei, move 20′ north–northeast to 6.5-magnitude Star a. Next move 40′ northeast to the roughly 8′-long, Y-shaped asterism (b). NGC 7510 is about 15′ due north of Asterism b.

The quick view

At 23× in the 4-inch, NGC 7510 displays a very bright linear core about 5′ long and oriented east–northeast to west–southwest. With averted vision, a 10′-wide a halo of dim suns materializes. At 72×, the cluster shows about a dozen bright suns in a very appealing asterism that looks like a fishhook, or a mouse with a thin tail curled up on its back. The cluster must be a glorious sight in larger telescopes because this tiny grouping actually contains about 75 members of 10th magnitude and fainter.

3. NGC 40 (H IV-58)

Type	Con	RA	Dec	Mag	Dim	Rating
Planetary nebula	Cepheus	00ʰ 13.0ᵐ	+72° 31′	12.3	38″× 35″	4

4′

Star charts for fourth night

General description

NGC 40 is a bright but unimposing planetary nebula about $5\frac{1}{2}°$ southeast of 3rd-magnitude Gamma (γ) Cephei. The planetary nebula's 11.6-magnitude central star is twice as bright as the surrounding nebula. So you are searching, in essence (at least at first), for a 12th-magnitude star.

Directions

Use Chart 68 to find Gamma Cephei, which marks the tip of the House asterism in Cepheus. Now look 3° southwest for 4.5-magnitude Pi (π) Cephei. Now use your binoculars to look for a string of two roughly 7th-magnitude stars to the east that attach to a 2°-long kite-shaped asterism comprising four 6th- to 7th-magnitude stars, the easternmost of which is 6.5-magnitude Star a. You want to center Star a in your telescope at low power, then switch to Chart 68f. From Star a, move about 25′ south–southeast to two 7.5-magnitude stars (b). Next, move equidistant to the southeast to 6.5-magnitude Star c. Now move 30′ to the south–southeast to two roughly 7.5-magnitude stars (d). Another 25′ hop further to the south–southeast brings you to a pair of 8.5-magnitude stars (e). NGC 40 is 30′ southeast of Star e, between two roughly 9.5-magnitude stars.

The quick view

At 23× in the 4-inch, NGC 40 is virtually stellar. At 72×, the nebula becomes apparent especially with averted vision. Magnifications of 200× and higher make the nebula stand out more clearly from the bright central star.

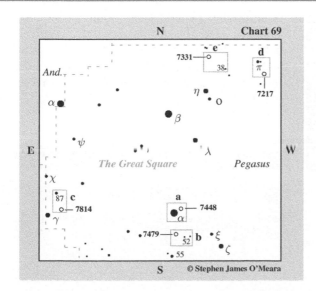

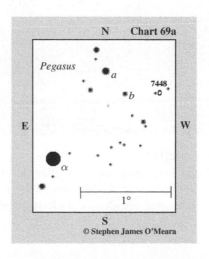

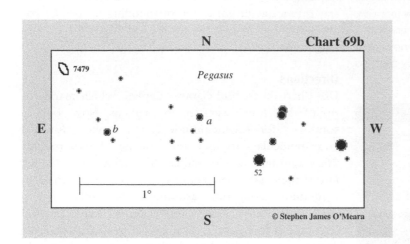

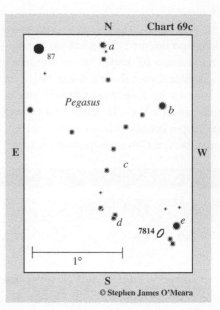

Fourth night

1. NGC 7448 (H II-251)

Type	Con	RA	Dec	Mag	Dim	Rating
Spiral galaxy	Pegasus	23^h 00.1^m	+15° 59′	11.7	2.5′× 1.0′	1.5

General description

NGC 7448 is a very small and dim galaxy a little less than $1\frac{1}{2}°$ northwest of 2.5-magnitude Alpha (α) Pegasi. Seeing it in a small telescope requires a dark sky and moderate

magnification. There are several dim stars and a grouping of three dim stars nearby, so you must be certain of the galaxy's position.

Directions

Use Chart 69 to find Alpha (α) Pegasi, the southwest corner star in the Great Square of Pegasus. Center Alpha Pegasi in your telescope at low power, then switch to Chart 69a. From Alpha Pegasi, make a careful $1\frac{1}{4}°$ sweep north–northwest to 6.5-magnitude Star a. Now drop 20′ southwest to 8th-magnitude Star b. NGC 7448 is about 25′ west of Star b and about 2′ west of a 10.5-magnitude star.

The quick view

In the 4-inch at 23× under a dark sky, the tiny 1′-wide core of NGC 7448 is visible as a roughly 12th-magnitude star 2′ west of an 11th-magnitude star. (Use the accompanying photograph to confirm the galaxy's location.) At 72×, the galaxy's slightly elliptical form is apparent, being oriented north–northwest to south–southeast. At 101×, the galaxy remains condensed and apparent with a small, slightly elliptical core surrounded by a 2′ elliptical halo.

Stop. Do not move the telescope.

2. NGC 7479 (H I-55)

Type	Con	RA	Dec	Mag	Dim	Rating
Barred spiral galaxy	Pegasus	23^h 05.0^m	+12° 19′	10.8	3.9′× 3.0′	3

General description

Under a dark sky, NGC 7479 is a nice barred spiral galaxy about 3° south of Alpha (α) Pegasi. It is a reasonable object for small telescopes users, but observing it requires much patience. When searching, look first for a streak of light – the galaxy's bright bar. Light pollution will affect the visibility of this object, whose surface brightness comes in at a weak 13.8.

Directions

Using Chart 69, from NGC 7448, return your gaze to Alpha Pegasi. Now use your unaided eyes or binoculars to find 6th-magnitude 52 Pegasi, which is $3\frac{1}{2}°$ to the south–southwest. Note that 52 Pegasi has a 6.5-magnitide companion 35' to the west–northwest. Center 52 Pegasi in your telescope at low power, and confirm the field with Chart 69b. From 52 Pegasi, move 30' northeast to 8th-magnitude Star a. Now make a roughly 40' hop east–southeast to similarly bright Star b. NGC 7479 is 30' northeast of Star b and 10' northeast of a 10th-magnitude star.

The quick view

At 23× in the 4-inch under a dark sky, NGC 7479 is a slender lens of light. With time, a fuzzy streak along its major axis, which is oriented north to south, can be seen. Give it more time and this bar will emerge from the hazy lens like a woman in white emerging from a fog. At 72×, the galaxy starts to reveal finer structures, such as some spiral structure near the limit of visibility sweeping from the bar and some knots embedded in the arms. But this requires more time than you might want to give at this moment, so plan to return to this object.

3. NGC 7814 (H II-240)

Type	Con	RA	Dec	Mag	Dim	Rating
Spiral galaxy	Pegasus	00ʰ 03.2ᵐ	+16° 09'	10.6	5.5' × 2.3'	2.5

General description

NGC 7814 is an almost exactly edge-on spiral galaxy $2\frac{1}{2}°$ west–northwest of 3rd-magnitude Gamma (γ) Pegasi. It is a very ghostly glow that will be affected greatly by any light pollution, especially if you're using a small telescope. You should focus in on searching for the galaxy's small but bright core. Moderate power will help. Also, be sure to take your time in searching for the object. Patience, in this case, should pay off.

Directions

Use Chart 69 to locate Gamma Pegasi, the southeast corner star in the Great Square of Pegasus. Now use your unaided eyes or binoculars to locate 5.5-magnitude 87 Pegasi, which is a little more than 3° to the north–northwest. Center 87 Pegasi in your telescope at low power, then switch to Chart 69c. From 87 Pegasi move about 45' west and slightly north to 7.5-magnitude Star a. Now move about 50' southwest to 6.5-magnitude Star b. Note that a $1\frac{1}{4}°$-long chain of seven roughly 8.5-magnitude stars (c) arcs southeast, then south, then southwest, before ending at a nice pair of stars (d). Follow Chain c until you get to Pair d, then swing your scope 40' west to 7th-magnitude Star e. NGC 7814 is 12' east–southeast of Star e.

The quick view

At 23× in the 4-inch, NGC 7814 is a 4'-long phantom of light with a stellarlike core. At 72×, that core turns into an extremely sharp star surrounded by a tight inner haze, which itself is surrounded by a dim oval glow oriented northwest to southeast.

Star charts for fifth night

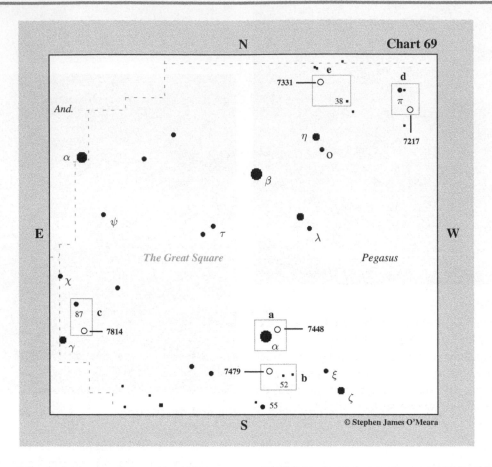

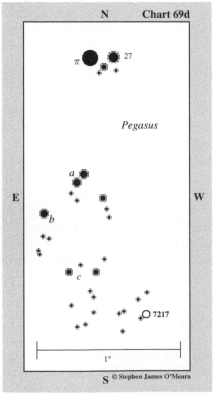

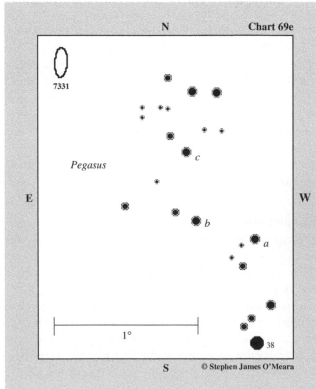

FIFTH NIGHT

1. NGC 7217 (H II-207)

Type	Con	RA	Dec	Mag	Dim	Rating
Spiral galaxy	Pegasus	22ʰ 07.9ᵐ	+31° 22′	10.1	3.5′ × 3.0′	3

4′

General description

NGC 7217 is a very small and somewhat condensed galaxy nearly 2° south–southwest of 4.5-magnitude Pi (π) Pegasi. You will be looking for a very small fuzzy "star." Be sure to follow the stars carefully to it. Take your time with this one.

Directions

Use Chart 69 to find Beta (β) Pegasi, the northwest corner star in the Great Square of Pegasus. Next look about $4\frac{1}{2}°$ northwest for 3rd-magnitude Eta (η) Pegasi. About $7\frac{1}{2}°$ further to the northwest will be 4th-magnitude Pi Pegasi, with 6th-magnitude 27 Pegasi 10′ to the west. Center Pi Pegasi in your telescope at low power, then switch to Chart 69d. From Pi Pegasi, move 50′ south and slightly east to a pair of 7.5-magnitude stars (a). Then move 20′ southeast to 7th-magnitude Star b. Now move 30′ southwest to Arc c, which consists of two 9.5-magnitude stars and one 10th-magnitude star. NGC 7217 is a little more than 30′ southwest of Arc c, immediately northwest of an 11th-magnitude star.

The quick view

In the 4-inch at 23× under a dark sky, NGC 7217 is very small (1′) and stellar in appearance; this "star" is the galaxy's core. With averted vision, a roughly 2′-wide halo can be discerned. At 72×, the galaxy is difficult to see with direct vision, but it becomes more obvious with averted vision, appearing as a circular glow with a bright inner core.

Stop. Do not move the telescope.

2. NGC 7331 (H I-53)

Type	Con	RA	Dec	Mag	Dim	Rating
Spiral galaxy	Pegasus	22ʰ 37.1ᵐ	+34° 25′	9.5	9.7′ × 4.5′	4

4′

General description

NGC 7331 is a bright spiral wonder $4\frac{1}{2}°$ north–northwest of Eta (η) Pegasi. Under a a dark sky it can be seen in 7×50 binoculars as a dim slash of light. Telescopically, the galaxy is striking in telescopes of all sizes.

Directions

Using Chart 69, from NGC 7217 return your gaze to Eta Pegasi. Now use your unaided eyes or binoculars to locate 6th-magnitude 38 Peg $3\frac{1}{2}°$ to the northwest. Center 38 Pegasi in your telescope at low power, then switch to Chart 69e. From 38 Pegasi move about 40′ north to 7.5-magnitude Star a. Next, move 25′ east–northeast to similarly bright Star b. Now move 25′ north–northeast to 8th-magnitude Star c. NGC 7331 is just a little more than 1° northeast of Star c.

The quick view

At 23× in the 4-inch, NGC 7331 is very obvious at first as a 5′ elongated glow, oriented north to south, in a rich star-field. The galaxy's outer halo swells to prominence with averted vision. At 72×, the galaxy's nucleus is extremely sharp and it lies in a complex mottled glow that gradually fades away from the nucleus. With averted vision, and concentration, the galaxy's dust lanes can be inferred, but these patches come and go fleetingly in a small telescope. Several galaxies lie along its eastern flank and they pop in and out of view like restless spirits. A definite object for further study.

Star charts for sixth night

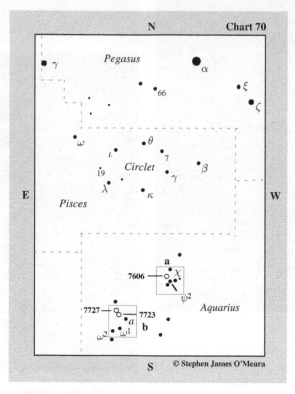

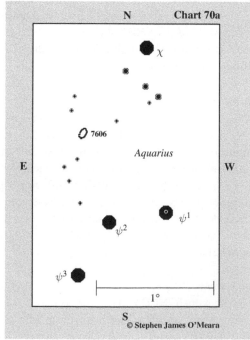

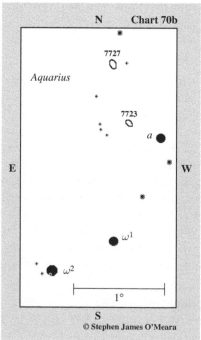

SIXTH NIGHT

1. NGC 7606 (H I-104)

Type	Con	RA	Dec	Mag	Dim	Rating
Barred spiral galaxy	Aquarius	23^h 19.1^m	$-08°$ $29'$	10.8	$4.4' \times$ $2.0'$	2

General description

NGC 7606 is a moderately small and very dim galaxy 45' northeast of 4.5-magnitude Psi² (ψ^2) Aquarii. The galaxy just stands above the background sky in a small telescope even under a dark sky, so being at a dark-sky site is probably a must for this object to be seen well. Light pollution of any sort will ruin the view. You will be looking mainly for the object's core.

Directions

Use Chart 70 to find Psi² Aquarii, which is about 12° south of Gamma (γ) Pisces, the westernmost star in the Circlet. Center this star in your telescope at low power, then switch to Chart 70a. Psi² Aquarii is also the centermost star in a 1°-long arc of similarly bright suns in the Water Bearer's jug – the other two stars being Psi¹ and Psi³ Aquarii. NGC 7606 is only 45' north–northeast of Psi² Aquarii.

The quick view

In the 4-inch at 23× NGC 7606 is a very dim amorphous glow, just ever-so-slightly visible against the background sky. It is better seen at 72×, when it appears as a 3'-wide oval glow, oriented northwest to southeast. The light is mostly uniform, though with time and averted vision the core does get somewhat brighter in the middle. It is a much better sight, appearing more concentrated, in larger telescopes.

Stop. Do not move the telescope.

2. NGC 7723 (H I-110)

Type	Con	RA	Dec	Mag	Dim	Rating
Barred spiral galaxy	Aquarius	23^h 38.9^m	$-12°$ $58'$	11.2	$2.8' \times$ $1.9'$	2.5

General description

NGC 7723 is another small and very dim galaxy in Aquarius, a little less than $1\frac{1}{2}°$ north–northwest of 5.5-magnitude Omega¹ (ω^1) Aquarii. It is also near the galaxy NGC 7727 (see below). Although it is fainter than NGC 7606 (see above) its light is more concentrated, making it appear brighter in small telescopes.

Directions

Using Chart 70, from NGC 7606, return your gaze to Psi² Aquarii. Now use your unaided eyes or binoculars to look 8° southeast for 5th-magnitude Omega² (ω^2) Aquarii. A little less than 50' northwest of Omega² Aquarii is Omega¹ (ω^1) Aquarii. Now look about $1\frac{1}{4}°$ north–northwest for 6th-magnitude Star a. Center Star a in your telescope at low power, then switch to Chart 70b. NGC 7723 is only 20' east–northeast of Star a.

The quick view

At 23× in the 4-inch under a dark sky, NGC 7723 is an amorphous 3' wide oval disk, oriented northeast to southwest, that hovers just above the background sky. Its intensity increases with averted vision and time. It is better seen at 72×, which shows the oval disk getting gradually brighter toward the middle to a fuzzy nucleus.

Stop. Do not move the telescope. Your next target is nearby!

3. NGC 7727 (H I-111)

Type	Con	RA	Dec	Mag	Dim	Rating
Galaxy	Aquarius	23^h 39.9^m	−12° 18′	10.6	5.6′ × 4.0′	3

4′

General description

NGC 7727 is a moderately bright and small galaxy only 45′ north–northeast of NGC 7723. Of the three galaxies listed here in Aquarius, NGC 7727 wins the visual-interest award.

Directions

Using Chart 70b, NGC 7727 is only 45′ north–northeast of NGC 7723.

The quick view

At 23× in the 4-inch, NGC 7727 is a small but reasonably obvious little galaxy that displays a tight core with dim asymmetrical extensions to the northeast and southwest. At 72×, the core is a stellar pip surrounded by a disheveled oval of light perhaps 1.5′ in extent. The extensions in a small telescope make it appear slightly lens shaped, as if a solid disk were here stretching out to 3′ or so, but, as the accompanying photograph shows, the bright inner disk is warped and in much turmoil; it has very faint sweeping arms, which may be seen in much larger instruments. The core takes power well, so do not be afraid to explore this intriguing galaxy with more power at a later date.

Star charts for seventh night

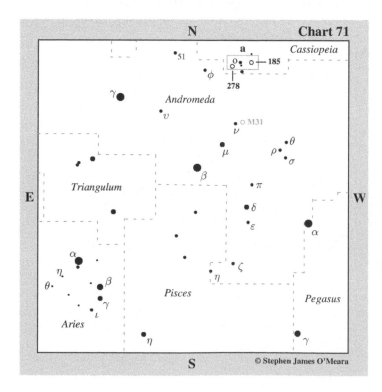

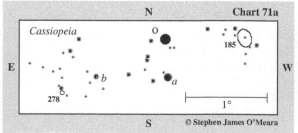

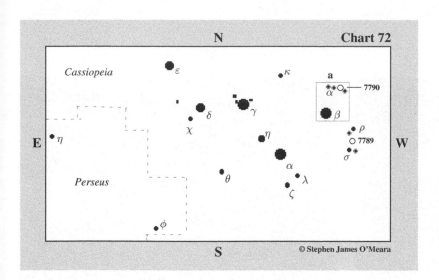

Chart 72

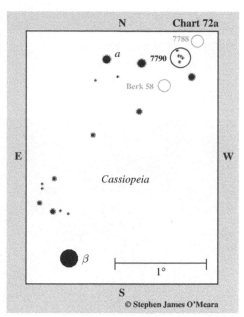

Chart 72a

SEVENTH NIGHT

1. NGC 185 (H II-707)

Type	Con	RA	Dec	Mag	Dim	Rating
Elliptical galaxy	Cassiopeia	00^h 39.0^m	+48° 20′	9.2	17.0′ × 14.3′	3.5

General description

NGC 185 is the brighter of two dwarf companion galaxies to the great Andromeda spiral M31, which lie some 7° to the north within the boundaries of Cassiopeia. NGC 185 is visible under a dark sky in binoculars – with effort. Its light is somewhat concentrated toward the core, so it looks like a ghostly version of M32 – M31's more concentrated and celebrated dwarf companion neighboring that great spiral immediately to the south. The galaxy requires a dark sky and patience to see.

Directions

Use Chart 71 to locate the great spiral galaxy M31 near 5th-magnitude Nu (ν) Andromedae. Now look 6° due north for 5th-magnitude Omicron (o) Cassiopeiae, which marks the northern end of a $1\frac{1}{2}$°-long chain of three roughly 5th- and 6th-magnitude suns. NGC 185 is a little less than 1° west of Omicron Cassiopeiae. You can confirm its position with Chart 71a.

The quick view

In the 4-inch at 23× under a dark sky, NGC 185 is relatively easy to see, though any light pollution will affect its visibility greatly. At a glance, with averted vision, it appears as a 10′ uniform glow, slightly oval in shape and oriented northeast to southwest. At 72×, the galaxy becomes gradually brighter toward the middle, though not to a stellar nucleus. The core measures about 1′ and appears mottled under higher powers. It also has a granular texture, making it look like a globular cluster near the limit of resolution.

Stop. Do not move the telescope. Your next object is nearby!

2. NGC 278 (H I-159)

Type	Con	RA	Dec	Mag	Dim	Rating
Mixed spiral galaxy	Cassiopeia	00ʰ 52.1ᵐ	+47° 33′	10.8	2.6′ × 2.6′	3.5

General description

NGC 278 is a very small and somewhat dim galaxy $1\frac{1}{2}°$ southeast of 5th-magnitude Omicron (o) Cassiopeiae. Fortunately, the galaxy's core is small and bright. The key to seeing this galaxy, then, in a small telescope, is to pinpoint the galaxy's location among the stars, then use magnification.

Directions

Using Chart 71a, from NGC 185, return to Omicron Cassiopeiae. Now move 25′ south–southwest to 6th-magnitude Star *a*. Next, move about 55′ east to 7th-magnitude Star *b*. NGC 278 is a little less than 30′ southeast of Star *b*, just south of a 9th-magnitude sun.

The quick view

At 23× in the 4-inch under a dark sky, NGC 278 is not apparent at a quick glance. If you know exactly where to look and use averted vision, its nucleus can be seen as a tiny pip of light. At 72×, the galaxy is not visible with a direct gazer but pops into view with averted vision, appearing simply as a small, highly condensed glow. The galaxy is best seen at 101×, when it looks like a small comet with a moderately condensed core.

3. NGC 7789 (H VI-30)

Type	Con	RA	Dec	Mag	Diam	Rating
Open cluster	Cassiopeia	23ʰ 57.5ᵐ	+56° 43′	6.6	25.0′	4

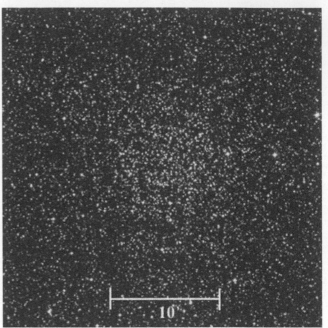

General description

Caroline Herschel discovered NGC 7789 in October, 1783. This beautiful open cluster lies just 3° southwest of Beta (β) Cassiopeiae and is at the limit of unaided vision. The cluster is a magnificent spectacle in 7×50 binoculars, being nearly as large as the full Moon and just as round. In fact, through binoculars, the cluster looks like a tailless comet crossing the rich star fields of the northern Milky Way. With any scrutiny, the cluster sparkles as the light of dim suns pop in and out of view in your peripheral vision. Through a telescope, it is one of the finest and richest open star clusters in the sky.

Directions

Use Chart 72 to locate Beta Cassiopeiae (the westernmost star in Cassiopeia's famous W asterism). Now use binoculars to look 3° southwest. NGC 7789 should be an obvious glow midway between Rho (ρ) and Sigma (σ) Cassiopeiae.

The quick view

At 23× in the 4-inch, NGC 7789 is at its best with some 60 or so suns of 9th magnitude and fainter sprinkled across 25′ of sky. With imagination, the central core of suns appears three-dimensional – like a fistful of diamond dust. At 72×, the number of visible suns increases to about 150. The view is splendid in larger scopes. The cluster sports some 580 stars, the brightest of which shine at a magnitude of 10.0; the faintest stars hover at around 18th magnitude!

Stop. Do not move the telescope.

4. NGC 7790 (H VII-56)

Type	Con	RA	Dec	Mag	Diam	Rating
Open cluster	Cassiopeia	23^h 58.4^m	+61° 12′	8.5	5.0′	3

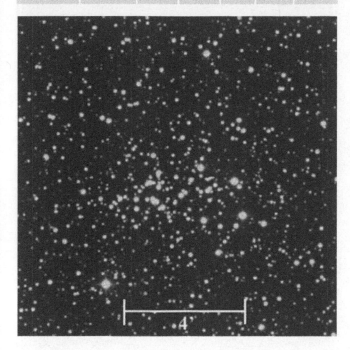

General description

NGC 7790 is a very small and dim open cluster about $2\frac{1}{2}°$ northwest of Beta (β) Cassiopeiae. In a small telescope just its brighter core stands out well against the rich Milky Way background. Beware, two dim clusters lie nearby. This cluster is best seen under a dark sky.

Directions

Using Chart 72, from NGC 7789, return your gaze to Beta Cassiopeiae. Now use your unaided eyes or binoculars to locate a pair of 6th-magnitude stars (a) a little more than 2° to the north–northwest. You want to center the westernmost star in Pair a in your telescope at low power, then confirm the field with Chart 72a. NGC 7790 is only about 25′ due west of that star, and about 15′ northeast of a magnitude 6.5 star.

The quick view

At 23× in the 4-inch, NGC 7790 is an extremely small and sideways V of starlight, perhaps 2′ across. At 72×, the cluster's core remains extremely tight and about a dozen suns are splashed across 5′ of sky in an east to west line. It's difficult to discern the other members of the cluster against the rich Milky Way background. But the cluster is well populated with some 134 members of 10th-magnitude and fainter in an area only 5′ across.

11 · November

Star charts for first night

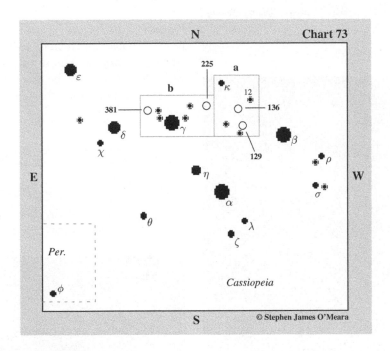

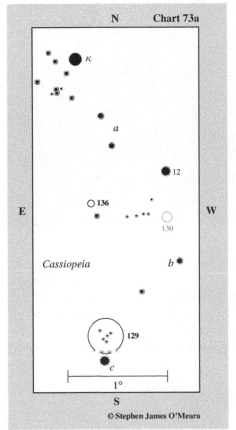

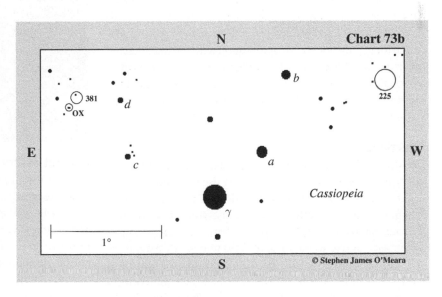

First Night

1. NGC 136 (H VI-35)

Type	Con	RA	Dec	Mag	Diam	Rating
Open cluster	Cassiopeia	00ʰ 31.5ᵐ	+61° 30′	~9th	1.5′	3

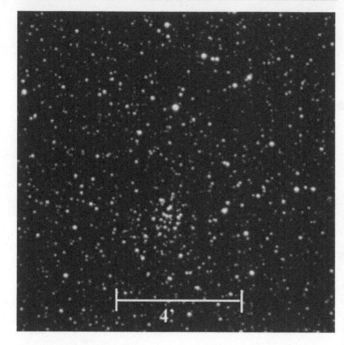

General description

NGC 136 is an extremely small and inconspicuous open star cluster a little less than $1\frac{1}{2}°$ south and a tad west of 4th-magnitude Kappa (κ) Cassiopeiae. You will need to make a careful search for it, and moderate magnification will help make it stand out from the Milky Way background, especially in a small telescope. Fortunately, the cluster has a bright core. No magnitudes for the cluster are given in any popular references, though visually it seems to shine at about 9th magnitude.

Directions

Use Chart 73 to find Kappa Cassiopeiae, which is almost 5° northeast of Beta (β) Cassiopeiae. Now use your unaided eyes or binoculars to locate 5.5-magnitude 12 Cassiopeiae, which is about $1\frac{1}{2}°$ to the southwest. Otherwise you can star-hop to it from Kappa Cassiopeiae using Chart 73a. From Kappa Cassiopeiae, move about 45′ southwest to two 7th-magnitude suns (a), separated by 20′ and oriented northeast to southwest. Now move about 40′ southwest of their midpoint to 12 Cassiopeiae. NGC 136 is about 50′ east–southeast of 12 Cassiopeiae, about 5′ northeast of a 9th-magnitude sun.

The quick view

At 23× in the 4-inch under a dark sky, NGC 136 is a very small (1.5′) and irregularly round glow that seems to scintillate with averted vision. At 72×, the cluster is a clump of dim suns (of 13th magnitude and fainter) with a spray of field stars surrounding it like jets emanating from the head of a comet. The cluster contains only 20 stars of 13th magnitude and fainter.

Field note

Be careful that you are not looking at any one of the several larger grouping of stars in this rich region. Magnification will help you identify the true cluster.

Stop. Do not move the telescope. Your next target is nearby!

2. NGC 129 (H VIII-79)

Type	Con	RA	Dec	Mag	Diam	Rating
Open cluster	Cassiopeia	00ʰ 29.9ᵐ	+60° 13′	6.5	12.0′	4

General description

NGC 129 is a moderately large and bright cluster about $2\frac{3}{4}°$ south–southwest of 4th-magnitude Kappa (κ) Cassiopeiae and about $1\frac{1}{4}°$ south–southwest of NGC 136 (see above). The cluster is visible in binoculars as a small round haze just north of a 6.5-magnitude star. Through a telescope, the cluster appears asymmetrical and scattered, with several nice star clumps.

Directions

Using Chart 73a, from NGC 136, return to 12 Cassiopeiae, then move about 50′ south and a tad west to 7th-magnitude Star b. (Note that another dim open cluster NGC 130 is about midway between and slightly east of 12 Cassiopeiae and Star b.)

NGC 129 is a nice 1° sweep southeast of Star *b*, just about 12′ north of 6.5-magnitude Star *c*.

The quick view

In the 4-inch at 23×, NGC 129 is an asymmetrical ellipse of fuzzy starlight (oriented north to south), with the southern section of the ellipse being the brightest. With a little concentration, an obvious triangular core of stars rests inside that bright southern section. At 72×, the cluster is patchy and scattered. About a dozen stars hover around 11th and 12th magnitude, and splashes of dimmer suns race through the cluster haphazardly for 10′ or so. The cluster is quite rich, having nearly 200 members of 11th magnitude and fainter, and is a splendid sight in larger telescopes.

3.   NGC 225 (H VIII-78)

Type	Con	RA	Dec	Mag	Diam	Rating
Open cluster	Cassiopeia	00ʰ 43.6ᵐ	+61° 46′	7.0	15.0′	4

General description

Caroline Herschel discovered NGC 225 in September 1783. It lies just 2° northwest of Gamma (γ) Cassiopeiae, the central gem of the celestial W. It also lies in a kidney-bean-shaped bay in the river Milky Way, exactly halfway between Gamma Cassiopeiae and 4th-magnitude Kappa (κ) Cassiopeiae. The 7th-magnitude glow is easily spied in 7×50 binoculars, if you know exactly where to look. It is a fine sight in small telescopes, being in an extremely rich Milky Way field. Look for a bright chevron of stars with a wavy line of a half-dozen or so stars to the west. The cluster is a "milkweed pod" of hazy suns just east of that wavy line.

Directions

Use Chart 73 to locate Gamma Cassiopeiae. Center that star in your telescope at low power, then switch to Chart 73b. From Gamma Cassiopeiae, move about 35′ northwest to 5th-magnitude Star *a*. Next, look for solitary 6th-magnitude Star *b* about 40′ to the north–northwest. NGC 225 is almost exactly 1° due west of Star *b*.

The quick view

At 23× in the 4-inch, NGC 225 is easily resolved in a 4-inch telescope. The brightest members shine at around 9th magnitude. The cluster's faintest suns dip sharply into the rich Milky Way background. At a glance NGC 225 looks like a broken valentine – a heart-shaped cluster with one lobe filled with stars of near uniform brightness (the "milkweed pod"), and the other lobe being little more than an empty shell. At 72×, the cluster's stars spread out and lose their appeal. Still, with some concentration, you should see that, though the cluster is loose and scattered, its stars are arranged in two distinct regions – a rich section to the southeast (like a lemon slice) and a smaller and sparser congregation to the northwest. If you defocus the view ever so slightly and tap the tube, you should see that these two regions are separated by a meandering lane of darkness.

Stop. Do not move your telescope.

4.   NGC 381 (H VIII-64)

Type	Con	RA	Dec	Mag	Diam	Rating
Open cluster	Cassiopeia	01ʰ 08.3ᵐ	+61° 35′	9.3	7.0′	3

General description

Contrary to popular belief, Caroline Herschel did not discover NGC 381, her brother William did. It is a dim and scattered cluster that lies about $1\frac{3}{4}°$ northeast of Gamma (γ) Cassiopeiae, making it a near mirror image of NGC 225 (see above). Under a dark sky, NGC 381 is barely – just hyperfine barely – visible in 7×50 binoculars. Seeing it in such a small instrument requires much time and effort (rest assured you'll get eye strain). Others have detected it clearly in 11×80 and 15×80 binoculars. The trick to seeing it in binoculars is to be relaxed, preferably lying back in a comfortable outdoor recliner with your arms supported, and to use gentle sweeping motions; slowly move back and forth across the sky from the nearby 6th-magnitude star to NGC 381. The cluster is situated in a rich Milky Way field, so look for a slight oval enhancement in the background glow. In small telescopes, it is a dim oval glow.

Directions

Using Chart 73b, from NGC 225, return to Gamma Cassiopeiae. Now move your scope nearly 1° northeast to 6th-magnitude Star c; it is a fine double star (Burnham 396) with a 6th-magnitude primary (c) and a 9th-magnitude companion 1.2″ to the northeast. Just 30′ to the north–northeast is solitary 6th-magnitude Star d. NGC 381 is only 30′ further east of Star d.

The quick view

At 23× in the 4-inch, NGC 381 is a very delicate display of dim suns. It looks as if a child had poured salt on a countertop, then spread a portion of it with his hand to the southeast. The cluster is the pile of salt and the arc is a fortuitous arrangement of field stars, one of which is OX Cassiopeiae. In a more fanciful way, the fuzzy cluster and its adjacent crooked "stick" of stars looks like a lollipop full of lint. Of the cluster's 50 stars, the brightest shines at 10th magnitude, and it goes downhill from there. At 72×, NGC 381 first seems to vanish; the cluster has little or no central concentration, so its most obvious members blend into the surrounding Milky Way. But if you use averted vision and breathe steadily with an occasional deep breath, a haze should suddenly appear in the cluster's position. This haze is the light of fainter, unresolved members just beginning to surface from the depths of the surrounding darkness. Now, if you can retain seeing the cluster's brighter stars at this power, look for an obvious double star at the cluster's center. Also note that the other bright members have a weak spiral structure. Of course, the view will be grander in larger instruments.

Star charts for second night

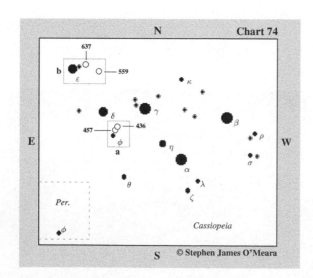

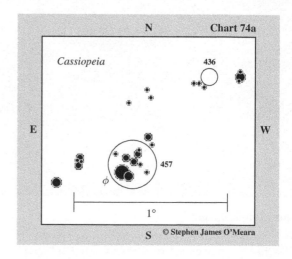

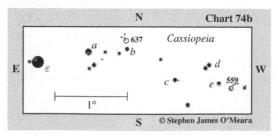

SECOND NIGHT

1. NGC 457 (H VII-42)

Type	Con	RA	Dec	Mag	Diam	Rating
Open cluster	Cassiopeia	01^h 19.5^m	+58° 17′	6.4	20.0′	4

General description
NGC 457, the E.T. Cluster, is one of the most popular non-Messier open clusters in the northern night skies. The T-shaped gathering of bright suns, with two prominent beacons as "eyes," looks remarkably like Spielberg's charismatic alien. The cluster lies 2° south–southwest of Delta (δ) Cassiopeiae and hides in the glare of 5th-magnitude Phi (φ) Cassiopeiae. It is visible in binoculars, appearing as a fragmentary comet tail blowing northwest of Phi Cassiopeiae. The cluster is a joy to see in telescopes of all sizes.

Directions
Use Chart 74 to find Delta Cassiopeiae in the celestial W. Now use your unaided eyes or binoculars to find Phi Cassiopeiae 2° to the south–southwest. Center Phi Cassiopeiae in your telescope and you will be at the cluster. You can confirm its placement relative to Phi Cassiopeiae using Chart 74a.

The quick view
At 23× in the 4-inch, NGC 457 looks like a skeleton of stars surrounded by smoke rising from the celestial fire of Phi Cassiopeiae. Here is the unresolved glow of the cluster's 200

or so members spread across 20′ of sky. Just relax with this cluster and use your imagination; it is known by many names other than E. T., such as the Owl, Stick Man, and Worry Doll Cluster.

Stop. Do not move the telescope. Your next target is nearby!

2. NGC 436 (H VII-45)

Type	Con	RA	Dec	Mag	Diam	Rating
Open cluster	Cassiopeia	01^h 15.9^m	+58° 49′	8.8	5.0′	3.5

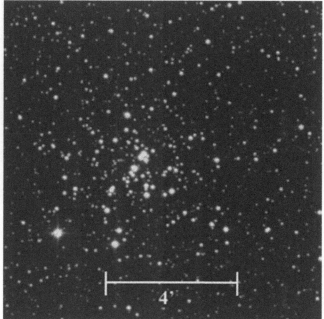

General description
NGC 436 is a small and dim open cluster near NGC 457. Although it is small, it is condensed and can just be seen as a small dim glow with 7×50 binoculars under a dark sky, if you know where to look. Telescopically, this cluster with its spread of dim suns looks like Hershey's Reese's Pieces® left out for our celestial E. T. (NGC 457).

Directions
Use Chart 74a. NGC 436 lies just 40′ northwest of NGC 457.

The quick view
In the 4-inch at 23×, NGC 436 is a nice concentration (5′) of 10th-magnitude and fainter suns at the northwest end of a little hook of stars. With averted vision the cluster looks round, like a ball balanced on a stick. At 72×, the cluster is more vivid. Its brightest stars trace out three sharp arcs that look like a crow's foot.

3. NGC 637 (H VII-49)

Type	Con	RA	Dec	Mag	Diam	Rating
Open cluster	Cassiopeia	01^h 43.1^m	+64° 02′	8.2	3.0′	3.5

4′

General description

NGC 637 is a small but condensed open cluster nearly 1$\frac{1}{2}$° west–northwest of Epsilon (ε) Cassiopeiae, the easternmost star in the celestial W. Although it is dim, the cluster is quite compact, and therefore noticeable in small telescopes.

Directions

Use Chart 74 to locate Epsilon Cassiopeiae. Center that star in your telescope at low power, then switch to Chart 74b. From Epsilon Cassiopeiae move about 45′ west–northwest to 6th-magnitude Star a. NGC 637 lies a little more than 40′ further to the west–northwest – just 5′ north of 8th-magnitude Star b, at the southwest end of a tiny tilted Y-shaped asterism of four 10th-magnitude stars.

The quick view

At 23× in the 4-inch, NGC 637 is a relatively bright speck of clustered starlight. With any concentration about a half dozen suns can be seen against a 3′-wide dim background haze of unresolved suns. At 72×, a significant pair of stars appears at center. These stars are part of a line of stars that zigzag from north to south. About a half dozen dimmer stars swim in and out of view as your eye tries to navigate this tiny space. The cluster has only 55 members of 8th magnitude and fainter.

Stop. Do not move your telescope. Your next target is nearby!

4. NGC 559 (H VII-48)

Type	Con	RA	Dec	Mag	Diam	Rating
Open cluster	Cassiopeia	01^h 29.5^m	+63° 18′	9.5	7.0′	3

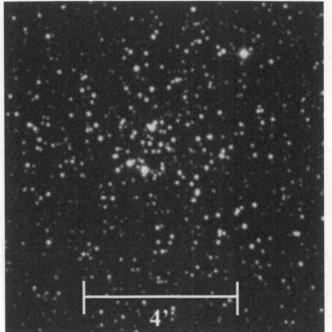

4′

General description

NGC 559 is another very small and dim open cluster in Cassiopeia, about 2$\frac{3}{4}$° west–southwest of Epsilon (ε) Cassiopeiae and a little more than 1$\frac{1}{2}$° southwest of NGC 637. It is a dim and somewhat difficult object that looks like a small fuzzy patch in small telescopes. It is a much better sight in larger telescopes, which show many of its 120 or so members splashed across 7′ of sky. The cluster's brightest stars shine at 9th magnitude.

Directions

Using Chart 74b, from NGC 637 return to Star b, then move about 50′ southwest to 7.5-magnitude Star c. Now move 30′ to west–northwest to a close pair of 7th-magnitude suns (d), which is oriented northwest to southeast. A little less than 20′ southwest of Pair d is 7.5-magnitude Star e. NGC 559 is a little more than 10′ west–southwest of Star e.

The quick view

At 23× in the 4-inch under a dark sky, NGC 559 is a faintly glowing orb 5′ in diameter that requires averted vision to see. At 72×, some stars start to sizzle out of the thin mist, mostly on the cluster's western flank. Most noticeable is a double star at the southwest edge. Increasing the magnification does little to enhance this dim concentration of suns.

Star charts for third night

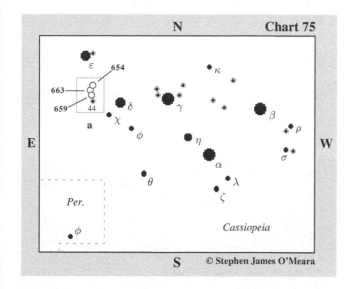

N   Chart 75

E

W

S   © Stephen James O'Meara

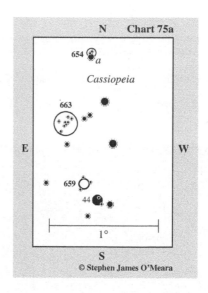

N   Chart 75a

Cassiopeia

E

W

1°

S   © Stephen James O'Meara

THIRD NIGHT

1. NGC 659 (H VIII-65)

Type	Con	RA	Dec	Mag	Diam	Rating
Open cluster	Cassiopeia	01^h 44.4^m	+60° 42′	8.2	6.0′	3.5

4′

General description

Caroline Herschel discovered NGC 659 in November, 1783. It lies only 2½° east–northeast of 3rd-magnitude Delta (δ)

Cassiopeiae in the constellation's celestial W asterism. It is one of five open star clusters forming an incomplete 2°-wide circle around a diamond of roughly 7th-magnitude stars. Under a dark sky, the cluster can be spotted in 7×50 binoculars, though only with difficulty; there is barely enough contrast to distinguish it from the Milky Way background. The task might actually be simpler from a suburban location, where the Milky Way background is subdued. Through a small telescope, it is a gasp of dim light with tiny sparkling jewels.

Directions

Use Chart 75 to find Delta Cassiopeiae. Now use your unaided eyes or binoculars to locate 5th-magnitude Chi (χ) Cassiopeiae, which is nearly 1½° to the southeast. Next look for 6th-magnitude 44 Cassiopeiae, which forms the northeast apex of a near-equilateral triangle with Delta and Chi Cassiopeiae. Center 44 Cassiopeiae in your telescope and confirm the field with Chart 75a. NGC 659 lies only about 10′ north–northeast of 44 Cassiopeiae.

The quick view

At 23× in the 4-inch under a dark sky, NGC 659 is a subtle glow that seems to hide in the shadow of 44 Cassiopeiae and its nearby stellar companions. It has a 3′-wide core that, with averted vision, is a tight, almost circular, collection of 10th- and 11th-magnitude suns; it is the main jewel in a necklace of four comparably bright suns that arc from the south to the northwest. The entire necklace is surrounded by a larger and fainter halo of suns, which glint in and out of view. At 101×, the cluster breaks down into individual packets of stars. The core, however, remains distinct. It is a rosette of a half dozen suns, one of which is a close 11th-magnitude double star that resolves further into a triple at higher magnifications.

Stop. Do not move the telescope. Your next target is nearby!

2. NGC 663 (H VI-31)

Type	Con	RA	Dec	Mag	Diam	Rating
Open cluster	Cassiopeia	01^h 46.3^m	+61° 13′	6.7	15.0′	4

General description

NGC 663 is a moderately large and bright cluster about $2\frac{2}{3}°$ east–northeast of Delta (δ) Cassiopeiae and about 30′ north–northeast of NGC 659 (see above). It is a wonderful object that rivals M103 in appearance. From a dark-sky site, NGC 663 can be glimpsed with the unaided eye with time and it is easy to see in 7×50 binoculars. Telescopically it is a moderately large horseshoe of stars with two bright pairs of stars.

Directions

Use Chart 75a. NGC 663 lies just about 30′ north–northeast of NGC 659.

The quick view

In the 4-inch at 23×, NGC 663 lies in a rich and patchy field of Milky Way. The cluster itself is beautiful, being a distinct east to west oriented ellipse with two striking chains of stars radiating northward from the cluster's core. Each arm ends to the north at a bright pair of suns. At 72×, the core is fractured into three distinct groupings – one to the east, one to the west, and one to the south – which gives the cluster

the appearance of a horseshoe or a lower-case Greek Nu (ν). A dark lane runs north to south through the cluster, dividing the east and west sections. The cluster is quite rich, having nearly 110 members of 9th magnitude and fainter.

Stop. Do not move the telescope. Your next target is nearby!

3. NGC 654 (H VII-46)

Type	Con	RA	Dec	Mag	Diam	Rating
Open cluster	Cassiopeia	01^h 44.0^m	+61° 53′	6.5	6.0′	4

General description

NGC 654 is a relatively small but well concentrated open cluster a little less than $2\frac{1}{2}°$ east–northeast of Delta (δ) Cassiopeiae and 40′ north–northwest of NGC 663. It sits just north and slightly west of a 7th-magnitude sun and can just be spied in binoculars from a dark-sky site. Telescopically it is a tiny and tight grouping of little gems.

Directions

Use Chart 75a. NGC 654 is only 40′ north–northwest of NGC 663, immediately north and slightly west of 7th-magnitude Star a.

The quick view

At 23× in the 4-inch, NGC 654 is a breath of circular light hugging a topaz-yellow 7th-magnitude sun. With averted vision the cluster's core pops into view as a 3′-wide C-shaped asterism with the open end toward the 7th-magnitude sun. At 72×, the C transforms into a 6′-wide inverted V-shaped asterism of irregularly bright suns. A stream of starlight drips from its V's northern tip like water from a needle.

Star charts for fourth night

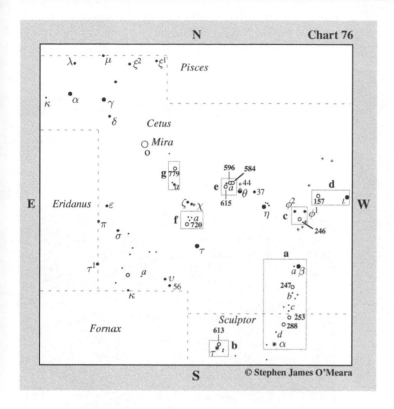

N   Chart 76

Pisces

λ   μ   ξ²   ξ¹

κ   α   γ

δ

Cetus

○ *Mira*
    o

g 779 ○
         a

                596   584
            e ○○○ •44
                   a θ •37
                  615

E   *Eridanus* •ε   ζ••χ   **d**
                         f ○ *a*   φ²   ○ 157 *ι* W
                π   **f** ○ 720   η   **c** ○ • φ¹
                σ   •   ○
                                                     246

            τ¹•   •τ   **a**
                       a   •v   •*a*•β
                 ○ •   •56   247○
                κ   b•:•
                                                     •:•c
                        Sculptor   ○ 253
                  613   ○ 288
            Fornax   ○ • **b**   •*d*
                          τ ○:   *∗α*

S   © Stephen James O'Meara

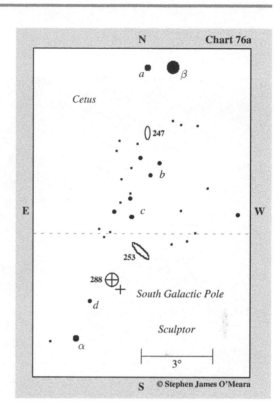

N   Chart 76a

                    a• ● β
        Cetus

                        ○ 247

                      •   • *b*

                          • *c*

E   • W

                        ⬭ 253

                288 ⊕
                   •d   + *South Galactic Pole*

                             Sculptor

            •
              ● α   ├────┤
                                        3°

S   © Stephen James O'Meara

FOURTH NIGHT

1. NGC 247 (H V-20)

Type	Con	RA	Dec	Mag	Dim	Rating
Mixed spiral galaxy	Cetus	00^h 47.1^m	$-20°$ $46'$	8.9	$22.2' \times$ $6.7'$	3.5

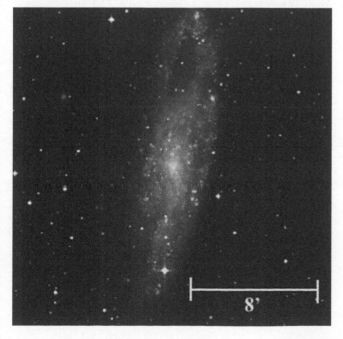

├────────── 8' ──────────┤

General description

NGC 247 is a fairly bright almost edge-on galaxy nearly 3° south–southeast of Beta (β) Ceti, the 2nd-magnitude star marking the celestial Whale's tail. From dark southerly locations, it is visible in binoculars as a thin thread of light extending from a 9.5-magnitude star near the galaxy's southern tip. Telescopically it is a thin narrow haze.

Directions

Use Chart 76 to find 2nd-magnitude Beta Ceti. Center that star in your telescope at low power, then switch to Chart 76a. From Beta Ceti, move 1° east to 6th-magnitude Star *a*. Now make a very slow and careful $2\frac{3}{4}°$ sweep south where you should intercept NGC 247. The galaxy is also about 1° north–northwest of the northeastern top of roughly 1°-wide Triangle *b*, which comprises three 6th-magnitude stars.

The quick view

At 23× in the 4-inch under a dark sky, NGC 247 is an anemic 15' elongated glow, oriented north to south. The core is round and bright and the north and south extensions swell and contract with averted and direct vision, respectively. At 72×, the galaxy is much the same, though averted vision begins to show some patchiness to the disk and a bright bead at the core of the nuclear region.

Stop. Do not move the telescope. Your next target is nearby!

2. NGC 253 (H V-1)

Type	Con	RA	Dec	Mag	Dim	Rating
Mixed spiral galaxy	Sculptor	00ʰ 47.6ᵐ	−25° 17′	7.1	25.8′ × 5.9′	4

General description

NGC 253, the Silver Coin Galaxy, was discovered by Caroline Herschel in October, 1783. It is located about $7\frac{1}{2}°$ south–southeast of 2nd-magnitude Beta (β) Ceti and has been spied with the unaided eye from southerly locals in the continental USA. The galaxy is a magnificent binocular object, mainly because it shares the field with equally magnificent globular cluster NGC 288 (see right). Through a telescope, NGC 253 is a marvel and one of the most popular non-Messier objects visible from the Northern Hemisphere.

Directions

Using Chart 76a, from NGC 247, move 1° south to Triangle b. If you center the southernmost star in triangle b, then move 1° southeast, you will arrive at the 6th-magnitude star marking the northern tip of another 1°-wide triangle comprising three 6th-magnitude stars (c). Center the south-westernmost star in Triangle c, then move about $1\frac{1}{4}°$ south–southwest to NGC 253.

The quick view

In the 4-inch at 23×, NGC 253 is a 20′-long spindle of moderately intense light trapped between a pair of roughly 9th-magnitude suns. The spindle is oriented northeast to southwest and shows structure almost immediately. The inner core is also elliptical and is punctuated by a bright starlike nucleus. At 72×, a weak spiral structure can be detected with time, especially a sweeping S-shape form – one arm of the S forms the outer northwest flank and the other forms the outer southeast flank. This galaxy is a must to study further on another night.

Stop. Do not move the telescope. Your next target is nearby!

3. NGC 288 (H VI-20)

Type	Con	RA	Dec	Mag	Diam	Rating
Globular cluster	Cassiopeia	00ʰ 52.8ᵐ	−26° 35′	7.9	13.0′	4

General description

NGC 288 lies only $1\frac{3}{4}°$ southeast of NGC 253, that's about $\frac{3}{4}°$ too far to be placed in the same low-power field as NGC 253 in many modern telescopes. From southerly locals in the continental USA it can be spotted in 7×50 binoculars together with NGC 253 (see above). In binoculars, the globular cluster looks like a smooth orb of light shining at 8th magnitude. It resides less than 9° southeast of 2nd-magnitude Beta (β) Ceti, 3° northwest of Alpha (α) Sculptoris, and just 40′ north and slightly east of the South Galactic Pole. Through the smallest of telescopes, NGC 288 will appear as a little ball of "gas" with a slightly condensed core. Averted vision will reveal a slight granular texture to the cluster's outer halo.

Directions

Using Chart 76a. From NGC 253, make a slow and careful $1\frac{3}{4}°$ sweep southeast to NGC 288. There will be no mistaking its large and swollen form.

The quick view

At 23× in the 4-inch, NGC 288 is a perfect sphere of stellar foam that seems to froth out of a squashed dipper asterism of surrounding starlight. A closer inspection shows the globular cluster to have an irregular border, one that is fractured into clumps of stars, many of which are immediately resolvable. NGC 288's core appears to be a large mottled region, perhaps 6′ across. With averted vision, this bright inner region seems to scintillate with frenetic energy, like bees swarming around a hive. At 72×, NGC 288 is well resolved, especially at the fringes, which appears patchy with clumps of starlight. The cluster's brightest members shine at a magnitude of 12.6, and its horizontal-branch magnitude is 15.3, so even a modest-sized amateur telescope can penetrate deep into the globular's inner sanctum. This feat is best accomplished at high magnifications, but be warned, even at 182×, the relatively low surface brightness of the cluster makes seeing any patterns difficult.

Star charts for fifth night

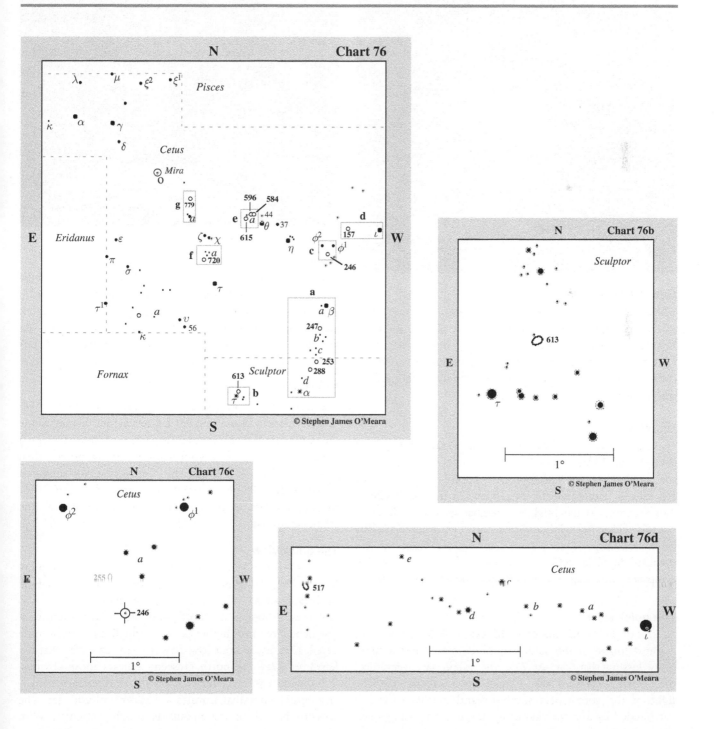

© Stephen James O'Meara

FIFTH NIGHT

1. NGC 613 (H I-281)

Type	Con	RA	Dec	Mag	Dim	Rating
Mixed spiral galaxy	Sculptor	01ʰ 34.3ᵐ	−29° 25′	10.0	5.2′ × 2.6′	3

General description

NGC 613 is a reasonably bright galaxy about 40′ northwest of 6th-magnitude Tau (τ) Sculptoris. Despite its southerly location, the galaxy has a high degree of central concentration, making it a good target for suburban locations. From a dark-sky site, it is visible in a small telescope under a first-quarter Moon.

Directions

Use Chart 76 to find 2nd-magnitude Beta (β) Ceti, the tail star of the celestial Whale. About 12° to the south–southeast is 4.5-magnitude Alpha (α) Sculptoris. Now use your unaided eyes or binoculars to locate Tau Sculptoris about 8° to the east–southeast. Tau Sculptoris is the brightest member of a 1°-wide acute triangle with two slightly dimmer suns. Center Tau Sculptoris in your telescope at low power, then switch to Chart 76b and confirm the field. NGC 613 is just 40′ northwest of Tau Sculptoris, immediately southwest of a 10th-magnitude star.

The quick view

At 23× in the 4-inch under a dark sky, NGC 613 looks like a ghost image of the 10th-magnitude star – just a fuzzy glow kissing that star. At 72×, the galaxy is a very nice, delicate elliptical glow with a lens-shaped core 2′-wide. The light of the core tapers southeast and northwest and is surrounded by a softer halo of light. Quite a pleasing view!

2. NGC 246 (H V-25)

Type	Con	RA	Dec	Mag	Dim	Rating
Planetary nebula	Cetus	00ʰ 47.0ᵐ	−11° 52′	10.9	4.6′ × 4.1′	3.5

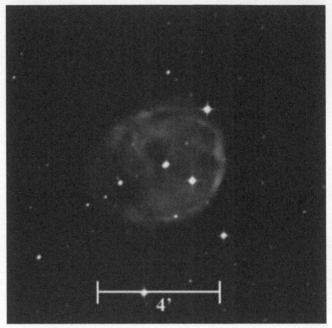

General description

NGC 246 is a moderately bright and large planetary nebula 6° north of 2nd-magnitude Beta (β) Ceti. It is involved with some bright stars so it is easy to sweep over if you are not careful. It is best to move to the object's location, then use averted vision to look for a tight and bright triangle of stars surrounded by a dim glow.

Directions

Use Chart 76 to find Beta Ceti. Now use your unaided eyes or binoculars to look about 7½° due north for 5th-magnitude Phi¹ (φ¹) Ceti and, 1½° farther east, 5th-magnitude Phi² (φ²) Ceti. Center Phi¹ Ceti in your telescope at low power, then switch to Chart 76c. From Phi¹ Ceti, move 50′ southeast to 20′-wide Triangle a, which is composed of three 8th-magnitude stars. NGC 246 lies just 40′ further to the south–southeast.

The quick view

In the 4-inch at 23×, NGC 246 is an elusive 5′-wide glow involved with a prominent, and equally spaced, triangle of bright suns. With attention, the nebula forms two crescents – one to the north and one to the southwest. The nebula requires higher power, though, to see it in more detail. (Try using your low-power eyepiece with a Barlow lens.) At 72×, the north crescent appears most obvious. The nebula also ends abruptly on the northeastern flank – at a bright knot that mimics a 13th-magnitude star. The western border of the nebula is roughly circular. With attention, voids, like eyes, can be seen within the glowing gas. The nebula's roughly 12th-magnitude central star looks offset to the east.

Stop. Do not move your telescope.

3. NGC 157 (H II-3)

Type	Con	RA	Dec	Mag	Dim	Rating
Mixed spiral galaxy	Cetus	00^h 34.8^m	−08° 24′	10.4	4.0′ × 2.4′	3

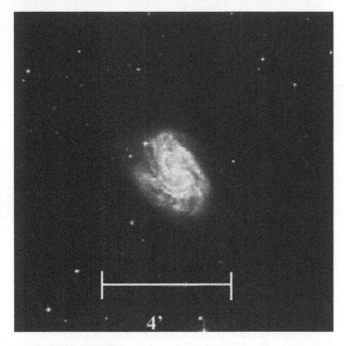

4′

General description

NGC 157 is a small and moderately bright galaxy in an empty recess of sky nearly 4° east of 4th-magnitude Iota (ι) Ceti. The galaxy is somewhat condensed so it should be a nice object from suburban locations in a moderate-sized telescope. Small-telescope users need to have patience and sweep slowly to find it. Once located in a small telescope, it can seen under a dark sky with a first-quarter Moon up. Be prepared to make a slow and careful search for this object.

Directions

Use Chart 76 to return to Phi^2 (φ^2) Ceti. Now look 6° northwest for Iota Ceti. Center Iota Ceti in your telescope at low power, then switch to Chart 76d. From Iota Ceti, move 35′ east–northeast to a 10′-wide arc of three 9th-magnitude suns (a). Now move another 35′ east–northeast to two 9th-magnitude suns (b), separated by 25′ and oriented east to west. Next, jump 30′ northeast to 7.5-magnitude Star c, which has a 10th-magnitude companion immediately to the southwest. Now make a 30′ hop south-east to 8th-magnitude Star d. Next, make a slow and careful sweep 1° northeast to 9th-magnitude Star e. NGC 157 lies a little more than 1° southeast of Star e; you'll find it nestled between two 9th-magnitude suns.

The quick view

At 23× in the 4-inch, under a dark sky NGC 157 hides between two 9th-magnitude stars. With averted vision, the galaxy pops into view and looks like a 4′-wide circular disk of even light. At first, the galaxy looks innocuous and some-what underwhelming. But with time, the disk is enticingly mottled. The view becomes more dramatic at higher power. At 72×, NGC 157 looks like a comet sailing between two stars. The mottling suspected at low power is due to some dim suns nearby. With time, the galaxy starts to dis-play some tightly wound spiral details. It is, overall, quite an interesting object and worthy of more time, so plan to return to it.

Star charts for sixth night

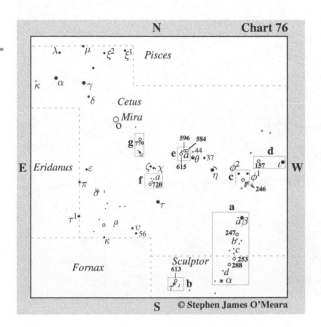

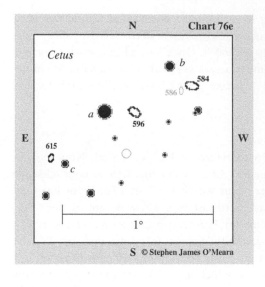

N   Chart 76e

Cetus

b

584
586

a
596

E   W

615
c

1°

S © Stephen James O'Meara

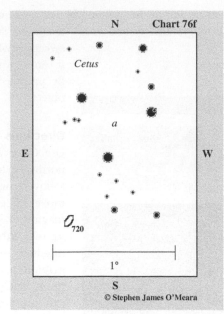

N   Chart 76f

Cetus

a

E   W

720

1°

S
© Stephen James O'Meara

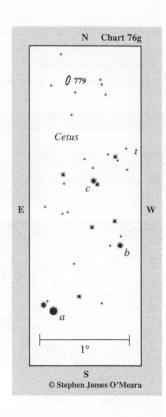

N   Chart 76g

779

Cetus

c

E   W

b

a

1°

S
© Stephen James O'Meara

SIXTH NIGHT

1. NGC 596 (H II-4)

Type	Con	RA	Dec	Mag	Dim	Rating
Mixed spiral galaxy	Cetus	01^h 32.9^m	−07° 02′	10.9	2.7′× 2.0′	3.5

4′

General description

NGC 596 is a small, though surprisingly nice galaxy about $2\frac{1}{2}°$ northeast of 3.6-magnitude Theta (θ) Ceti, a fine golden star with a 6th-magnitude neighbor (44 Ceti) about 10′ due north; the galaxy is almost hidden in the glare of 6th-magnitude Star *a*, which is only about 12′ to its east. NGC 596 has been seen with an aperture of $2\frac{1}{2}$-inches, and it is a cinch to see in anything larger − at least from a dark sky. The galaxy makes an attractive companion to NGC 584, which is 25′ to the northwest (see below).

Directions

Use Chart 76 to find Theta Ceti. Now use your unaided eyes or binoculars to locate 6th-magnitude Star *a*, which is a little more than $2\frac{1}{2}°$ to the northeast. Center Star *a* in your telescope at low power, then switch to Chart 76e. NGC 596 lies only about 12′ to the west of Star *a*.

The quick view

At 23× in the 4-inch, NGC 596 is a fine sight − a condensed 2.5′-wide circular glow with an intense starlike core; the galaxy is a bit difficult to appreciate fully at this power because of the proximity of the 6th-magnitude star. At 72×, NGC 596 has a bright circular core in a slightly swollen elliptical halo that appears just slightly tilted northeast to southwest with averted vision.

Stop. Do not move the telescope. Your next target is nearby!

2. NGC 584 (H I-100)

Type	Con	RA	Dec	Mag	Dim	Rating
Lenticular galaxy	Cetus	01ʰ 31.3ᵐ	−06° 52′	10.4	3.8′ × 2.4′	4

General description
NGC 584 is a stunningly small, and deceivingly bright, lens-shaped galaxy about 25′ northwest of NGC 596. It is a fine example of how a tiny galaxy with a moderately faint listed magnitude (10.4) is quite easy to see. I have glimpsed it at the limit of vision in an antique 1¼-inch telescope under a dark sky. It is a beautiful object in all larger telescopes.

Directions
Use Chart 76e. NGC 584 is only 25′ northwest of NGC 596, about 12′ southwest of 7.5-magnitude Star *b*.

The quick view
In the 4-inch at 23×, NGC 584 is an almost stellar smear of light. The galaxy is so small and its core so bright and stellar that the system is easy to sweep over. The galaxy's fuzzy nature shows up better at 72×, but I find 101× provides the most comfortable view in the 4-inch. At that power, the galaxy displays an extremely-sharp core, an egg-shaped inner lens and a strong and equally sharp needle of light that slices through the major axis of the outer lens, which fades to insignificance under higher powers. NGC 586 is quite easily visible at 101× in the 4-inch just 4.3′ southeast of NGC 584. This non-interacting galaxy has a listed magnitude of 13.2, but that must be in error by a magnitude or more.

Stop. Do not move the telescope. Your next target is nearby!

3. NGC 615 (H II-282)

Type	Con	RA	Dec	Mag	Dim	Rating
Barred spiral galaxy	Cetus	01ʰ 35.1ᵐ	−07° 20′	11.6	2.5′ × 1.3′	3

General description
NGC 615 is nearly 3° northeast of Theta (θ) Ceti, a little less than 30′ southeast of Star *a*. It is a nearly edge-on spiral galaxy that, despite its dim listed magnitude (11.6) is not too difficult to see in a small telescope because its central lens is highly condensed. Remember, you're looking for something small and starlike.

Directions
Using Chart 76e, from NGC 584, return to Star *a*. NGC 615 lies just 30′ southeast of Star *a* and just 6′ east–northeast of 8.5-magnitude Star *c*.

The quick view
At 23× in the 4-inch, NGC 615 is a tiny fleck of elongated light with a starlike core. At 72×, the inner lens, which measures only about 1′ in length, appears elongated. It is surrounded by a slightly dimmer halo that extends its length by a factor of two, especially with averted vision.

Stop. Do not move the telescope.

4. NGC 720 (H I-105)

Type	Con	RA	Dec	Mag	Dim	Rating
Elliptical galaxy	Cetus	01^h 53.0^m	−13° 44′	10.2	4.3′ × 2.0′	3.5

General description
NGC 720 is a moderately bright elliptical galaxy about $3\frac{1}{2}°$ south and slightly east of 4th-magnitude Zeta (ζ) Ceti. As with many elliptical galaxies, NGC 720 is a very nice object for small telescopes, especially at higher powers. At low power, in your search, imagine that you are looking for a little planetary nebula rather than a galaxy.

Directions
Use Chart 76 to locate Zeta Ceti, which is about $6\frac{1}{2}°$ southeast of Theta (θ) Ceti. Zeta Ceti is the brighter star in a close pair with Chi (χ) Ceti to the southwest, which is a nice binocular double star. Now center Zeta and Chi Ceti in your binoculars, then drop $2\frac{1}{2}°$ south to 30′-wide Triangle a, which comprises three 7th-magnitude stars. You want to take the time to center Triangle a in your telescope at low power, then confirm its appearance with Chart 76f. Now center the southernmost star in Triangle a. NGC 720 is only about 35′ south–southeast of that star.

The quick view
In the 4-inch at 23×, NGC 720 is an almost stellar object that swells into a roundish or slightly elliptical form – like a 2′-wide, egg-shaped planetary nebula – with averted vision. At 72×, the galaxy is definitely elongated northwest to southeast with a pretty lens-shaped core

that brightens to a starlike nucleus. The cigar-shaped halo lengthens the object to about 4′ when viewed with averted vision.

Stop. Do not move the telescope.

5. NGC 779 (H I-101)

Type	Con	RA	Dec	Mag	Dim	Rating
Spiral galaxy	Cetus	01^h 59.7^m	−05° 58′	11.2	3.4′ × 1.2′	3

General description
NGC 779 is a very dim and small spiral galaxy about 5° northeast of Zeta (ζ) Ceti. In a small telescope it is best seen at moderate magnification. It is also in a relatively lonely part of the sky, so you must take some time to star hop to it. Fortunately the galaxy's core is condensed; it can be seen under the light of a first-quarter Moon.

Directions
Use Chart 76 to return to Zeta Ceti. Now use your unaided eyes or binoculars to look about 3° northeast for 5.5-magnitude Star a, which has a 7th-magnitude companion about 8′ to the northeast. Center Star a in your telescope at low power and confirm it with Chart 76g. From Star a, make a careful 1° sweep northwest to 7.5-magnitude Star b, which marks the southwestern apex of a 20′-wide triangle with two 8th-magnitude stars. From Star b, move a little less than 50′ north–northeast to 7th-magnitude Star c. NGC 779 is a little more than 1° north–northeast of Star c in a relatively lonely field. So make sure this sweep is a slow one.

The quick view

At 23× in the 4-inch under a dark sky, NGC 779 is a small (2′) long cloud of light elongated north–northwest to south–southeast. At 72×, the galaxy is quite nice, having a highly condensed oval core surrounded by a uniformly bright ellipse of light. Still, the galaxy is tiny.

Star charts for seventh night

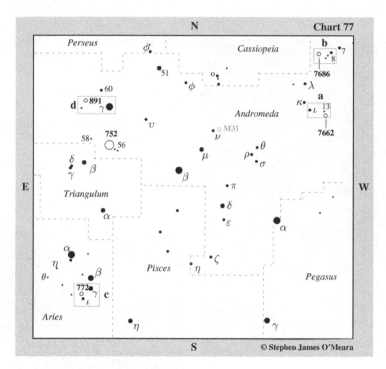

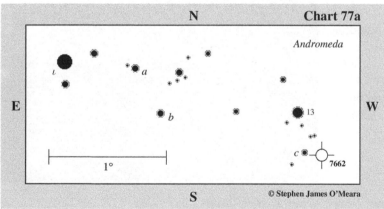

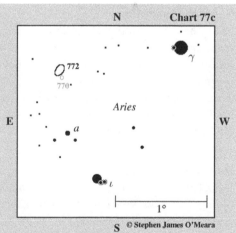

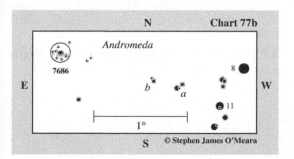

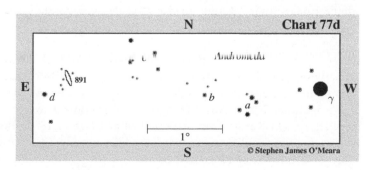

SEVENTH NIGHT

1. NGC 7662 (H IV-18)

Type	Con	RA	Dec	Mag	Dim	Rating
Planetary nebula	Andromeda	23^h 25.9^m	+42° 32'	8.3	32"× 28"	4

General description

NGC 7662, the famous Light Blue Snowball, is a beautiful planetary nebula just 25' southwest of 5th-magnitude 13 Andromedae. It is visible in 7×50 binoculars as an 8th-magnitude "star" and is a popular non-Messier object among amateurs using telescopes of all sizes. Its location in the far northern reaches of Andromeda, however, keeps it out of the mainstream limelight.

Directions

Use Chart 77 to find Alpha (α) Andromedae, which marks the northeast corner of the Great Square of Pegasus. Now look 15° north–northwest for 4th-magnitude Iota (ι) Andromedae. Now use your unaided eyes or binoculars to find 6th-magnitude 13 Andromedae 2° to the west–southwest. If you center 13 Andromedae in your telescope at low power, NGC 7662 will be just 25' to the southwest, just about 8' west of 8th-magnitude Star c (see Chart 77a). Otherwise, you can use Chart 77a to star-hop to it from Iota Andromedae. From Iota Andromedae move 35' west to 7.5-magnitude Star a. Now dip nearly 30' southwest to similarly bright Star b. Now make a slow and careful 1¼° sweep due west to 13 Andromedae. Again, your target is only 25' southwest of 13 Andromedae.

The quick view

At 23× in the 4-inch, NGC 7662 looks simply as an 8th-magnitude star. It requires magnification to see the disk well. At 72×, the nebula is very apparent, appearing as a pale aquamarine ellipse, oriented northeast to southwest with a sharp core and dense inner ring. At 189×, the sharp core is composed of two bright arcs of nebulosity – one on either side of the nebula's major axis. These arcs surround the annulus's cavity. The nebula's central star may be variable, with a brightness that ranges from a magnitude of 11.5 to 13.0, with 12.5 being the average magnitude.

Stop. Do not move the telescope.

2. NGC 7686 (H VIII-69)

Type	Con	RA	Dec	Mag	Dim	Rating
Open cluster	Andromeda	23^h 30.1^m	+49° 08'	5.6	15.0'	4

General description

NGC 7686 is a large and bright open cluster in far northern Andromeda, just 6° north–northwest of 4th-magnitude Iota (ι) Andromedae. The cluster is centered on a 6.5-magnitude star with an 8th-magnitude companion about 5' to the southwest. These stars can be easily spied in binoculars. And the surrounding cluster can be spied as a fuzzy glow in binoculars under a dark sky.

Directions

Using Chart 77, from NGC 7662, return your gaze to Iota Andromedae. Next look 3° north for 3.5-magnitude Lambda

(λ) Andromedae. Now use your unaided eyes or binoculars to look $4\frac{1}{2}°$ northwest for the 5th-magnitude pair of stars 7 and 8 Andromedae, which has a 6th-magnitude companion (11 Andromedae) 30′ to the southeast. Center 8 Andromedae in your telescope at low power, then switch to Chart 77b. Center 11 Andromedae and move 30′ northeast to 7.5-magnitude Star *a*. Now hop about 18′ east–northeast to 9th-magnitude Star *b*. NGC 7686 is 1° further to the east–northeast.

The quick view

In the 4-inch at 23×, NGC 7686 is a large (15′) and scattered splash of suns tickling a central double star, which has a dimmer companion to the south–southeast, forming a clean central equilateral triangle. The bright, 6th-magnitude member shines with a crisp golden hue. The field is filled with a diffuse, asymmetrical light that transforms into tiny stars that jump about with averted vision. At 72×, the cluster's core is distinctly V-shaped with faint outliers rounding out the cluster's edges, but mostly to the north and west. Though large, the cluster has only about 80 true members shining at 7th-magnitude and fainter.

3. NGC 772 (H I-112)

Type	Con	RA	Dec	Mag	Dim	Rating
Spiral galaxy	Aries	01ʰ 59.3ᵐ	+19° 00′	9.9	7.1′× 4.7′	3.5

4′

General description

NGC 772 is a moderately sized and somewhat bright spiral galaxy $1\frac{1}{2}°$ east–southeast of 3.5-magnitude Gamma (γ) Arietis. The galaxy's core is most prominent in small apertures at moderate magnifications. So you'll be looking for a "star" with a very dim sheen around it. As with its phantom neighbor, M74, which is only about $6\frac{1}{2}°$ to the southwest, light pollution will affect the visibility of NGC 772.

Directions

Use Chart 77 to locate Gamma Arietis, the First Star in Aries, which is about $1\frac{1}{2}°$ west–southwest of 3rd-magnitude Beta (β) Arietis. NGC 772 is a clean sweep about $1\frac{1}{2}°$ to the east–southeast of Gamma Arietis. But it is easier to use Chart 77c. First locate 5.5-magnitude Iota (ι) Arietis, which is $1\frac{3}{4}°$ south-southeast of Gamma Arietis. From Iota Arietis, move about 40′ northeast to 7th-magnitude Star *a*. NGC 772 is about 40′ north and slightly east of Star *a*.

Field note

Gamma Arietis is an attractive double star. Here are two 4.8-magnitude beacons separated by 7.8″, looking like a car's headlights blazing forth on a dark highway.

The quick view

At 23× in the 4-inch, NGC 772 is a ghostly 5′-wide glow in a barren field. It has an ever-so-slightly condensed core surrounded by a weak halo oriented northwest to southeast. But this pale ghost glows more strongly with time and averted vision. At 72×, the galaxy's core is a white egg inside a mottled nest of light, which tapers into a whisper of light toward the west. The "apostrophe" shape of NGC 772 is most apparent at 101×. The southeast flank is remarkably rounded and smoothed, while the long western arm is but a slim arc of light with a slight mottled texture. Also at this power, the galaxy's elliptical companion, NGC 770, just snaps into view; NGC 770, then, must be brighter than its listed magnitude (12.9).

4. NGC 752 (H VII-32)

Type	Con	RA	Dec	Mag	Diam	Rating
Open cluster	Andromeda	01ʰ 57.6ᵐ	+37° 50′	5.7	75.0′	4

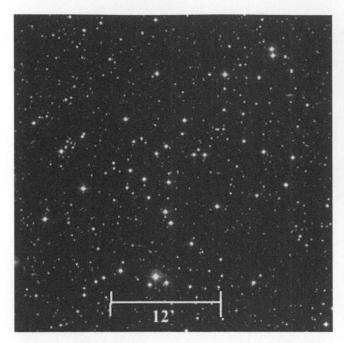

General description

NGC 752 is a bright, though very large open cluster about 5°
south–southwest of 2nd-magnitude Gamma (γ) Andromedae.
Under a dark sky, it is visible with the unaided eye as a soft
glow. It is a grand binocular object appearing as a scattering of
8th- to 12th-magnitude suns spread across 75′ of sky! The
view is best in binoculars under a dark sky, or with a rich field
telescope.

Directions

Use Chart 77 to locate Gamma Andromedae. Now use your
unaided eyes or binoculars to look 5° south-southwest for
NGC 752.

The Quick View

In the 4-inch at 23×, NGC 752 fills the field of view. It is
large and scattered with no obvious central concentration or
boundaries! Again, this is a better object to hunt down in
binoculars. In his description of this object, of which he
must have seen only a part, William Herschel notes that the
cluster looks like a "nebulous star to the naked eye." So, even
if you see this object only with your unaided eye, you have
seen it as Herschel did.

5. NGC 891 (H V-19)

Type	Con	RA	Dec	Mag	Dim	Rating
Spiral galaxy	Andro-meda	02^h 22.6^m	+42° 21′	9.9	12.2′ × 3.0′	3.5

General description

NGC 891 was discovered by William, not Caroline Herschel.
It is a highly popular edge-on galaxy, but it is only moderately large and somewhat dim (owing to the galaxy's thick
dust lane), making it a challenge to see in a small telescope
unless you are under a dark sky. Under a dark sky, it has been
seen in a 2½-inch telescope.

Directions

Use Chart 77 to locate 2nd-magnitude Gamma (γ)
Andromedae, then center it in your telescope at low power
and switch to Chart 77d. From Gamma Andromedae, make
a slow and careful 1° sweep east–southeast to 10′-wide
Trapezoid *a*, which comprises 7.5- to 9th-magnitude
suns. Now hop about 35′ east–northeast to 8th-magnitude
Star *b*. Now make another careful 1° sweep, this time to the
northeast, where you will find a 30′-long diamond (*c*) of
four 7th- to 8th-magnitude suns, oriented northeast to
southwest. NGC 891 is another 1° sweep to the east–
southeast, just 20′ northwest of 7th-magnitude Star *d*.

The quick view

At 23× in the 4-inch under a dark sky, NGC 891 is a ghostly,
10′-long sliver of light (oriented northeast to southwest)
that vanishes when direct vision is used. At 72×, the galaxy is
more clearly defined. With averted vision, the length of the
galaxy is peppered with faint suns that pop in and out of
view as the eye tries to trace the galaxy from end to end.
Most distinct is the galaxy's oval hub that, with attention,
can be seen divided by dust.

12 · December

Star charts for first night

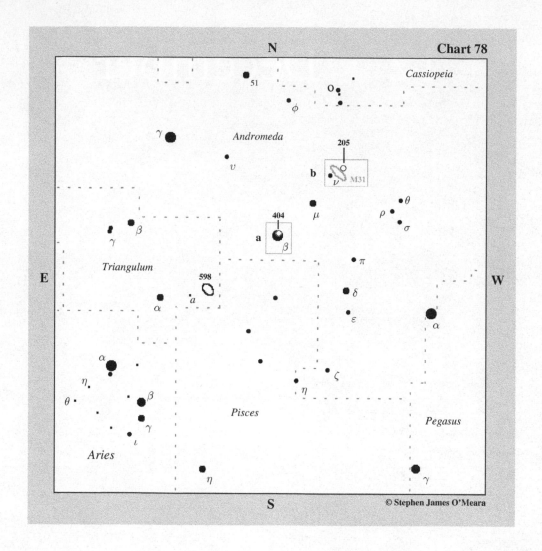

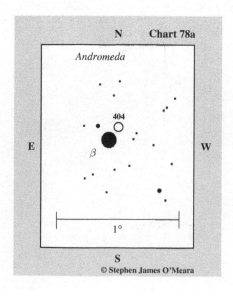

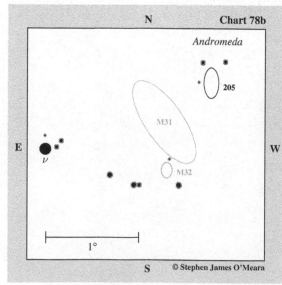

FIRST NIGHT

1. NGC 598 (H V-17) = M33

Type	Con	RA	Dec	Mag	Dim	Rating
Spiral galaxy	Trian-gulum	01^h 33.9^m	+30° 39'	5.7	71.0'× 42.0'	5

General description

Out of respect for Charles Messier, William Herschel did not include known Messier objects in his catalog. He did, however, make exceptions, and NGC 598 (M33) is one of them. Herschel catalogued M33 as H V-17 to differentiate it from a new "nebula" (H III-150 [NGC 604]) he found on September 11, 1784, which was associated with H V-17. We now recognize the new "nebula" as a large H II region in M33. M33, the Great Triangulum Spiral, is visible to the unaided eye under a dark sky about 4$\frac{1}{4}$° west–northwest of 3rd-magnitude Alpha (α) Trianguli. It appears as a distinct oval glow in 7×50 binoculars. It is a wonderful sight in telescopes of all sizes.

Directions

Use Chart 78 to find Alpha Trianguli. Now use your unaided eyes or binoculars to look 2$\frac{1}{2}$° west-northwest for 6th-magnitude Star a. Center Star a in your telescope at low power, then make a slow and careful sweep about 1$\frac{3}{4}$° further to the west–northwest, where you'll find M33.

The quick view

At 23× in the 4-inch under a dark sky, M33 is a compressed oval disk nearly 1° in extent, oriented north–northeast to south–southwest. With attention, spiral structure can be seen sweeping away from the galaxy's tight, lens-shaped nuclear region, but it is dim. The galaxy is best appreciated in an eyepiece that provides both low power and a wide field of view. A magnification of 144× is great for studying the tight inner region, but save that for another night.

2. NGC 404 (H II-224)

Type	Con	RA	Dec	Mag	Dim	Rating
Dwarf lenticular galaxy	Andro-meda	01^h 09.4^m	+35° 43'	9.8	6.6'× 6.6'	4

General description

NGC 404 lies a little more than 6' northwest of 2nd-magnitude Beta (β) Andromedae. Many observers have found NGC 404 by chance, mistaking it for an undiscovered comet. The galaxy is just visible in a 2.4 inch refractor as a small but bright polished pearl. It is a beautiful, though simple, sight in larger telescopes.

Directions

Use Chart 78 to locate Beta Andromedae. Center that star in your telescope, then switch to Chart 78a, which will help you pinpoint NGC 404's position.

The quick view

In the 4 inch at 23×, NGC 404 is a condensed 6' round glow with a milky-smooth texture and stellar core. At 72× (with Beta Andromedae out of the field of view), the galaxy has an enticingly mottled appearance, as if it were breaking up into individual starlight. That view is shattered, though, at 101×, when it becomes apparent that the mottled nature is due, in part, to some superimposed suns. One star lies just beyond the galaxy's north flank; it's enough to make one question whether it is a supernova, so be careful. The galaxy takes magnification surprisingly well.

3. NGC 205 (H V-18) = M110

Type	Con	RA	Dec	Mag	Dim	Rating
Elliptical galaxy	Andromeda	00^h 40.4^m	+41° 41′	8.0	21.9′ × 11.0′	4

4′

Star charts for second night

General description
Caroline Herschel discovered NGC 205 on July 1, 1783. Although it had been discovered by Messier in 1773, the French comet hunter did not publish that discovery; so when Caroline independently found it, it remained unknown to her and her brother, who catalogued it as H V-18. NGC 205 (M110) is the larger (and more diffuse) of the two elliptical companions flanking majestic M31. Although it can be seen in 7×50 binoculars under a dark sky, it may prove difficult in a telescope under light-polluted skies because it has a low surface brightness.

Directions
Use Chart 78 to locate the bright, 3rd-magnitude spiral galaxy M31, which is a beautiful 3°-long cocoon of light 1° west of 4.5-magnitude Nu (ν) Andromedae in the Chained Maiden's belt. Center M31 in your telescope at low power, then switch to Chart 78b. NGC 205 (M110) is about 35′ northwest of M31's core.

The quick view
At 23× in the 4-inch under a dark sky, M110 is a 20′-long fuzzy-oval glow that gradually, then suddenly brightens toward the center to a nucleus. At 72×, the galaxy's disk appears rather mottled, but this is undoubtedly due to some dim superimposed stars.

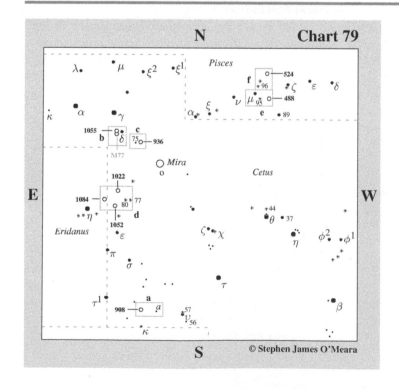

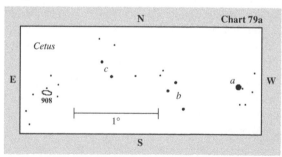

© Stephen James O'Meara

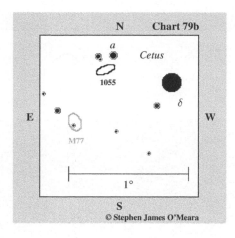

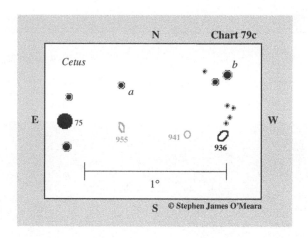

SECOND NIGHT

1. NGC 908 (H I-153)

Type	Con	RA	Dec	Mag	Dim	Rating
Spiral galaxy	Cetus	02^h 23.1^m	−21° 14′	10.4	6.1′× 2.7′	3.5

4'

General description

NGC 908 is a moderately bright starburst galaxy about 5½° west of 4th-magnitude Upsilon (υ) Ceti. The galaxy's core is so bright that it is *barely* visible in 7×50 binoculars under a dark sky, just southeast of a little triangle of roughly 10th-magnitude suns. Through a telescope the galaxy is an elliptical haze with an intense stellar nucleus.

Directions

Use Chart 79 to find Upsilon Ceti, which may be a difficult naked-eye catch from a suburban sky; if so, use binoculars to locate it. It lies in the far southeastern quadrant of the constellation and is the brightest star in the region for many degrees. You'll know when you have it, because 5th-magnitude 56 Ceti will be about 1½° to the southwest and, closer in, 5th-magnitude 57 Ceti, will be 15′ to the north–northwest. Once you're certain you have Upsilon, use your unaided eyes or binoculars to look a little more than 3° due east, where you'll find solitary 6th-magnitude Star *a*. Center Star *a* in your telescope at low power, then switch to Chart 79a. Note that Star *a* is bordered to the southwest by a 10′-wide trapezoid of 8th- and 9th-magnitude suns. From Star *a*, move 45′ due east, you'll encounter a 20′-long acute triangle of 8th- to 9th-magnitude stars (*b*). Center the northernmost star (the faintest of the three) in Triangle *b*, then move 50′ east–northeast to a pair of roughly 9th-magnitude stars (*c*) oriented northeast to southwest and separated by about 12′. The galaxy is about 50′ southeast of Pair *c*. Look for a 10th-magnitude cigar-shaped glow (6′-long) with a stellar core.

The quick view

At 23× in the 4-inch, NGC 908 is immediately obvious as a slightly elongated haze surrounded by a trapezoid of suns. With averted vision, the galaxy has a stellar nucleus within a dense core of light. An outer envelope tapers to the east and west. At 72× and 101×, the galaxy breaks down into finer details. It is symmetrical along the east–west axis. The dense core is no longer entirely stellar. Instead, it appears to consist of an extremely sharp stellar nucleus surrounded by a fuzzy elliptical core, which is at the center of a broad lens; that lens is slightly tilted with respect to the galaxy's major axis.

2. NGC 1055 (H II-6)

Type	Con	RA	Dec	Mag	Dim	Rating
Spiral galaxy	Cetus	02^h 41.8^m	+00° 26′	10.6	7.3′ × 3.3′	2.5

3. NGC 936 (H IV-23)

Type	Con	RA	Dec	Mag	Dim	Rating
Barred spiral galaxy	Cetus	02^h 27.6^m	−01° 09′	10.2	5.7′ × 4.6′	4

General description

NGC 1055 is a moderately large but dim nearly edge-on galaxy about 35′ east and slightly north of 4th-magnitude Delta (δ) Ceti and 30′ northwest of the bright Seyfert galaxy M77. Small-telescope users should be prepared to use averted vision and low power to see its dim form. The galaxy's major axis is infiltrated with light-obscuring dust, which makes it a challenge to see.

Directions

Use Chart 79 to locate Delta Ceti. Center that star in your telescope at low power, then switch to Chart 79b. From Delta Ceti, move 35′ east–northeast to 7th-magnitude Star a, which has an 8th-magnitude companion about 6′ to the east. NGC 1055 lies about 6′ south of the midpoint between those two stars.

The quick view

In the 4-inch at 23×, NGC 1055 is an elongated haze immediately southeast of a roughly 12th-magnitude star. The galaxy is elongated west–northwest to east–southeast, and it swells with averted vision. Seen together with M77, NGC 1055 appears much larger, though much dimmer. At 72×, NGC 1055 is more difficult to see, as just the central lens is readily visible.

Stop. Do not move the telescope.

General description

NGC 936 is a small but obvious galaxy about 1$\frac{1}{4}$° west of 5.5-magnitude 75 Ceti. The galaxy's core is extremely bright, so is a good object for suburban observers.

Directions

Use Chart 79 to locate 75 Ceti, which is about 2$\frac{1}{4}$° southwest of Delta Ceti. Center that star in your telescope at low power, then switch to Chart 79c. From 75 Ceti, move a little less than 30′ northwest to 9th-magnitude Star a. Now move about 45′ west–northwest to 7th-magnitude Star b, which has a 9th-magnitude companion about 6′ to the south–southeast. NGC 936 is about 25′ south of that pair of stars.

The quick view

At 23× in the 4-inch, NGC 936 is a 3′-wide oval glow, very condensed and immediately bright. At 72×, the galaxy's core is quite stellar and intense. It is oriented roughly east to west. Larger scopes should show the galaxy's larger, though dimmer oval disk, which is oriented northwest to southeast.

Star charts for third night

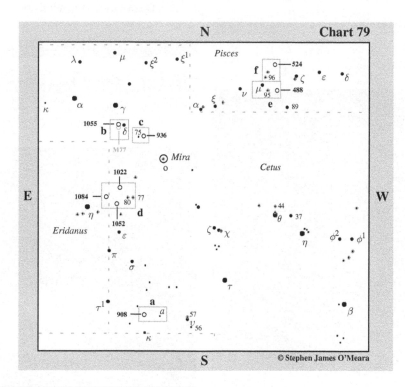

Chart 79

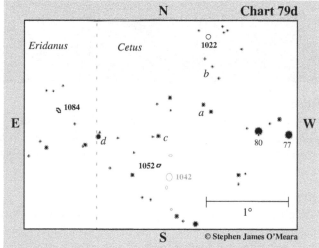

Chart 79d

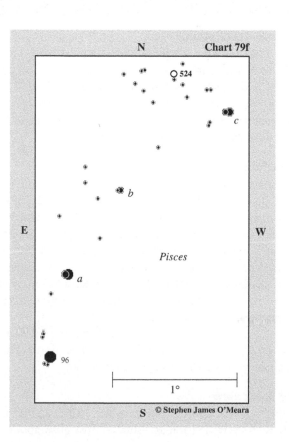

Chart 79f

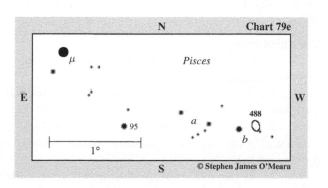

Chart 79e

THIRD NIGHT

1. NGC 1022 (H I-102)

Type	Con	RA	Dec	Mag	Dim	Rating
Barred spiral galaxy	Cetus	02ʰ 38.5ᵐ	−06° 40′	11.3	2.7′ × 2.7′	2

General description

NGC 1022 is a very small and dim galaxy about $1\frac{1}{4}°$ northwest of 5.5-magnitude 80 Ceti. It is in a very sparse and lonely field, making it all the more difficult to find. It is not an object for a small telescope in a city or suburb. It is more diffuse than concentrated, adding to its difficulty. You'll need to make a slow and careful search for it.

Directions

Use Chart 79 to locate 4th-magnitude Eta (η) Eridani, which is 10° (a fist) south–southwest of Delta (δ) Ceti. Now use your unaided eyes or binoculars to find 80 Ceti, which is 5° to the west–northwest. You'll know when you have this star because similarly bright 77 Ceti lies only about 20′ to its west. Center 80 Ceti in your telescope at low power and switch to Chart 79d. (Note that 80 Ceti has a 9.5-magnitude companion immediately to its south.) From 80 Ceti, move 40′ northeast to a pair of 8.5-magnitude stars (a), which are oriented northeast to southwest and separated by about 5′. NGC 1022 is 55′ due north of Pair a, and 20′ north of a 20′-long line of three 10.5-magnitude suns (b).

The quick view

In the 4-inch at 23× under a dark sky, NGC 1022 is visible only with averted vision as a small diffuse glow about 1′ in diameter; it shines just above the background sky, so any light pollution will wash it out. At 72×, the galaxy expands to about 1.5′ and appears as a little diffuse oval with a slight and circular central condensation.

Stop. Do not move the telescope. Your next target is nearby!

2. NGC 1052 (H I-63)

Type	Con	RA	Dec	Mag	Dim	Rating
Spiral galaxy	Cetus	02ʰ 41.1ᵐ	−08° 15′	10.5	2.5′ × 2.0′	4

General description

NGC 1052 is a small but highly condensed galaxy about $1\frac{1}{4}°$ east–southeast of 80 Ceti. It's a fairly obvious object in a small telescope under a dark sky and is likely an easy target from the suburbs.

Directions

Using Chart 79d, from NGC 1022, return to Pair a. Now make a careful sweep about 45′ southeast to 8.5-magnitude Star c, which has an 11th-magnitude companion about 4′ to the east–southeast. NGC 1052 is a little more than 20′ south of Star c.

The quick view

At 23× in the 4-inch, NGC 1052 stands out with direct vision as a small starlike nucleus surrounded by a small haze. With averted vision, the galaxy appears as a fairly bright and concentrated oval disk about 2′ in extent and oriented northwest to southeast. At 72×, the galaxy is a cometlike glow, with a starlike nucleus in a bright inner lens surrounded by a diffuse oval halo.

Stop. Do not move the telescope. Your next target is nearby!

3. NGC 1084 (H I-64)

Type	Con	RA	Dec	Mag	Dim	Rating
Barred spiral galaxy	Eridanus	02^h 46.0^m	−07° 35′	10.7	3.2′ × 1.9′	4

General description

NGC 1084 is a small but bright galaxy about 3° west–north-west of 4th-magnitude Eta (η) Eridani, just east of the Cetus border, and a little less than $1\frac{1}{2}$° east–northeast of NGC 1052. The galaxy is very visible in a small telescope, being highly condensed.

Directions

Using Chart 79d, from NGC 1052, move a little less than 50′ east–northeast to 7th-magnitude Star d. NGC 1084 is about 35′ northeast of Star d.

The quick view

At 23× in the 4-inch, NGC 1084 is easily seen with direct vision as a highly condensed oval glow about 2′ in diameter and oriented northeast to southwest. At 72×, the galaxy appears much the same as it does at low power (though more obvious). Averted vision shows the core to be slightly brighter and elongated.

General description

NGC 488 is a very small but bright galaxy about $2\frac{1}{4}$° west–southwest of 5th-magnitude Mu (μ) Piscium. It is yet another good target for small telescopes, even from a suburban location.

Directions

Use Chart 79 first to locate 3.5-magnitude Alpha (α) Piscium, which is 10° due west of 4th-magnitude Gamma (γ) Ceti. Now use your unaided eyes or binoculars to find Mu Piscium about 8° further to the northwest in the Fishes's cord. Center Mu Piscium in your telescope at low power, then switch to Chart 79e. From Mu Piscium, move 1° south–southwest to 7th-magnitude 95 Piscium. Now move about 50′ west to a wide pair of 8.5-magnitude stars (a) oriented east–northeast to west–southwest and separated by about 20′. Next move about 35′ west–southwest to 7.5-magnitude Star b. NGC 488 is only 10′ west of Star b.

The quick view

At 23× in the 4-inch, NGC 488 can be seen as a small (1′) but very condensed glow just northeast of a roughly 12th-magnitude star. It can be seen even with direct vision under the light of a first-quarter Moon. The galaxy swells to about 2′ with averted vision. At 72×, the galaxy's dim halo diffuses making it less distinct, though the core remains tight and bright. The view in a small telescope is better at low power. Stop. Do not move the telescope.

4. NGC 488 (H III-252)

Type	Con	RA	Dec	Mag	Dim	Rating
Spiral galaxy	Pisces	01^h 21.8^m	+05° 15′	10.3	5.5′ × 4.0′	4

5. NGC 524 (H I-151)

Type	Con	RA	Dec	Mag	Dim	Rating
Spiral galaxy	Pisces	01^h 24.8^m	+09° 32′	10.2	3.5′ × 3.5′	2.5

General description

NGC 524 is a small and somewhat difficult galaxy about $3\frac{3}{4}°$ north–northwest of Mu (μ) Piscium. It best seen at moderate magnification in a small telescope.

Directions

Using Chart 79, from NGC 488, return to Mu Piscium. Now use binoculars to locate 6.5-magnitude 96 Piscium $1\frac{1°}{4}$ to the north–northwest. Center 96 Piscium in your telescope at low power, then switch to Chart 79f. From 96 Piscium, move about 40′ north–northwest to 6.5-magnitude Star *a*, which has an 8.5-magnitude companion immediately to its east. Now make a careful 45′ sweep northwest to a faint double star (*b*), with a 9.5-magnitude primary and an 11th-magnitude companion to the east. Next, move a little more than 1° further to the northwest, where you'll find yet another double star (*c*), with a 7.5-magnitude primary and a 9.5-magnitude companion, also to the east. NGC 524 is nearly 35′ northeast of Double Star *c*.

The quick view

At 23× in the 4-inch (under the light of a first-quarter Moon), NGC 524 requires averted vision to see. It lies in a little "rectangle" comprising five dim suns. The galaxy is very round and tiny, appearing perhaps 1′ in diameter. At 72×, NGC 524 is a tiny (though somewhat concentrated) ball of light with a slight outer halo.

Star charts for fourth night

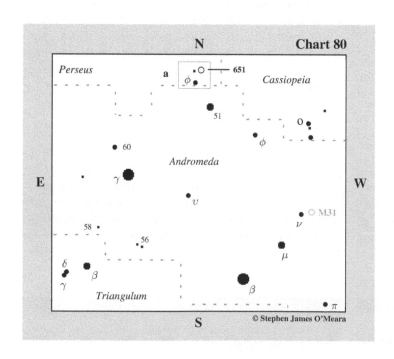

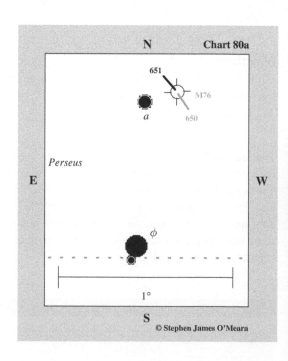

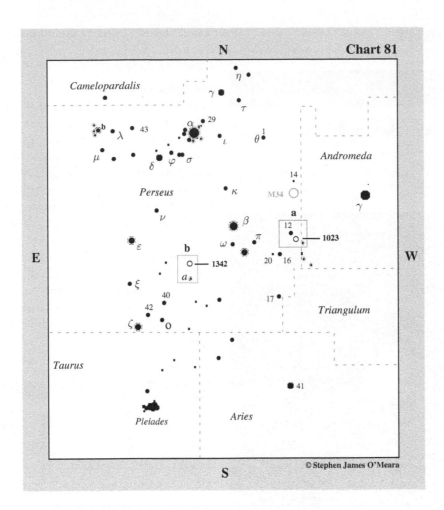

Chart 81

N

Camelopardalis

η
γ
τ
b
α 29
λ 43
ι
θ 1
μ
δ φ σ

Andromeda

Perseus
κ
14
M34
γ
ν
a
12
β
1023
b
ω π
1342
20 16
E
ε
a
W
ξ
17
40
Triangulum
42
ζ o

Taurus

41

Pleiades

Aries

S

© Stephen James O'Meara

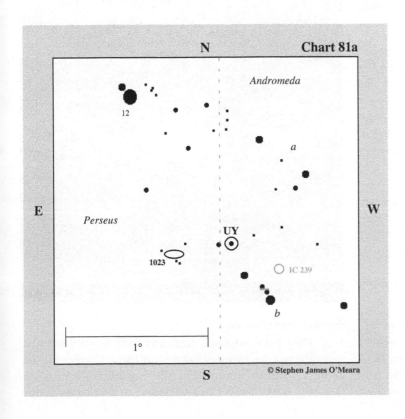

Chart 81a

N

Andromeda

12

a

Perseus

UY

1023

IC 239

E

W

b

1°

S

© Stephen James O'Meara

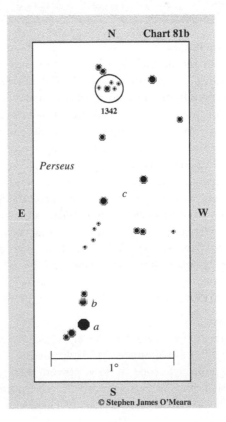

Chart 81b

N

1342

Perseus

c

E

W

b

a

1°

S

© Stephen James O'Meara

Fourth Night

1. NGC 651 (H I-193) = NORTHEAST HALF OF M76

Type	Con	RA	Dec	Mag	Diam	Rating
Planetary nebula	Perseus	01ʰ 42.4ᵐ	+51° 34′	10.1	~35″ (northeast half)	3.5

4′

General description

NGC 651 is the northeastern half of the Little Dumbbell Nebula (M76), near 4th-magnitude Phi (φ) Persei. The planetary nebula can just be glimpsed under a dark sky with 7×50 binoculars as a 10th-magnitude "star." Pierre Mechain discovered the nebula in 1780, and Messier listed it as the 76th object in his famous catalog. When William Herschel swept it up in 1787, he saw not one but two nebulae in contact, so he decided to give each nebula a separate number in his catalog. Today we know that Herschel's two nebulae are part of one ring-type planetary nebula whose central taurus is seen nearly edge on. To see H I-193, then, you must perceive the nebula's dumbbell shape and focus in on its northeast end.

Directions

Use Chart 80 to locate Phi Persei, which is about $2\frac{1}{4}°$ northeast of 3.6-magnitude 51 Andromedae. Center Phi Persei in your telescope at low power, then switch to Chart 80a. From Phi Persei, move 50′ due north to 6.5-magnitude Star *a*. M76 is about 12′ west–northwest of

Star *a*, and NGC 651 is the northeastern half of the Little Dumbbell.

The quick view

In the 4-inch at 23×, M76 is a 1′-wide opalescent disk of light in a rich field of stars, so you must look with averted vision at its plotted position to see the nebula swell against its stellar background. At 72×, the dumbbell impression starts to emerge. With concentration, the nebula appears as a patchy, hourglass-shaped orb of light oriented northeast to southwest; two semicircular arcs border the hourglass – one to the southeast, the other to the northwest. At 101×, NGC 651, the northern half of the hourglass or dumbbell stands out well (remember, seeing it as a distinct nebulous glow is your goal); It tapers to the southwest where its ends at a distinct dark lane running northwest to southeast. The dark lane separates NGC 651 from the southwestern half of the hourglass (NGC 650).

2. NGC 1023 (H I-156)

Type	Con	RA	Dec	Mag	Dim	Rating
Barred lenticular galaxy	Perseus	02ʰ 40.4ᵐ	+39° 04′	9.3	7.5′× 3.0′	4

4′

General description

NGC 1023 is a moderately large and fairly bright lenticular galaxy about $5\frac{1}{2}°$ west–southwest of Beta (β) Persei and about $1\frac{1}{4}°$ south–southwest of 5th-magnitude 12 Persei. It is a nice target for a small telescope, appearing as a tiny sliver of light nestled inside a triangle of roughly 9th-magnitude stars. It is a bright, yet simple-looking ellipse in larger telescopes with a southward kink on the eastern flank.

Directions

Use Chart 81. NGC 1023 lies $3\frac{3}{4}°$ almost due south of M34. If you have an equatorial mount with setting circles, just center that bright Messier open cluster with low power and swing your scope $3\frac{3}{4}°$ to the south. Otherwise, use your unaided eyes or binoculars to find 5th-magnitude 12 Persei, which is $2\frac{1}{2}°$ south of M34, or just south of the midpoint between the 2nd-magnitude stars Beta Persei and Gamma (γ) Andromedae. There is no mistaking 12 Persei, since it is the brightest star in the region. Center 12 Persei in your telescope at low power, then switch to Chart 81a. Note that through a telescope, 12 Persei is a close pair of stars, and the fainter companion is itself a double. From 12 Persei, move $1\frac{1}{4}°$ southwest, to a pair of 7th-magnitude suns oriented northeast to southwest (a). Next, center the southwestern member of that pair (which has an 8th-magnitude companion about $10'$ to the southeast). Now sweep $1°$ south–southeast to 6.5-magnitude Star b, which marks the southern end of a $15'$-long arc of slightly fainter suns. NGC 1023 is about $45'$ northeast of Star b.

The quick view

At 23× in the 4-inch under a dark sky, NGC 1023 is a tiny ellipse of light with a bright core, nestled among a splash of field stars. (Under suburban skies, the object might look more like an oval planetary nebula with a bright central star.) At 72×, NGC 1023's core is quite brilliant and takes magnification well. The surrounding lens fades gradually away from this center, however, you must sacrifice one to see the other well. But it is through such studies at various powers that you can see the most detail. At 101×, the galaxy is a very puzzling sight. The sharp center is surrounded by a bright lens of light, punctuated on the west, northwest, and east by nebulous knots or stars.

4'

General description

NGC 1342 is a fairly bright, though loosely packed and scattered, open cluster about $5\frac{1}{2}°$ west–southwest of 3rd-magnitude Epsilon (ε) Persei. Under a dark sky it is visible in 7×50 binoculars.

Directions

Use Chart 81 to find Epsilon Persei and Xi (ξ) Persei 4° roughly due south. NGC 1342 marks the western apex of an equilateral triangle with those stars. Just raise your binoculars to that location and look for what appears to be a diffuse, 7th-magnitude "comet" immediately south–southwest of an 8.5-magnitude star. Otherwise, from Xi Persei, look 5° west for 6th-magnitude Star a, which, in binoculars should be a nice triple. Center Star a in your telescope at low power, then switch to Chart 81b. From Star a, hop about $10'$ north to 7.5-magnitude Star b. Next, move about $55'$ north–northwest to a wide pair of 8th-magnitude stars (c), oriented northwest to southeast and separated by about $20'$. NGC 1342 is $50'$ due north of the southeastern 8th-magnitude star in Pair c.

The quick view

At 23× in the 4-inch, NGC 1342 is a jagged, $10'$-long line of some half dozen suns of mixed magnitudes (oriented east to west) that shine brightly against an oval haze of dim suns. The background haze scintillates with averted vision. At 72×, the cluster's two dozen or so brightest stars form a sideways broken T against a fuzzy backdrop of fainter suns. The cluster contains nearly 100 members of 8th magnitude and fainter.

3.	NGC 1342 (H VIII-88)					

Type	Con	RA	Dec	Mag	Diam	Rating
Open cluster	Perseus	03^h 31.7^m	+37° 22'	6.7	17.0'	4

Star charts for fifth night

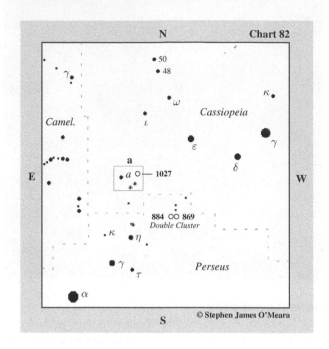

Chart 82

© Stephen James O'Meara

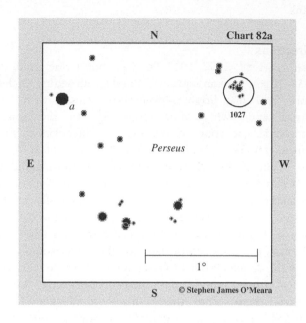

Chart 82a

© Stephen James O'Meara

FIFTH NIGHT

1. NGC 869 (H VI-33)

Type	Con	RA	Dec	Mag	Diam	Rating
Open cluster	Perseus	02^h 19.0^m	+57° 08′	4.5	~18′	5

General description

NGC 869 is the western (and brighter) member of the famous Double Cluster in Perseus. Although the Double Cluster has been known since antiquity as a single nebula, William Herschel was the first to catalog the two individual clusters that comprise it. Under dark skies the Double Cluster can be seen with the unaided eyes as a 4th-magnitude glow in the rich Milky Way about halfway between Delta (δ) Cassiopeiae and Gamma (γ) Persei. The light of NGC 869 dominates that naked-eye view. The Double Cluster is one of the wonders of the binocular sky, being well resolved and a joy to see. In 7×50 binoculars NGC 869 is a sharp cache of stars with a stunning cross of stars at the core, which is punctuated by a tight pair of suns. It is a delight to see in any size telescope.

Directions

Use Chart 82 to locate Gamma Persei and Delta Cassiopeiae, then use either your unaided eyes or binoculars to look for the Double Cluster which is about halfway between these stars. You want to point your telescope at the Double Cluster at low power and zoom in on the western member of the pair.

The quick view

In the 4-inch at 23×, NGC 869 has a tight core of suns that forms an island of starlight around a piercing sun. A myriad of irregularly bright suns populates the island and plays with the imagination. This cluster has more than 300 stars of 6.5

magnitude and fainter. This stellar island is bordered to the east by a dark Y-shaped rift.

Stop. Do not move the telescope. Your next target is nearby!

2. NGC 884 (H VI-34)

Type	Con	RA	Dec	Mag	Diam	Rating
Open cluster	Perseus	02^h 22.0^m	+57° 08'	5.7	18.0'	5

General description

NGC 884 is the eastern member of the famous Double Cluster in Perseus. It is just visible with the unaided eye under a dark sky and is a delight to see in binoculars, appearing as a translucent diamond surrounded by white smoke.

Directions

Use Chart 82. NGC 884 is only about 25' east of NGC 869.

The quick view

At 23× in the 4-inch, NGC 884 resembles a hollow rib cage of stars surrounded by a ruby heart – the semiregular variable star RS Persei, which pulses between 7.8 magnitude and 10.0 magnitude about every 224 days. NGC 884 also contains some 300 members of 6th magnitude and fainter. If you haven't spent time with NGC 884 and 869, be sure to return to them once you complete the Herschel 400. Rich star clusters like this are best admired without words – only thoughts! But now it's time to move on.

3. NGC 1027 (H VIII-66)

Type	Con	RA	Dec	Mag	Diam	Rating
Open cluster	Cassiopeia	02^h 42.6^m	+61° 36'	6.7	15.0'	4

General description

NGC 1027 is a bright open cluster about $5\frac{3}{4}°$ east–southeast of Epsilon (ε) Cassiopeiae, the easternmost star in the celestial W, or about $5\frac{3}{4}°$ north–northwest of Eta (η) Persei. It is easy to see in binoculars, especially because a 7th-magnitude star is near its core. Telescopically it is a sweet mix of bright and faint suns swarming around the bright central star.

Directions

Use Chart 82 to find Eta Persei. Now take the time to use your unaided eyes or binoculars to locate 6th-magnitude Star a, $5\frac{1}{2}°$ to the north–northeast. It is the brightest star in the region for some degrees. NGC 1027 is about $1\frac{1}{2}°$ west of Star a. Try to see it first in binoculars. If you can't, center Star a in your telescope at low power, then switch to Chart 82a, confirm the field, then make the $1\frac{1}{2}°$ sweep from Star a to the cluster.

The quick view

At 23× in the 4-inch, NGC 1027 is an intriguing sight, consisting of a 7th-magnitude sun surrounded by a milky haze that swells (and resolves) with averted vision into a myriad of irregularly bright stars spanning an area of about $\frac{1}{4}°$. At 72×, the cluster loses its luster. There are some two dozen suns scattered in front of an unresolved background of light. NGC 1027 is actually quite rich, containing more than 150 stars of 9th magnitude and fainter.

Star charts for sixth night

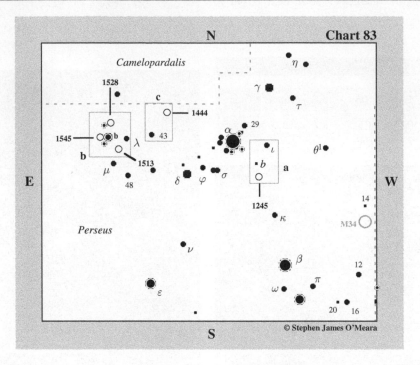

Chart 83

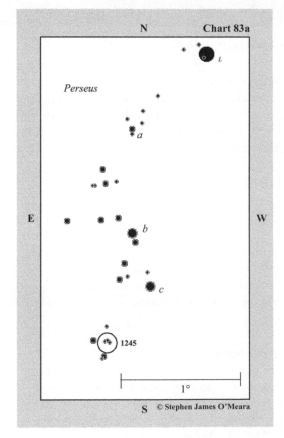

Chart 83a

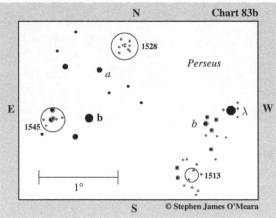

Chart 83b

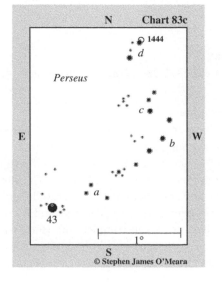

Chart 83c

SIXTH NIGHT

1. NGC 1245 (H VI-25)

Type	Con	RA	Dec	Mag	Diam	Rating
Open cluster	Perseus	03^h 14.7^m	+47° 14'	8.4	10.0'	3.5

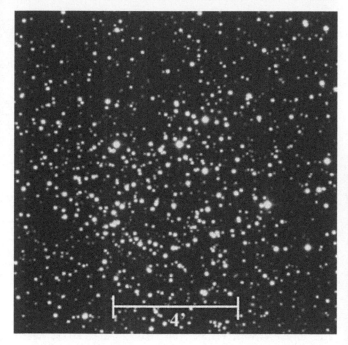

General description

NGC 1245 is a relatively small, dim, but rich open cluster a little more than 3° southwest of 2nd-magnitude Alpha (α) Persei. In small telescopes, it displays a starlike skeleton of moderately bright suns surrounded by a granular background comprising countless dim suns. It is quite an interesting object. Light pollution will greatly affect the view.

Directions

Use Chart 83 to locate Alpha Persei. Now look about 2½° west for 4th-magnitude Iota (ι) Persei. Center Iota Persei in your telescope at low power, then switch to Chart 83a. From Iota Persei, make a slow sweep 50' southeast to 8th-magnitude Star a. Now make another sweep 50' south to 6th-magnitude Star b, then slide about 30' south–southwest to 6.5-magnitude Star c. NGC 1245 is about 35' southeast of Star c.

The quick view

In the 4-inch at 23× under a dark sky, NGC 1245 is quite impressive, being a tiny packet of bright and faint suns in a starfish pattern. The little suns look like droplets of dew on a flower. At 72×, some 40 to 50 bright suns are scattered

about, forming an irregular oval with loose arms. The core, at a glance, has an east to west oriented river of about eight similarly bright suns running along the northern border. The cluster's granular texture is most apparent between the cluster's two brightest stars. The cluster has some 200 members of 12th magnitude and fainter packed into an area only 10' across. So it is a fine spectacle in larger telescopes. In the 4-inch, the best overall view is at 41×.

Stop. Do not move the telescope.

2. NGC 1513 (H VII-60)

Type	Con	RA	Dec	Mag	Diam	Rating
Open cluster	Perseus	04^h 09.9^m	+49° 31'	8.4	12.0'	3

General description

NGC 1513 is a rather dim and loose open cluster about 1° southeast of 4.5-magnitude Lambda (λ) Persei. Lambda Persei marks the end of the hook of that extends from the constellation's prominent spine of bright suns known as the Segment of Perseus. This graceful arc follows a very rich part of the Milky Way. The brightest part of the Segment measures about 12° in length and consists of the stars Eta (η), Gamma (γ), Alpha (α), Sigma (σ), Phi (φ), and Delta (δ) Persei. A fainter extension curves to the northeast from Delta and includes the stars 48, Mu (μ), and Lambda Persei. So the overall shape of the Segment is a J or fishhook. NGC 1513 is a delicate object and is best seen in small telescopes at moderate magnification. It will be difficult to appreciate under any light pollution, and it is better appreciated in large telescopes.

Directions

Use Chart 83 to follow the Segment of Perseus to Lambda Persei, then center the star in your telescope at low power and switch to Chart 83b. From Lambda Persei, hop about 25′ southeast to 7th-magnitude Star *b*, which marks the southeastern tip of a 10′-wide kite-shaped asterism of stars. NGC 1513 is 40′ south–southeast of Star *b*, and is on the southwest edge of a 20′-wide circlet of 8.5- to 10.5-magnitude suns.

The quick view

At 23× in the 4-inch under a dark sky, NGC 1513 is at first only a dim glow. With averted vision, the cluster's brighter members pop in and out of view, forming a semicolon of twinkling starlight against a uniform background glow. At 72×, the cluster is a circlet of about a dozen 11th-magnitude and fainter suns that look like a mix of salt and pepper with averted vision. The cluster contains only 50 members.

Stop. Do not move the telescope. Your next target is nearby!

3. NGC 1545 (H VIII-85)

Type	Con	RA	Dec	Mag	Diam	Rating
Open cluster	Perseus	04ʰ 20.9ᵐ	+50° 15′	6.2	12.0′	4

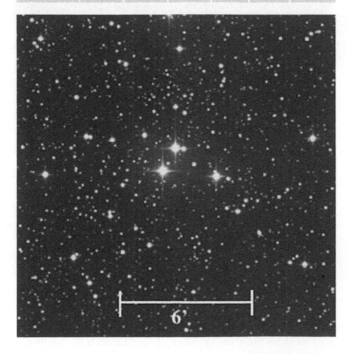

6′

General description

NGC 1545 is a bright open cluster a little less than 30′ east of 5th-magnitude b Persei, which is nearly 2° east of Lambda Persei. The cluster is condensed and borders on the limit of naked-eye visibility for an extended object. In 7×50 binoculars, only three stars of 7th to 8th magnitude are immediately apparent. But in a small telescope, NGC 1545 is a fascinating object.

Directions

Using Chart 83b, from Lambda Persei, look about 2° east for b Persei. Center that star in your telescope at low power. NGC 1545 is only $\frac{1}{2}$° to the east.

The quick view

At 23× in the 4-inch under dark skies, a clear pattern of about a dozen suns emerges. What appears to be a close pair of stars to the south is actually a tight triangle of suns, which lies at the center of a five-pointed-star-shaped pattern like a pentagram. The pentagram in NGC 1545 is comprised of about a dozen 10th- to 11th-magnitude suns, many of which are in pairs. These stars are only the first tier of bright stars that can be seen in small telescopes with direct vision. It's amazing how many stars suddenly appear with averted vision. The pentagram appears to be surrounded by a ghostly mist that wafts in and out of view with averted and direct vision, respectively. At 72× and 101×, nearly three dozen additional dim suns can be counted inside an imaginary pentacle surrounding the pentagram. By alternating between averted and direct gazes, the cluster's crisp background stars seem to appear and disappear like lights blinking on a Christmas tree. Seen another way, the stars look like the stick figure of a man running. The cluster contains 65 members of 9th magnitude and fainter, but it lies in a rich Milky Way field.

Stop. Do not move the telescope. Your next target is nearby!

4. NGC 1528 (H VII-61)

Type	Con	RA	Dec	Mag	Diam	Rating
Open cluster	Perseus	04ʰ 15.3ᵐ	+51° 12′	6.2	18.0′	4

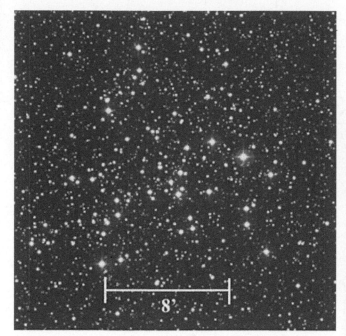

5. NGC 1444 (H VIII-80)

Type	Con	RA	Dec	Mag	Diam	Rating
Open cluster	Perseus	03^h 49.1^m	+52° 39′	6.6	4.0′	4

General description

NGC 1528 is 1° north–northwest of 5th-magnitude b Persei in the Champion's knee, and just 30′ northwest of a 7.5-magnitude star. NGC 1528 shines at a magnitude of 6.2 and makes a decent naked-eye challenge for anyone under dark skies. It is a fine object for binoculars, though, and its brightest members will start to appear to the dedicated observer in even the smallest pair. In 7×50 binoculars the cluster is round and well formed. Its brightest stars appear unmistakably in a small antique telescope against the cluster's unresolved background, which shines with a cometlike luster. It is a marvel in telescopes of all sizes.

Directions

Using Chart 83b, from NGC 1545 return to b Persei. NGC 1528 is just 1° to the north–northwest.

The quick view

At 23× in the 4-inch, NGC 1528 reveals a veil of well-resolved suns scintillating in front of a fuzzy backdrop of fainter stars. The brightest ones are curved in rows, like waves. NGC 1528's brightest stars form a teardop-shaped asterism that is elongated along a northwest-to-southeast axis; its narrow tip points to the southeast. A dark U-shaped bay can be seen at the cluster's western end, where it is outlined by seven bright stars. That U can also be extended a bit further to the southwest, making the asterism look like the Sickle of Leo. NGC 1528 comprises a healthy dose of mini-asterisms, all of which look hazy at low power but resolve into distinct geometrical patterns under 101×. The cluster has some 165 stars of 9th magnitude and fainter.

Stop. Do not move the telescope.

General description

NGC 1444 is a small and curious open cluster about 2$\frac{1}{4}$° northwest of 5.5-magnitude 43 Persei. It is dominated by a 7th-magnitude star that is easy to see in binoculars. Through a telescope, that star seems to "occult" a scattering of fainter cluster members to the northwest. Your main goal is to get that 7th-magnitude sun in your telescope.

Directions

Use Chart 83 to locate 43 Persei 1$\frac{1}{2}$° west–northwest of Lambda (λ) Persei. Center 43 Persei in your telescope at low power, then switch to Chart 83c. From 43 Persei, move 40′ west–northwest to a 20′-long triangle of 8th-magnitude suns (a). Next make a slow and careful sweep 1° northwest to a 25′-long arc of three 7.5-magnitude suns (b), then hop 15′ east–northeast of the northernmost star in Arc b to 7.5-magnitude Star c. Now move 45′ north–northeast to a pair of 7th-magnitude suns (d). NGC 1444 is kissing the northernmost star in Pair d.

The quick view

At 23× in the 4-inch NGC 1444 is a 7th-magnitude star abutting a curious milky haze, which, with averted vision, begins to resolve into patches of dim starlight. At 72×, the cluster is an insignificant agglomeration of about a dozen suns, some in pairs. The cluster has nearly 60 members, though of 7th magnitude and fainter, so it is better appreciated in large telescopes.

Star charts for seventh night

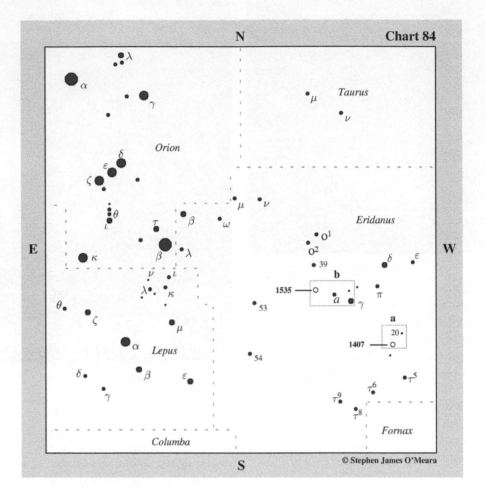

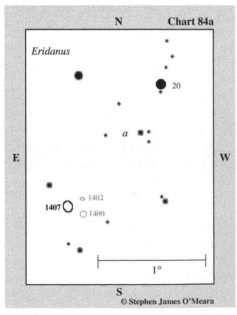

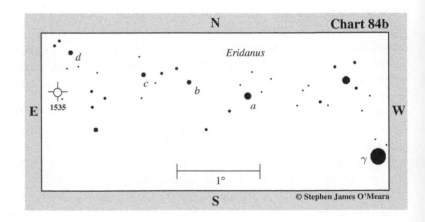

SEVENTH NIGHT

1.   NGC 1407 (H I-107)

Type	Con	RA	Dec	Mag	Dim	Rating
Elliptical galaxy	Eridanus	03^h 40.2^m	−18° 35′	9.7	6.0′× 5.8′	4

General description

NGC 1407 is a very nice and bright elliptical galaxy 1$\frac{1}{2}$° southeast of 5.5-magnitude 20 Eridani, which is 6$\frac{1}{2}$° south-west of 3rd-magnitude Gamma (γ) Eridani, which is 20° (two fists) west–southwest of Beta (β) Orionis (Rigel). The galaxy is a fine, yet simple, ellipse of light in telescopes of all sizes.

Directions

Use Chart 84 to locate 20 Eridani. Center that star in your telescope at low power, then switch to Chart 84a. From 20 Eridani, move a little less than 30′ south–southeast to 8th-magnitude Star a. NGC 1407 is 1° to the southeast.

The quick view

In the 4-inch at 23×, NGC 1047 is a bright elliptical haze, oriented north–northeast to south–southwest, about 1′ in extent. Its core is bright and starlike. The surrounding disk swells to nearly 2′ with averted vision. At 72×, the galaxy is much the same, though more obvious. Two fainter galaxies flank it – one to the northwest, the other to the southwest. A very subtle yet beautiful sight.

Stop. Do not move the telescope.

2.   NGC 1535 (H IV-26)

Type	Con	RA	Dec	Mag	Dim	Rating
Planetary nebula	Eridanus	04^h 14.3^m	−12° 44′	9.1	48″× 42″	4

General description

NGC 1535, Cleopatra's Eye, is one of the brightest planetary nebulae. If you observe under a dark sky, the nebula should be visible as a dim star in 7×50 binoculars. It is a remark-able object for telescopes of all sizes. Its visual appearance consists of a bright inner annulus with a much fainter, though uniformly bright, crown. An 11.6-magnitude star burns at the center of the shell.

Directions

Using Chart 83, from NGC 1407, return your gaze to Gamma (γ) Eridani. Center that star in your telescope at low power, then switch to Chart 83b. There is no mistaking Gamma Eridani, because it is the brightest star in the region. Train your binoculars on it, and you will see another elegant curve of three 6th-magnitude stars just north of it (note that only the two easternmost stars in that arc are shown on the chart). If you center the easternmost star a in that curve in your telescope, you will be within striking distance of NGC 1535, which is almost 2$\frac{1}{2}$° due east. Start by moving 45′ east–northeast to a solitary 8.5-magnitude Star b. Then hop 35′ east–northeast to 8th-magnitude Star c. A greater 55′ hop to the northeast will bring you to 8th-magnitude Star d. You'll find the 9th-magnitude planetary nebula just $\frac{1}{2}$° south–southeast of that sun.

The quick view

At 23× in the 4-inch, NGC 1535 is a tiny blue-gray disk, which is so nearly stellar that it's easy to pass over it in a sweep. With averted vision, it looks slightly fuzzy. At 72×, the nebula's round appearance and pale blue color are reminiscent of Neptune's disk, which is why I also call it the Ghost of Neptune Nebula. At 101× and higher powers, the structure in the annulus is complex. It's difficult to separate the bright central star from the equally bright inner annulus, which extends to about 20″ in diameter. So be sure to use very high power when you have the time to study this celestial wonder.

Stephen James O'Meara's

Herschel 400 Award

This Certifies that

_____

has observed all of the objects in the Herschel 400 list and
achieved the status of an advanced deep-sky observer

Date: _____

Stephen James O'Meara

Appendix A Herschel 400 observing list[a]

NGC	Type[b]	Con	RA	Dec	Mag	Size (')	Rating	Herschel No.
7814	GX	Peg	00^h 03.2^m	+16° 09′	10.6	5.5 × 2.3	2.5	H II-240
40	PN	Cep	00^h 13.0^m	+72° 31′	12.3	38″ × 35″	4	H IV-58
129	OC	Cas	00^h 29.9^m	+60° 13′	6.5	12.0	4	H VIII-79
136	OC	Cas	00^h 31.5^m	+61° 30′	~9th	1.5	3	H VI-35
157	GX	Cet	00^h 34.8^m	−08° 24′	10.4	4.0 × 2.4	3	H II-3
185	GX	Cas	00^h 39.0^m	+48° 20′	9.2	17.0 × 14.3	3.5	H II-707
205	GX	And	00^h 40.4^m	+41° 41′	8.0	21.9 × 11.0	4	H V-18 = M110
225	OC	Cas	00^h 43.6^m	+61° 46′	7.0	15.0	4	H VIII-78
246	PN	Cet	00^h 47.0^m	−11° 52′	10.9	4.6 × 4.1	3.5	H V-25
247	GX	Cet	00^h 47.1^m	−20° 46′	8.9	22.2 × 6.7	3.5	H V-20
253	GX	Scl	00^h 47.6^m	−25° 17′	7.1	25.8 × 5.9	4	H V-1
278	GX	Cas	00^h 52.1^m	+47° 33′	10.8	2.6 × 2.6	3.5	H I-159
288	GC	Scl	00^h 52.8^m	−26° 35′	7.9	13.0	4	H VI-20
381	OC	Cas	01^h 08.3^m	+61° 35′	9.3	7.0	3	H VIII-64
404	GX	And	01^h 09.4^m	+35° 43′	9.8	6.6 × 6.6	4	H II-224
436	OC	Cas	01^h 15.9^m	+58° 49′	8.8	5.0	3.5	H VII-45
457	OC	Cas	01^h 19.5^m	+58° 17′	6.4	20.0	4	H VII-42
488	GX	Pis	01^h 21.8^m	+05° 15′	10.3	5.5 × 4.0	4	H III-252
524	GX	Pis	01^h 24.8^m	+09° 32′	10.2	3.5 × 3.5	2.5	H I-151
559	OC	Cas	01^h 29.5^m	+63° 18′	9.5	7.0	3	H VII-48
584	GX	Cet	01^h 31.3^m	−06° 52′	10.4	3.8 × 2.4	4	H I-100
596	GX	Cet	01^h 32.9^m	−07° 02′	10.9	2.7 × 2.0	3.5	H II-4
598	GX	Tri	01^h 33.9^m	+30° 39′	5.7	71 × 42	5	H V-17 = M33
613	GX	Scl	01^h 34.3^m	−29° 25′	10.0	5.2 × 2.6	3	H I-281
615	GX	Cet	01^h 35.1^m	−07° 20′	11.6	2.5 × 1.3	3	H II-282
651	PN	Pcr	01^h 42.4^m	+51° 34′	10.1	~35″	3.5	H I-193 = M76
637	OC	Cas	01^h 43.1^m	+64° 02′	8.2	3.0	3.5	H VII-49
654	OC	Cas	01^h 44.0^m	+61° 53′	6.5	6.0	4	H VII-46
659	OC	Cas	01^h 44.4^m	+60° 42′	8.2	6.0	3.5	H VIII-65
663	OC	Cas	01^h 46.3^m	+61° 13′	6.7	15.0	4	H VI-31

NGC	Type[b]	Con	RA	Dec	Mag	Size (')	Rating	Herschel No.
720	GX	Cet	$01^h 53.0^m$	$-13° 44'$	10.2	4.3×2.0	3.5	H I-105
752	OC	And	$01^h 57.6^m$	$+37° 50'$	5.7	75.0	4	H VII-32
772	GX	Ari	$01^h 59.3^m$	$+19° 00'$	9.9	7.1×4.7	3.5	H I-112
779	GX	Cet	$01^h 59.7^m$	$-05° 58'$	11.2	3.4×1.2	3	H I-101
869	OC	Per	$02^h 19.0^m$	$+57° 08'$	4.5	~18	5	H VI-33
884	OC	Per	$02^h 22.0^m$	$+57° 08'$	5.7	18.0	5	H VI-34
891	GX	And	$02^h 22.6^m$	$+42° 21'$	9.9	12.2×3.0	3.5	H V-19
908	GX	Cet	$02^h 23.1^m$	$-21° 14'$	10.4	6.1×2.7	3.5	H I-153
936	GX	Cet	$02^h 27.6^m$	$-01° 09'$	10.2	5.7×4.6	4	H IV-23
1022	GX	Cet	$02^h 38.5^m$	$-06° 40'$	11.3	2.7×2.7	2	H I-102
1023	GX	Per	$02^h 40.4^m$	$+39° 04'$	9.3	7.5×3.0	4	H I-156
1052	GX	Cet	$02^h 41.1^m$	$-08° 15'$	10.5	2.5×2.0	4	H I-63
1055	GX	Cet	$02^h 41.8^m$	$+00° 26'$	10.6	7.3×3.3	2.5	H II-6
1027	OC	Cas	$02^h 42.6^m$	$+61° 36'$	6.7	15.0	4	H VIII-66
1084	GX	Eri	$02^h 46.0^m$	$-07° 35'$	10.7	3.2×1.9	4	H I-64
1245	OC	Per	$03^h 14.7^m$	$+47° 14'$	8.4	10.0	3.5	H VI-25
1342	OC	Per	$03^h 31.7^m$	$+37° 22'$	6.7	17.0	4	H VIII-88
1407	GX	Eri	$03^h 40.2^m$	$-18° 35'$	9.7	6.0×5.8	4	H I-107
1444	OC	Per	$03^h 49.1^m$	$+52° 39'$	6.6	4.0	4	H VIII-80
1501	PN	Cam	$04^h 07.0^m$	$+60° 55'$	10.6	$56'' \times 48''$	4	H IV-53
1502	OC	Cam	$04^h 07.8^m$	$+62° 20'$	6.0	20.0	5	H VII-47
1513	OC	Per	$04^h 09.9^m$	$+49° 31'$	8.4	12.0	3	H VII-60
1535	PN	Eri	$04^h 14.3^m$	$-12° 44'$	9.1	$48'' \times 42''$	4	H IV-26
1528	OC	Per	$04^h 15.3^m$	$+51° 12'$	6.2	18.0	4	H VII-61
1545	OC	Per	$04^h 20.9^m$	$+50° 15'$	6.2	12.0	4	H VIII-85
1647	OC	Tau	$04^h 45.7^m$	$+19° 07'$	6.2	40.0	4	H VIII-8
1664	OC	Aur	$04^h 51.1^m$	$+43° 41'$	7.6	18.0	3	H VIII-59
1788	BN	Ori	$05^h 06.9^m$	$-03° 21'$	–	5×3	3	H V-32
1817	OC	Tau	$05^h 12.4^m$	$+16° 41'$	7.7	20.0	3	H VII-4
1857	OC	Aur	$05^h 20.1^m$	$+39° 21'$	7.0	10.0	?	H VII-33
1907	OC	Aur	$05^h 28.1^m$	$+35° 19.5'$	8.2	5.0	3	H VII-39
1931	BN	Aur	$05^h 31.4^m$	$+34° 15'$	–	4.0	3	H I-261
1964	GX	Lep	$05^h 33.4^m$	$-21° 57'$	10.7	5.0×2.1	3	H IV-21
1980	BN	Ori	$05^h 35.4^m$	$-05° 54'$	–	14×14	3	H V-31

NGC	Type[b]	Con	RA	Dec	Mag	Size (')	Rating	Herschel No.
1999	BN	Ori	$05^h\,36.5^m$	$-06°\,42'$	9.5	2×2	3	H IV-33
2024	BN	Ori	$05^h\,41.9^m$	$-01°\,51'$	7th	30×30	4	H V-28
1961	GX	Cam	$05^h\,42.1^m$	$+69°\,23'$	11.0	4.2×4.0	1	H III-747
2022	PN	Ori	$05^h\,42.1^m$	$+09°\,05'$	11.9	$28'' \times 27''$	2	H IV-34
2129	OC	Gem	$06^h\,01.1^m$	$+23°\,19'$	6.7	6.0	4	H VIII-26
2126	OC	Aur	$06^h\,02.6^m$	$+49°\,52'$	10.2	6.0	2	H VIII-68
2158	OC	Gem	$06^h\,07.4^m$	$+24°\,06'$	8.6	5.0	3	H VI-17
2169	OC	Ori	$06^h\,08.4^m$	$+13°\,58'$	5.9	6.0	4	H VIII-24
2185	BN	Mon	$06^h\,11.1^m$	$-06°\,13'$	–	1.5×1.5	2	H IV-20
2186	OC	Ori	$06^h\,12.1^m$	$+05°\,27.5'$	8.7	5.0	2	H VII-25
2194	OC	Ori	$06^h\,13.8^m$	$+12°\,48'$	8.5	9.0	2.5	H VI-5
2204	OC	CMa	$06^h\,15.5^m$	$-18°\,40'$	8.6	10.0	2	H VII-13
2215	OC	Mon	$06^h\,20.8^m$	$-07°\,17'$	8.4	8.0	3	H VII-20
2232	OC	Mon	$06^h\,27.2^m$	$-04°\,45'$	4.2	53.0	5	H VIII-25
2244	OC	Mon	$06^h\,32.3^m$	$+04°\,51'$	4.8	30.0	5	H VII-2
2251	OC	Mon	$06^h\,34.6^m$	$+08°\,22'$	7.3	10.0	3	H VIII-3
2264	OC	Mon	$06^h\,41.0^m$	$+09°\,54'$	4.4	40.0	5	H VIII-5
2266	OC	Gem	$06^h\,43.3^m$	$+26°\,58'$	9.5	5.0	3	H VI-21
2286	OC	Mon	$06^h\,47.7^m$	$-03°\,09'$	7.5	15.0	3	H VIII-31
2281	OC	Aur	$06^h\,48.3^m$	$+41°\,05'$	5.4	25.0	4	H VIII-71
2301	OC	Mon	$06^h\,51.8^m$	$+00°\,28'$	6.0	15.0	4	H VI-27
2304	OC	Gem	$06^h\,55.2^m$	$+17°\,59'$	10.0	3.0	2	H VI-2
2311	OC	Mon	$06^h\,57.8^m$	$-04°\,37'$	9.6	7.0	3	H VIII-60
2324	OC	Mon	$07^h\,04.1^m$	$+01°\,03'$	8.4	8.0	2	H VII-38
2335	OC	Mon	$07^h\,06.8^m$	$-10°\,02'$	7.2	7.0	4	H VIII-32
2343	OC	Mon	$07^h\,08.1^m$	$-10°\,37'$	6.7	6.0	4	H VIII-33
2354	OC	CMa	$07^h\,14.3^m$	$-25°\,41'$	6.5	18.0	3	H VII-16
2353	OC	Mon	$07^h\,14.5^m$	$-10°\,16'$	7.1	18.0	5	H VIII-34
2355	OC	Gem	$07^h\,17.0^m$	$+13°\,45'$	9.7	8.0	3	H VI-6
2360	OC	CMa	$07^h\,17.7^m$	$-15°\,38'$	7.2	14.0	4	H VII-12
2362	OC	CMa	$07^h\,18.7^m$	$-24°\,57'$	3.8	6.0	4	H VII-17
2371–2	PN	Gem	$07^h\,25.6^m$	$+29°\,29'$	11.3	$>55''$	1	H II-316/17
2395	OC	Gem	$07^h\,27.2^m$	$+13°\,36'$	8.0	15.0	3	H VIII-11
2392	PN	Gem	$07^h\,29.2^m$	$+20°\,55'$	9.2	$47'' \times 43''$	5	H IV-45

NGC	Type[b]	Con	RA	Dec	Mag	Size (')	Rating	Herschel No.
2421	OC	Pup	07^h 36.2^m	−20° 37′	8.3	8.0	3	H VII-67
2422	OC	Pup	07^h 36.6^m	−14° 29′	4.4	25.0	5	H VIII-38 = M47
2403	GX	Cam	07^h 36.9^m	+65° 36′	8.5	25.5 × 13.0	4	H V-44
2423	OC	Pup	07^h 37.1^m	−13° 52′	6.7	12.0	4	H VII-28
2419	GC	Lyn	07^h 38.1^m	+38° 52′	10.3	4.6	2	H I-218
2420	OC	Gem	07^h 38.4^m	+21° 34′	8.3	6.0	2	H VI-1
2438	PN	Pup	07^h 41.8^m	−14° 44′	11.0	>66″	4	H IV-39
2440	PN	Pup	07^h 41.9^m	−18° 12′	9.1	74″ × 42″	3.5	H IV-64
2479	OC	Pup	07^h 55.1^m	−17° 42′	9.6	11.0	3	H VII-58
2482	OC	Pup	07^h 55.2^m	−24° 15′	7.3	10.0	3	H VII-10
2489	OC	Pup	07^h 56.2^m	−30° 04′	7.9	5.0	4	H VII-23
2506	OC	Mon	08^h 00.0^m	−10° 46′	7.6	12.0	4	H VI-37
2509	OC	Pup	08^h 00.8^m	−19° 03′	9.3	12.0	3	H VIII-1
2527	OC	Pup	08^h 04.9^m	−28° 08′	6.5	10.0	4	H VIII-30
2539	OC	Pup	08^h 10.6^m	−12° 49′	6.5	15.0	4	H VII-11
2548	OC	Hya	08^h 13.7^m	−05° 45′	5.8	30.0	5	H VI-22 = M48
2567	OC	Pup	08^h 18.5^m	−30° 39′	7.4	11.0	3	H VII-64
2571	OC	Pup	08^h 18.9^m	−29° 45′	7.0	7.0	3	H VI-39
2613	GX	Pyx	08^h 33.4^m	−22° 58′	10.5	7.6 × 1.9	3	H II-266
2627	OC	Pyx	08^h 37.2^m	−29° 57′	8.4	9.0	3	H VII-63
2683	GX	Lyn	08^h 52.7^m	+33° 25′	9.7	9.1 × 2.7	5	H I-200
2681	GX	UMa	08^h 53.5^m	+51° 19′	10.3	3.5 × 3.5	3	H I-242
2655	GX	Cam	08^h 55.6^m	+78° 13′	10.2	5.9 × 5.3	4	H I-288
2742	GX	UMa	09^h 07.6^m	+60° 29′	11.4	3.0 × 1.6	2	H I-249
2775	GX	Can	09^h 10.3^m	+07° 02′	10.1	4.6 × 3.7	4	H I-2
2768	GX	UMa	09^h 11.6^m	+60° 02′	9.9	6.6 × 3.2	4	H I-250
2782	GX	Lyn	09^h 14.1^m	+40° 07′	11.6	3.8 × 2.9	2	H I-167
2811	GX	Hya	09^h 16.2^m	−16° 19′	11.3	1.9 × 0.6	2	H II-505
2787	GX	UMa	09^h 19.3^m	+69° 12′	10.8	3.5 × 2.4	2	H I-216
2841	GX	UMa	09^h 22.0^m	+50° 59′	9.0	6.6 × 3.1	5	H I-205
2859	GX	LMi	09^h 24.3^m	+34° 31′	10.9	4.6 × 4.1	2	H I-137
2903	GX	Leo	09^h 32.2^m	+21° 30′	9.0	11.6 × 5.7	4	H I-56
2950	GX	UMa	09^h 42.6^m	+58° 51′	10.9	3.3 × 2.4	3	H IV-68
2974	GX	Sex	09^h 42.6^m	−03° 42′	10.9	3.0 × 1.7	2	H I-61

NGC	Type[b]	Con	RA	Dec	Mag	Size (')	Rating	Herschel No.
2964	GX	Leo	09^h 42.9^m	+31° 51′	11.3	3.2 × 1.8	2	H I-114
2976	GX	UMa	09^h 47.3^m	+67° 55′	10.2	5.0 × 2.8	2	H I-285
2985	GX	UMa	09^h 50.4^m	+72° 17′	10.4	3.9 × 3.0	3	H I-78
3034	GX	UMa	09^h 55.8^m	+69° 41′	8.4	11.2 × 4.3	5	H IV-79
3079	GX	UMa	10^h 02.0^m	+55° 41′	10.9	8.0 × 1.5	3	H V-47
3077	GX	UMa	10^h 03.3^m	+68° 44′	9.8	5.5 × 4.1	4	H I-286
3115	GX	Sex	10^h 05.2^m	−07° 43′	8.9	6.9 × 3.4	5	H I-163
3166	GX	Sex	10^h 13.8^m	+03° 26′	10.4	4.6 × 2.6	4	H I-3
3169	GX	Sex	10^h 14.2^m	+03° 28′	10.2	5.0 × 2.8	3	H I-4
3147	GX	Dra	10^h 16.9^m	+73° 24′	10.6	4.3 × 3.7	2	H I-79
3190	GX	Leo	10^h 18.1^m	+21° 50′	11.2	4.1 × 1.6	2	H II-44
3184	GX	UMa	10^h 18.3^m	+41° 25′	9.4	7.5 × 7.0	4	H I-168
3193	GX	Leo	10^h 18.4^m	+21° 54′	10.9	2.5 × 2.5	2	H II-45
3198	GX	UMa	10^h 19.9^m	+45° 33′	10.3	9.2 × 3.5	3	H I-199
3226	GX	Leo	10^h 23.4^m	+19° 54′	11.4	2.5 × 2.2	3	H II-28
3227	GX	Leo	10^h 23.5^m	+19° 52′	10.3	6.9 × 5.4	4	H II-29
3242	PN	Hya	10^h 24.8^m	−18° 38′	7.3	45″ × 36″	5	H IV-27
3245	GX	LMi	10^h 27.3^m	+28° 30′	10.8	2.9 × 2.0	3	H I-86
3277	GX	LMi	10^h 32.9^m	+28° 31′	11.7	2.2 × 2.0	2	H II-359
3294	GX	LMi	10^h 36.3^m	+37° 20′	11.8	3.5 × 1.7	2	H I-164
3310	GX	UMa	10^h 38.7^m	+53° 30′	10.8	3.5 × 3.2	4	H IV-60
3344	GX	LMi	10^h 43.5^m	+24° 55′	9.3	6.7 × 6.3	4	H I-81
3377	GX	Leo	10^h 47.7^m	+13° 59′	10.4	4.1 × 2.6	3	H II-99
3379	GX	Leo	10^h 47.8^m	+12° 35′	9.3	3.9 × 3.9	5	H I-17 = M105
3384	GX	Leo	10^h 48.3^m	+12° 38′	9.9	5.5 × 2.9	4	H I-18
3395	GX	LMi	10^h 49.8^m	+32° 59′	12.1	1.6 × 0.9	2	H I-116
3412	GX	Leo	10^h 50.9^m	+13° 25′	10.5	3.3 × 2.0	3	H I-27
3414	GX	LMi	10^h 51.3^m	+27° 59′	11.0	3.0 × 2.7	2	H II-362
3432	GX	LMi	10^h 52.5^m	+36° 37′	11.2	6.9 × 1.9	4	H I-172
3489	GX	Leo	11^h 00.3^m	+13° 54′	10.3	3.2 × 2.0	3	H II-101
3486	GX	LMi	11^h 00.4^m	+28° 58′	10.5	6.6 × 4.7	4	H I-87
3504	GX	LMi	11^h 03.2^m	+27° 58′	10.9	2.3 × 2.3	3	H I-88
3521	GX	Leo	11^h 05.8^m	−00° 02′	9.1	11.7 × 6.5	4	H I-13
3556	GX	UMa	11^h 11.5^m	+55° 40′	10.0	8.1 × 2.1	4	H V-46 = M108

NGC	Type[b]	Con	RA	Dec	Mag	Size (')	Rating	Herschel No.
3593	GX	Leo	$11^h 14.6^m$	$+12° 49'$	10.9	5.3×2.2	3	H I-29
3607	GX	Leo	$11^h 16.9^m$	$+18° 03'$	9.9	4.6×4.1	4	H II-50
3608	GX	Leo	$11^h 17.0^m$	$+18° 09'$	10.8	2.7×2.3	3	H II-51
3621	GX	Hya	$11^h 18.3^m$	$-32° 49'$	8.5	14.9×7.4	4	H I-241
3610	GX	UMa	$11^h 18.4^m$	$+58° 47'$	10.8	3.2×3.2	4	H I-270
3613	GX	UMa	$11^h 18.6^m$	$+58° 00'$	10.9	3.4×1.9	3	H I-271
3619	GX	UMa	$11^h 19.4^m$	$+57° 46'$	11.5	3.7×2.8	2	H I-244
3626	GX	Leo	$11^h 20.1^m$	$+18° 21'$	11.0	2.6×1.8	2	H II-52
3628	GX	Leo	$11^h 20.3^m$	$+13° 35'$	9.5	14.8×3.3	4	H V-8
3631	GX	UMa	$11^h 21.0^m$	$+53° 10'$	10.4	5.5×4.6	2	H I-226
3640	GX	Leo	$11^h 21.1^m$	$+03° 14'$	10.4	4.6×4.1	4	H II-33
3655	GX	Leo	$11^h 22.9^m$	$+16° 35'$	11.7	1.5×0.9	2	H I-5
3665	GX	UMa	$11^h 24.7^m$	$+38° 46'$	10.8	3.5×3.1	3	H I-219
3675	GX	UMa	$11^h 26.1^m$	$+43° 35'$	10.2	6.2×3.2	4	H I-194
3686	GX	Leo	$11^h 27.7^m$	$+17° 13'$	11.3	2.8×2.3	1	H II-160
3726	GX	UMa	$11^h 33.3^m$	$+47° 02'$	10.4	5.6×3.8	3	H II-730
3729	GX	UMa	$11^h 33.8^m$	$+53° 08'$	11.4	3.1×2.2	2	H I-222
3810	GX	Leo	$11^h 41.0^m$	$+11° 28'$	10.8	3.8×2.6	3	H I-21
3813	GX	UMa	$11^h 41.3^m$	$+36° 33'$	11.7	1.9×1.1	2	H I-94
3877	GX	UMa	$11^h 46.1^m$	$+47° 30'$	11.0	5.1×1.1	2	H I-201
3893	GX	UMa	$11^h 48.6^m$	$+48° 43'$	10.5	4.2×2.3	4	H II-738
3898	GX	UMa	$11^h 49.2^m$	$+56° 05'$	10.7	3.3×1.9	2	H I-228
3900	GX	Leo	$11^h 49.2^m$	$+27° 01'$	11.3	2.9×1.5	1.5	H I-82
3912	GX	Leo	$11^h 50.1^m$	$+26° 29'$	12.4	1.6×0.9	1	H II-342
3938	GX	UMa	$11^h 52.8^m$	$+44° 07'$	10.4	4.9×4.7	3	H I-203
3941	GX	UMa	$11^h 52.9^m$	$+36° 59'$	10.3	3.7×2.6	4	H I-173
3945	GX	UMa	$11^h 53.2^m$	$+60° 41'$	10.8	5.9×3.7	3	H I-251
3949	GX	UMa	$11^h 53.7^m$	$+47° 52'$	11.1	2.6×1.6	2	H I-202
3953	GX	UMa	$11^h 53.8^m$	$+52° 20'$	10.1	6.0×3.2	4	H V-45
3962	GX	Crt	$11^h 54.7^m$	$-13° 58'$	10.7	2.6×2.2	2.5	III 67
3982	GX	UMa	$11^h 56.5^m$	$+55° 08'$	11.0	2.2×2.0	2	H IV-62
3992	GX	UMa	$11^h 57.6^m$	$+53° 23'$	9.8	7.6×4.3	4	H IV-61 = M109
3998	GX	UMa	$11^h 57.9^m$	$+55° 27'$	10.7	3.0×2.6	3	H I-229
4026	GX	UMa	$11^h 59.4^m$	$+50° 58'$	10.8	4.6×1.2	2	H I-223

NGC	Type[b]	Con	RA	Dec	Mag	Size (')	Rating	Herschel No.
4027	GX	Crv	$11^h 59.5^m$	$-19° 16'$	11.2	3.8×2.3	2	H II-296
4030	GX	Vir	$12^h 00.4^m$	$-01° 06'$	10.6	3.8×2.9	3	H I-121
4036	GX	UMa	$12^h 01.4^m$	$+61° 54'$	10.7	3.8×1.9	3	H I-253
4038	GX	Crv	$12^h 01.9^m$	$-18° 52'$	10.5	11.2×5.9	3.5	H IV-28
4041	GX	UMa	$12^h 02.2^m$	$+62° 08'$	11.3	2.6×2.6	2	H I-252
4051	GX	UMa	$12^h 03.2^m$	$+44° 32'$	10.2	5.5×4.6	4	H IV-56
4085	GX	UMa	$12^h 05.4^m$	$+50° 21'$	12.4	2.5×0.8	2	H I-224
4088	GX	UMa	$12^h 05.6^m$	$+50° 33'$	10.6	5.4×2.1	2	H I-206
4102	GX	UMa	$12^h 06.4^m$	$+52° 43'$	11.2	2.9×1.8	3	H I-225
4111	GX	CVn	$12^h 07.1^m$	$+43° 04'$	10.7	4.4×0.9	3	H I-195
4143	GX	CVn	$12^h 09.6^m$	$+42° 32'$	10.7	2.9×1.9	2.5	H IV-54
4147	GC	Com	$12^h 10.1^m$	$+18° 32'$	10.4	4.0	3	H I-19
4151	GX	CVn	$12^h 10.5^m$	$+39° 24'$	10.8	6.4×5.5	3	H I-165
4150	GX	Com	$12^h 10.6^m$	$+30° 24'$	11.6	2.1×1.5	2.5	H I-73
4179	GX	Vir	$12^h 12.9^m$	$+01° 18'$	11.0	3.9×1.1	2.5	H I-9
4203	GX	Com	$12^h 15.1^m$	$+33° 12'$	10.9	3.5×3.4	2	H I-175
4214	GX	CVn	$12^h 15.6^m$	$+36° 20'$	9.1	9.6×8.1	4	H I-95
4216	GX	Vir	$12^h 15.9^m$	$+13° 09'$	10.0	7.9×1.7	4	H I-35
4245	GX	Com	$12^h 17.6^m$	$+29° 36'$	11.4	3.2×3.0	2	H I-74
4251	GX	Com	$12^h 18.1^m$	$+28° 10'$	10.7	3.7×2.1	3	H I-89
4258	GX	CVn	$12^h 19.0^m$	$+47° 18'$	8.4	20.0×8.4	4	H V-43 = M106
4261	GX	Vir	$12^h 19.4^m$	$+05° 49'$	10.4	3.5×3.1	2	H II-139
4274	GX	Com	$12^h 19.8^m$	$+29° 37'$	10.4	6.7×2.5	3	H I-75
4273	GX	Vir	$12^h 19.9^m$	$+05° 21'$	11.9	2.3×1.1	2	H II-569
4278	GX	Com	$12^h 20.1^m$	$+29° 17'$	10.2	3.5×3.5	3.5	H I-90
4281	GX	Vir	$12^h 20.4^m$	$+05° 23'$	11.3	2.5×1.3	2	H II-573
4293	GX	Com	$12^h 21.2^m$	$+18° 23'$	10.4	5.3×3.1	3.5	H V-5
4303	GX	Vir	$12^h 21.9^m$	$+04° 28'$	9.7	6.0×5.9	4	H I-139 = M61
4314	GX	Com	$12^h 22.6^m$	$+29° 53'$	10.6	4.2×4.1	2.5	H I-76
4346	GX	CVn	$12^h 23.5^m$	$+47° 00'$	11.1	3.2×1.4	3	H I-210
4350	GX	Com	$12^h 24.0^m$	$+16° 42'$	11.0	2.5×1.0	3.5	H II-86
4365	GX	Vir	$12^h 24.5^m$	$+07° 19'$	9.6	5.6×4.6	3.5	H I-30
4361	PN	Crv	$12^h 24.5^m$	$-18° 47'$	10.2	1.9×1.9	4	H I-65
4371	GX	Vir	$12^h 24.9^m$	$+11° 42'$	10.8	4.6×2.2	2.5	H I-22

NGC	Type[b]	Con	RA	Dec	Mag	Size (')	Rating	Herschel No.
4394	GX	Com	$12^h 25.9^m$	$+18° 13'$	10.9	3.3×3.1	3	H II-55
4414	GX	Com	$12^h 26.4^m$	$+31° 13'$	10.1	4.4×3.0	4	H I-77
4419	GX	Com	$12^h 26.9^m$	$+15° 03'$	11.2	2.8×0.9	2	H II-113
4429	GX	Vir	$12^h 27.4^m$	$+11° 07'$	10.0	5.6×2.6	3.5	H II-65
4435	GX	Vir	$12^h 27.7^m$	$+13° 05'$	10.8	3.2×2.0	3.5	H I-28,1
4438	GX	Vir	$12^h 27.8^m$	$+13° 01'$	10.2	8.9×3.6	4	H I-28,2
4442	GX	Vir	$12^h 28.1^m$	$+09° 48'$	10.4	4.6×1.9	3.5	H II-156
4449	GX	CVn	$12^h 28.2^m$	$+44° 06'$	9.6	5.5×4.1	4	H I-213
4448	GX	Com	$12^h 28.2^m$	$+28° 37'$	11.1	3.7×1.4	3	H I-91
4450	GX	Com	$12^h 28.5^m$	$+17° 05'$	10.1	5.0×3.4	4	H II-56
4459	GX	Com	$12^h 29.0^m$	$+13° 59'$	10.4	3.5×2.8	3.5	H I-169
4473	GX	Com	$12^h 29.8^m$	$+13° 26'$	10.2	3.7×2.4	4	H II-114
4477	GX	Com	$12^h 30.0^m$	$+13° 38'$	10.4	3.9×3.6	4	H II-115
4478	GX	Vir	$12^h 30.3^m$	$+12° 20'$	11.4	1.7×1.4	2.5	H II-124
4485	GX	CVn	$12^h 30.5^m$	$+41° 42'$	11.9	2.7×2.3	3	H I-197
4490	GX	CVn	$12^h 30.6^m$	$+41° 39'$	9.5	5.6×2.8	4	H I-198
4494	GX	Com	$12^h 31.4^m$	$+25° 47'$	9.8	4.6×4.4	3	H I-83
4526	GX	Vir	$12^h 34.0^m$	$+07° 42'$	9.9	7.4×2.7	4	H I-31 = H I-38
4527	GX	Vir	$12^h 34.1^m$	$+02° 39'$	10.5	6.0×2.1	2.5	H II-37
4535	GX	Vir	$12^h 34.3^m$	$+08° 12'$	10.0	7.1×6.4	3.5	H II-500
4548	GX	Com	$12^h 35.4^m$	$+14° 30'$	10.2	5.0×4.1	4	H II-120 = M91
4536	GX	Vir	$12^h 34.5^m$	$+02° 11'$	10.6	6.4×2.6	3	H V-2
4550	GX	Vir	$12^h 35.5^m$	$+12° 13'$	11.7	3.3×1.0	1.5	H I-36
4546	GX	Vir	$12^h 35.5^m$	$-03° 48'$	10.3	3.2×1.4	3.5	H I-160
4559	GX	Com	$12^h 36.0^m$	$+27° 58'$	10.0	11.3×5.0	4	H I-92
4565	GX	Com	$12^h 36.3^m$	$+25° 59'$	9.6	16.2×2.3	4	H V-24
4570	GX	Vir	$12^h 36.9^m$	$+07° 15'$	10.9	4.3×1.3	2	H I-32
4596	GX	Vir	$12^h 39.9^m$	$+10° 11'$	10.4	4.6×4.1	3.5	H I-24
4594	GX	Vir	$12^h 40.0^m$	$-11° 37'$	8.0	7.1×4.4	4	H I-43 = M104
4618	GX	CVn	$12^h 41.5^m$	$+41° 09'$	10.8	4.1×3.2	3	H I-178
4631	GX	CVn	$12^h 42.1^m$	$+32° 32'$	9.2	14.7×3.5	4	H V-42
4636	GX	Vir	$12^h 42.8^m$	$+02° 41'$	9.5	7.1×5.2	4	H II-38
4643	GX	Vir	$12^h 43.3^m$	$+01° 59'$	10.8	3.0×3.0	3	H I-10
4656	GX	CVn	$12^h 44.0^m$	$+32° 10'$	10.5	18.8×3.2	3	H I-176

NGC	Type[b]	Con	RA	Dec	Mag	Size (′)	Rating	Herschel No.
4654	GX	Vir	$12^h\,44.0^m$	$+13°\,08′$	10.5	4.9×2.7	3	H II-126
4660	GX	Vir	$12^h\,44.5^m$	$+11°\,11′$	11.2	2.4×2.1	2.5	H II-71
4665	GX	Vir	$12^h\,45.1^m$	$+03°\,03′$	10.5	4.1×4.1	3.5	H I-142
4666	GX	Vir	$12^h\,45.1^m$	$-00°\,28′$	10.7	4.1×1.3	3.5	H I-15
4689	GX	Com	$12^h\,47.8^m$	$+13°\,46′$	10.9	3.7×3.2	3	H II-128
4698	GX	Vir	$12^h\,48.4^m$	$+08°\,29′$	10.6	3.2×1.7	3.5	H I-8
4697	GX	Vir	$12^h\,48.6^m$	$-05°\,48′$	9.0	7.1×5.4	4	H I-39
4699	GX	Vir	$12^h\,49.0^m$	$-08°\,40′$	9.5	3.1×2.5	4	H I-129
4725	GX	Com	$12^h\,50.4^m$	$+25°\,30′$	9.2	10.5×8.1	4	H I-84
4754	GX	Vir	$12^h\,52.3^m$	$+11°\,19′$	10.6	4.6×2.6	3.5	H I-25
4753	GX	Vir	$12^h\,52.4^m$	$-01°\,12′$	9.9	4.1×2.3	4	H I-16
4762	GX	Vir	$12^h\,52.9^m$	$+11°\,14′$	10.3	9.1×2.2	3.5	H II-75
4781	GX	Vir	$12^h\,54.4^m$	$-10°\,32′$	11.1	2.9×1.3	2	H I-134
4800	GX	CVn	$12^h\,54.6^m$	$+46°\,32′$	11.5	1.6×1.1	2.5	H I-211
4845	GX	Vir	$12^h\,58.0^m$	$+01°\,35′$	11.2	4.8×1.2	1.5	H II-536
4856	GX	Vir	$12^h\,59.3^m$	$-15°\,02′$	10.5	3.1×0.9	3	H I-68
4866	GX	Vir	$12^h\,59.5^m$	$+14°\,10′$	11.2	5.5×1.2	3.5	H I-162
4900	GX	Vir	$13^h\,00.6^m$	$+02°\,30′$	11.4	2.3×2.3	2.5	H I-143
4958	GX	Vir	$13^h\,05.8^m$	$-08°\,01′$	10.7	3.6×1.4	3	H I-130
4995	GX	Vir	$13^h\,09.7^m$	$-07°\,50′$	11.1	2.5×1.8	2	H I-42
5005	GX	CVn	$13^h\,10.9^m$	$+37°\,03′$	9.8	5.8×2.8	4	H I-96
5033	GX	CVn	$13^h\,13.4^m$	$+36°\,36′$	10.2	10.5×5.1	3.5	H I-97
5054	GX	Vir	$13^h\,17.0^m$	$-16°\,38′$	10.9	4.8×2.8	2.5	H II-513
5195	GX	CVn	$13^h\,30.0^m$	$+47°\,16′$	9.6	6.4×4.6	4	H I-186
5248	GX	Boo	$13^h\,37.5^m$	$+08°\,53′$	10.3	6.2×4.6	3.5	H I-34
5273	GX	CVn	$13^h\,42.1^m$	$+35°\,39′$	11.6	2.8×2.4	2	H I-98
5322	GX	UMa	$13^h\,49.3^m$	$+60°\,12′$	10.2	6.1×4.1	3.5	H I-256
5363	GX	Vir	$13^h\,56.1^m$	$+05°\,15′$	10.1	4.7×3.2	4	H I-6
5364	GX	Vir	$13^h\,56.2^m$	$+05°\,01′$	10.5	6.6×5.1	3	H II-534
5473	GX	UMa	$14^h\,04.7^m$	$+54°\,54′$	11.4	2.2×1.7	2	H I-231
5474	GX	UMa	$14^h\,05.0^m$	$+53°\,40′$	10.8	6.0×4.9	2	H I-214
5466	GC	Boo	$14^h\,05.5^m$	$+28°\,32′$	9.2	9.0	3.5	H VI-9
5557	GX	Boo	$14^h\,18.4^m$	$+36°\,30′$	11.0	2.2×2.0	2	H I-99
5566	GX	Vir	$14^h\,20.3^m$	$+03°\,56′$	10.6	5.7×2.1	3.5	H I-144

NGC	Type[b]	Con	RA	Dec	Mag	Size (′)	Rating	Herschel No.
5576	GX	Vir	$14^h\,21.1^m$	$+03°\,16'$	11.0	3.0×2.4	2.5	H I-146
5631	GX	UMa	$14^h\,26.6^m$	$+56°\,35'$	11.5	1.8×1.8	2	H I-236
5634	GC	Vir	$14^h\,29.6^m$	$-05°\,59'$	9.5	5.5	3.5	H I-70
5676	GX	Boo	$14^h\,32.8^m$	$+49°\,28'$	11.2	3.7×1.6	2	H I-189
5689	GX	Boo	$14^h\,35.5^m$	$+48°\,45'$	11.9	3.7×1.0	1.5	H I-188
5694	GC	Hya	$14^h\,39.6^m$	$-26°\,32'$	10.2	4.3	3	H II-196
5746	GX	Vir	$14^h\,44.9^m$	$+01°\,57'$	9.3	8.1×1.4	3.5	H I-126
5846	GX	Vir	$15^h\,06.4^m$	$+01°\,36'$	10.0	3.0×3.0	3	H I-128
5866	GX	Dra	$15^h\,06.5^m$	$+55°\,46'$	9.8	7.3×3.5	4	H I-215
5907	GX	Dra	$15^h\,15.9^m$	$+56°\,20'$	10.3	11.5×1.7	3	H II-759
5897	GC	Lib	$15^h\,17.4^m$	$-21°\,01'$	8.4	11.0	3.5	H VI-19
5982	GX	Dra	$15^h\,38.7^m$	$+59°\,21'$	11.1	3.0×2.2	1.5	H II-764
6118	GX	Ser	$16^h\,21.8^m$	$-02°\,17'$	11.7	4.6×1.9	1	H II-402
6144	GC	Sco	$16^h\,27.2^m$	$-26°\,01'$	9.0	7.4	3.5	H VI-10
6171	GC	Oph	$16^h\,32.5^m$	$-13°\,03'$	7.8	13.0	4	H VI-40 = M107
6217	GX	UMi	$16^h\,32.6^m$	$+78°\,12'$	11.2	3.3×3.3	2.5	H I-280
6207	GX	Her	$16^h\,43.1^m$	$+36°\,50'$	11.6	3.0×1.1	3	H II-701
6229	GC	Her	$16^h\,46.9^m$	$+47°\,32'$	9.4	4.5	3.5	H IV-50
6235	GC	Oph	$16^h\,53.4^m$	$-22°\,11'$	8.9	5.0	2.5	H II-584
6284	GC	Oph	$17^h\,04.5^m$	$-24°\,46'$	8.9	6.2	3	H VI-11
6287	GC	Oph	$17^h\,05.1^m$	$-22°\,42'$	9.3	4.8	2.5	H II-195
6293	GC	Oph	$17^h\,10.2^m$	$-26°\,35'$	8.3	8.2	3	H VI-12
6304	GC	Oph	$17^h\,14.5^m$	$-29°\,28'$	8.3	8.0	2	H I-147
6316	GC	Oph	$17^h\,16.6^m$	$-28°\,08'$	8.1	5.4	3	H I-45
6342	GC	Oph	$17^h\,21.2^m$	$-19°\,35'$	9.5	4.4	2	H I-149
6356	GC	Oph	$17^h\,23.6^m$	$-17°\,49'$	8.2	10.0	4	H I-48
6355	GC	Oph	$17^h\,24.0^m$	$-26°\,21'$	8.6	4.2	3.5	H I-46
6369	PN	Oph	$17^h\,29.3^m$	$-23°\,46'$	11.4	$58'' \times 34''$	4	H IV-11
6401	GC	Oph	$17^h\,38.6^m$	$-23°\,54'$	7.4	4.8	3.5	H I-44
6426	GC	Oph	$17^h\,44.9^m$	$+03°\,10'$	10.9	4.2	1.5	H II-587
6440	GC	Sgr	$17^h\,48.9^m$	$-20°\,21'$	9.3	4.4	4	H I-150
6445	PN	Sgr	$17^h\,49.2^m$	$-20°\,01'$	11.2	3×1	4	H II-586
6451	OC	Sco	$17^h\,50.7^m$	$-30°\,13'$	8.2	8.0	3.5	H VI-13
6543	PN	Dra	$17^h\,58.6^m$	$+66°\,38'$	8.1	$23'' \times 17''$	4	H IV-37

NGC	Type[b]	Con	RA	Dec	Mag	Size (')	Rating	Herschel No.
6517	GC	Oph	$18^h\,01.8^m$	$-08°\,57'$	10.1	4.0	1.5	H II-199
6514	BN	Sgr	$18^h\,02.5^m$	$-23°\,02'$	6.3	20×20	5	H IV-41 = H V-10,11,12
6520	OC	Sgr	$18^h\,03.4^m$	$-27°\,53'$	7.6	5.0	4	H VII-7
6522	GC	Sgr	$18^h\,03.6^m$	$-30°\,02'$	9.9	9.4	3.5	H I-49
6528	GC	Sgr	$18^h\,04.8^m$	$-30°\,03'$	9.6	5.0	3.5	H II-200
6540	GC	Sgr	$18^h\,06.1^m$	$-27°\,46'$	~10	~5	3	H II-198
6544	GC	Sgr	$18^h\,07.3^m$	$-25°\,00'$	7.5	9.2	4	H II-197
6553	GC	Sgr	$18^h\,09.3^m$	$-25°\,54'$	8.3	9.2	3.5	H IV-12
6568	OC	Sgr	$18^h\,12.7^m$	$-21°\,35'$	8.6	12.0	3.5	H VII-30
6569	GC	Sgr	$18^h\,13.6^m$	$-31°\,49'$	8.4	6.4	3.5	H II-201
6583	OC	Sgr	$18^h\,15.8^m$	$-22°\,08'$	10.0	5.0	2.5	H VII-31
6624	GC	Sgr	$18^h\,23.7^m$	$-30°\,22'$	7.6	8.8	4	H I-50
6629	PN	Sgr	$18^h\,25.7^m$	$-23°\,12'$	11.3	15''	3	H II-204
6633	OC	Oph	$18^h\,27.2^m$	$+06°\,30'$	4.3	20.0	4	H VIII-72
6638	GC	Sgr	$18^h\,30.9^m$	$-25°\,30'$	9.2	7.3	3.5	H I-51
6642	GC	Sgr	$18^h\,31.9^m$	$-23°\,28'$	8.9	5.8	3	H II-205
6645	OC	Sgr	$18^h\,32.6^m$	$-16°\,53'$	8.5	15.0	3.5	H VI-23
6664	OC	Sct	$18^h\,36.5^m$	$-08°\,11'$	7.8	12.0	3.5	H VIII-12
6712	GC	Sct	$18^h\,53.1^m$	$-08°\,42'$	8.3	9.8	4	H I-47
6755	OC	Aql	$19^h\,07.8^m$	$+04°\,16'$	7.5	15.0	3.5	H VII-19
6756	OC	Aql	$19^h\,08.7^m$	$+04°\,42'$	10.6	4.0	2.5	H VII-62
6781	PN	Aql	$19^h\,18.5^m$	$+06°\,32'$	11.4	1.9×1.8	4	H III-743
6802	OC	Vul	$19^h\,30.6^m$	$+20°\,16'$	8.8	5.0	3	H VI-14
6823	OC	Vul	$19^h\,43.2^m$	$+23°\,18'$	7.1	7.0	3.5	H VII-18
6818	PN	Sgr	$19^h\,43.9^m$	$-14°\,09'$	9.3	$22'' \times 15''$	4	H IV-51
6826	PN	Cyg	$19^h\,44.8^m$	$+50°\,31'$	8.5	$27'' \times 24''$	4	H IV-73
6830	OC	Vul	$19^h\,51.0^m$	$+23°\,06'$	7.9	6.0	3.5	H VII-9
6834	OC	Cyg	$19^h\,52.2^m$	$+29°\,24'$	7.8	6.0	2.5	H VIII-16
6866	OC	Cyg	$20^h\,03.9^m$	$+44°\,09'$	7.6	15.0	4	H VII-59
6882	OC	Cyg	$20^h\,03.9^m$	$+44°\,09'$	7.6	15.0	4	H VIII-22 = 6885
6885	OC	Vul	$20^h\,11.6^m$	$+26°\,28'$	8.1	20	4	H VIII-20
6905	PN	Del	$20^h\,22.4^m$	$+20°\,06'$	11.1	$42'' \times 35''$	3	H IV-16
6910	OC	Cyg	$20^h\,23.2^m$	$+40°\,47'$	7.4	10.0	4	H VIII-56

NGC	Type[b]	Con	RA	Dec	Mag	Size (')	Rating	Herschel No.
6939	OC	Cep	$20^h 31.5^m$	$+60° 40'$	7.8	10.0	4	H VI-42
6934	GC	Del	$20^h 34.2^m$	$+07° 24'$	8.9	7.1	3.5	H I-103
6940	OC	Vul	$20^h 34.5^m$	$+28° 17'$	6.3	25.0	4	H VII-8
6946	GX	Cep	$20^h 34.8^m$	$+60° 09'$	8.8	13.0×13.0	4	H IV-76
7000	BN	Cyg	$\sim21^h 00.0^m$	$\sim +43° 35'$	–	~7	3.5	H V-37
7008	PN	Cyg	$21^h 00.5^m$	$+54° 33'$	9.9	$98'' \times 75''$	4	H I-192
7006	GC	Del	$21^h 01.5^m$	$+16° 11'$	10.6	3.6	3	H I-52
7009	PN	Aqr	$21^h 04.2^m$	$-11° 22'$	8.0	$44'' \times 23''$	4	H IV-1
7044	OC	Cyg	$21^h 13.1^m$	$+42° 29'$	12.0	7.0	1.5	H VI-24
7062	OC	Del	$21^h 23.4^m$	$+46° 23'$	8.3	5.0	3.5	H VII-51
7086	OC	Cyg	$21^h 30.5^m$	$+51° 36'$	8.4	12.0	3.5	H VI-32
7128	OC	Cyg	$21^h 43.9^m$	$+53° 43'$	9.7	4.0	2.5	H VII-40
7142	OC	Cep	$21^h 45.1^m$	$+65° 46'$	9.3	12.0	3	H VII-66
7160	OC	Cep	$21^h 53.7^m$	$+62° 36'$	6.1	5.0	4	H VIII-67
7209	OC	Lac	$22^h 05.1^m$	$+46° 29'$	7.7	15.0	4	H VII-53
7217	GX	Peg	$22^h 07.9^m$	$+31° 22'$	10.1	3.5×3.0	3	H II-207
7243	OC	Lac	$22^h 15.0^m$	$+49° 54'$	6.4	30.0	4	H VIII-75
7296	OC	Lac	$22^h 28.0^m$	$+52° 19'$	9.7	3.0	3.5	H VII-41
7331	GX	Peg	$22^h 37.1^m$	$+34° 25'$	9.5	9.7×4.5	4	H I-53
7380	OC	Cep	$22^h 47.3^m$	$+58° 08'$	7.4	20.0	4	H VIII-77
7448	GX	Peg	$23^h 00.1^m$	$+15° 59'$	11.7	2.5×1.0	1.5	H II-251
7479	GX	Peg	$23^h 05.0^m$	$+12° 19'$	10.8	3.9×3.0	3	H I-55
7510	OC	Cep	$23^h 11.1^m$	$+60° 34'$	7.9	7.0	3.5	H VII-44
7606	GX	Aqr	$23^h 19.1^m$	$-08° 29'$	10.8	4.4×2.0	2	H I-104
7662	PN	And	$23^h 25.9^m$	$+42° 32'$	8.3	$32'' \times 28''$	4	H IV-18
7686	OC	And	$23^h 30.1^m$	$+49° 08'$	5.6	15.0	4	H VIII-69
7723	GX	Aqr	$23^h 38.9^m$	$-12° 58'$	11.2	2.8×1.9	2.5	H I-110
7727	GX	Aqr	$23^h 39.9^m$	$-12° 18'$	10.6	5.6×4.0	3	H I-111
7789	OC	Cas	$23^h 57.5^m$	$+56° 43'$	6.6	25.0	4	H VI-30
7790	OC	Cas	$23^h 58.4^m$	$+61° 12'$	8.5	5.0	3	H VII-56

[a] In order of increasing right ascension.
[b] Type: GX, galaxy; BN, bright nebula; PN, planetary nebula; OC, open cluster; GC, globular cluster.

Appendix B Herschel 400 checklist[a]

	NGC	Type[b]	Con	Date Observed	Observe Again?	Telescope & Magnification	Notes
(1)	40	PN	Cep				
(2)	129	OC	Cas				
(3)	136	OC	Cas				
(4)	157	GX	Cet				
(5)	185	GX	Cas				
(6)	205	GX	And				
(7)	225	OC	Cas				
(8)	246	PN	Cet				
(9)	247	GX	Cet				
(10)	253	GX	Scl				
(11)	278	GX	Cas				
(12)	288	GC	Scl				
(13)	381	OC	Cas				
(14)	404	GX	And				
(15)	436	OC	Cas				
(16)	457	OC	Cas				
(17)	488	GX	Pis				
(18)	524	GX	Pis				
(19)	559	OC	Cas				
(20)	584	GX	Cet				
(21)	596	GX	Cet				
(22)	598	GX	Tri				
(23)	613	GX	Scl				
(24)	615	GX	Cet				
(25)	637	OC	Cas				
(26)	651	PN	Per				
(27)	654	OC	Cas				
(28)	659	OC	Cas				
(29)	663	OC	Cas				
(30)	720	GX	Cet				

	NGC	Type[b]	Con	Date Observed	Observe Again?	Telescope & Magnification	Notes
(31)	752	OC	And				
(32)	772	GX	Ari				
(33)	779	GX	Cet				
(34)	869	OC	Per				
(35)	884	OC	Per				
(36)	891	GX	And				
(37)	908	GX	Cet				
(38)	936	GX	Cet				
(39)	1022	GX	Cet				
(40)	1023	GX	Per				
(41)	1027	OC	Cas				
(42)	1052	GX	Cet				
(43)	1055	GX	Cet				
(44)	1084	GX	Eri				
(45)	1245	OC	Per				
(46)	1342	OC	Per				
(47)	1407	GX	Eri				
(48)	1444	OC	Per				
(49)	1501	PN	Cam				
(50)	1502	OC	Cam				
(51)	1513	OC	Per				
(52)	1528	OC	Per				
(53)	1535	PN	Eri				
(54)	1545	OC	Per				
(55)	1647	OC	Tau				
(56)	1664	OC	Aur				
(57)	1788	BN	Ori				
(58)	1817	OC	Tau				
(59)	1857	OC	Aur				
(60)	1907	OC	Aur				
(61)	1931	BN	Aur				
(62)	1961	GX	Cam				
(63)	1964	GX	Lep				
(64)	1980	BN	Ori				

	NGC	Type[b]	Con	Date Observed	Observe Again?	Telescope & Magnification	Notes
(65)	1999	BN	Ori				
(66)	2022	PN	Ori				
(67)	2024	BN	Ori				
(68)	2126	OC	Aur				
(69)	2129	OC	Gem				
(70)	2158	OC	Gem				
(71)	2169	OC	Ori				
(72)	2185	OC	Mon				
(73)	2186	OC	Ori				
(74)	2194	OC	Ori				
(75)	2204	OC	CMa				
(76)	2215	OC	Mon				
(77)	2232	OC	Mon				
(78)	2244	OC	Mon				
(79)	2251	OC	Mon				
(80)	2264	OC	Mon				
(81)	2266	OC	Gem				
(82)	2281	OC	Aur				
(83)	2286	OC	Mon				
(84)	2301	OC	Mon				
(85)	2304	OC	Gem				
(86)	2311	OC	Mon				
(87)	2324	OC	Mon				
(88)	2335	OC	Mon				
(89)	2343	OC	Mon				
(90)	2353	OC	Mon				
(91)	2354	OC	CMa				
(92)	2355	OC	Gem				
(93)	2360	OC	CMa				
(94)	2362	OC	CMa				
(95)	2371	$\frac{1}{2}$PN	Gem				
(96)	2372	$\frac{1}{2}$PN	Gem				
(97)	2392	PN	Gem				
(98)	2395	OC	Gem				

	NGC	Type[b]	Con	Date Observed	Observe Again?	Telescope & Magnification	Notes
(99)	2403	GX	Cam				
(100)	2419	GC	Lyn				
(101)	2420	OC	Gem				
(102)	2421	OC	Pup				
(103)	2422	OC	Pup				
(104)	2423	OC	Pup				
(105)	2438	PN	Pup				
(106)	2440	OC	Pup				
(107)	2479	OC	Pup				
(108)	2482	OC	Pup				
(109)	2489	OC	Pup				
(110)	2506	OC	Mon				
(111)	2509	OC	Pup				
(112)	2527	OC	Pup				
(113)	2539	OC	Pup				
(114)	2548	OC	Hya				
(115)	2567	OC	Pup				
(116)	2571	OC	Pup				
(117)	2613	GX	Pyx				
(118)	2627	OC	Pyx				
(119)	2655	GX	Cam				
(120)	2681	GX	UMa				
(121)	2683	GX	Lyn				
(122)	2742	GX	UMa				
(123)	2768	GX	UMa				
(124)	2775	GX	Can				
(125)	2782	GX	Lyn				
(126)	2787	GX	UMa				
(127)	2811	GX	Hya				
(128)	2841	GX	UMa				
(129)	2859	GX	LMi				
(130)	2903	GX	Leo				
(131)	2950	GX	UMa				
(132)	2964	GX	Leo				

	NGC	Type[b]	Con	Date Observed	Observe Again?	Telescope & Magnification	Notes
(133)	2974	GX	Sex				
(134)	2976	GX	UMa				
(135)	2985	GX	UMa				
(136)	3034	GX	UMa				
(137)	3077	GX	UMa				
(138)	3079	GX	UMa				
(139)	3115	GX	Sex				
(140)	3147	GX	Dra				
(141)	3166	GX	Sex				
(142)	3169	GX	Sex				
(143)	3184	GX	UMa				
(144)	3190	GX	Leo				
(145)	3193	GX	Leo				
(146)	3198	GX	UMa				
(147)	3226	GX	Leo				
(148)	3227	GX	Leo				
(149)	3242	PN	Hya				
(150)	3245	GX	LMi				
(151)	3277	GX	LMi				
(152)	3294	GX	LMi				
(153)	3310	GX	UMa				
(154)	3344	GX	LMi				
(155)	3377	GX	Leo				
(156)	3379	GX	Leo				
(157)	3384	GX	Leo				
(158)	3395	GX	LMi				
(159)	3412	GX	Leo				
(160)	3414	GX	LMi				
(161)	3432	GX	LMi				
(162)	3486	GX	LMi				
(163)	3489	GX	Leo				
(164)	3504	GX	LMi				
(165)	3521	GX	Leo				
(166)	3556	GX	UMa				

	NGC	Type[b]	Con	Date Observed	Observe Again?	Telescope & Magnification	Notes
(167)	3593	GX	Leo				
(168)	3607	GX	Leo				
(169)	3608	GX	Leo				
(170)	3610	GX	UMa				
(171)	3613	GX	UMa				
(172)	3619	GX	UMa				
(173)	3621	GX	Hya				
(174)	3626	GX	Leo				
(175)	3628	GX	Leo				
(176)	3631	GX	UMa				
(177)	3640	GX	Leo				
(178)	3655	GX	Leo				
(179)	3665	GX	UMa				
(180)	3675	GX	UMa				
(181)	3686	GX	Leo				
(182)	3726	GX	UMa				
(183)	3729	GX	UMa				
(184)	3810	GX	Leo				
(185)	3813	GX	UMa				
(186)	3877	GX	UMa				
(187)	3893	GX	UMa				
(188)	3898	GX	UMa				
(189)	3900	GX	Leo				
(190)	3912	GX	Leo				
(191)	3938	GX	UMa				
(192)	3941	GX	UMa				
(193)	3945	GX	UMa				
(194)	3949	GX	UMa				
(195)	3953	GX	UMa				
(196)	3962	GX	Crt				
(197)	3982	GX	UMa				
(198)	3992	GX	UMa				
(199)	3998	GX	UMa				
(200)	4026	GX	UMa				

	NGC	Type^b	Con	Date Observed	Observe Again?	Telescope & Magnification	Notes
(201)	4027	GX	Crv				
(202)	4030	GX	Vir				
(203)	4036	GX	UMa				
(204)	4038	GX	Crv				
(205)	4041	GX	UMa				
(206)	4051	GX	UMa				
(207)	4085	GX	UMa				
(208)	4088	GX	UMa				
(209)	4102	GX	UMa				
(210)	4111	GX	CVn				
(211)	4143	GX	CVn				
(212)	4147	GC	Com				
(213)	4150	GX	Com				
(214)	4151	GX	CVn				
(215)	4179	GX	Vir				
(216)	4203	GX	Com				
(217)	4214	GX	CVn				
(218)	4216	GX	Vir				
(219)	4245	GX	Com				
(220)	4251	GX	Com				
(221)	4258	GX	CVn				
(222)	4261	GX	Vir				
(223)	4273	GX	Vir				
(224)	4274	GX	Com				
(225)	4278	GX	Com				
(226)	4281	GX	Vir				
(227)	4293	GX	Com				
(228)	4303	GX	Vir				
(229)	4314	GX	Com				
(230)	4346	GX	CVn				
(231)	4350	GX	Com				
(232)	4361	PN	Crv				
(233)	4365	GX	Vir				
(234)	4371	GX	Vir				

	NGC	Type[b]	Con	Date Observed	Observe Again?	Telescope & Magnification	Notes
(235)	4394	GX	Com				
(236)	4414	GX	Com				
(237)	4419	GX	Com				
(238)	4429	GX	Vir				
(239)	4435	GX	Vir				
(240)	4438	GX	Vir				
(241)	4442	GX	Vir				
(242)	4448	GX	Com				
(243)	4449	GX	CVn				
(244)	4450	GX	Com				
(245)	4459	GX	Com				
(246)	4473	GX	Com				
(247)	4477	GX	Com				
(248)	4478	GX	Vir				
(249)	4485	GX	CVn				
(250)	4490	GX	CVn				
(251)	4494	GX	Com				
(252)	4526	GX	Vir				
(253)	4527	GX	Vir				
(254)	4535	GX	Vir				
(255)	4536	GX	Vir				
(256)	4546	GX	Vir				
(257)	4548	GX	Com				
(258)	4550	GX	Vir				
(259)	4559	GX	Com				
(260)	4565	GX	Com				
(261)	4570	GX	Vir				
(262)	4594	GX	Vir				
(263)	4596	GX	Vir				
(264)	4618	GX	CVn				
(265)	4631	GX	CVn				
(266)	4636	GX	Vir				
(267)	4643	GX	Vir				
(268)	4654	GX	Vir				

	NGC	Type[b]	Con	Date Observed	Observe Again?	Telescope & Magnification	Notes
(269)	4656	GX	CVn				
(270)	4660	GX	Vir				
(271)	4665	GX	Vir				
(272)	4666	GX	Com				
(273)	4689	GX	Vir				
(274)	4697	GX	Vir				
(275)	4698	GX	Vir				
(276)	4699	GX	Vir				
(277)	4725	GX	Com				
(278)	4753	GX	Vir				
(279)	4754	GX	Vir				
(280)	4762	GX	Vir				
(281)	4781	GX	Vir				
(282)	4800	GX	CVn				
(283)	4845	GX	Vir				
(284)	4856	GX	Vir				
(285)	4866	GX	Vir				
(286)	4900	GX	Vir				
(287)	4958	GX	Vir				
(288)	4995	GX	Vir				
(289)	5005	GX	CVn				
(290)	5033	GX	CVn				
(291)	5054	GX	Vir				
(292)	5195	GX	CVn				
(293)	5248	GX	Boo				
(294)	5273	GX	CVn				
(295)	5322	GX	UMa				
(296)	5363	GX	Vir				
(297)	5364	GX	Vir				
(298)	5466	GC	Boo				
(299)	5473	GX	UMa				
(300)	5474	GX	UMa				
(301)	5557	GX	Boo				
(302)	5566	GX	Vir				

	NGC	Type[b]	Con	Date Observed	Observe Again?	Telescope & Magnification	Notes
(303)	5576	GX	Vir				
(304)	5631	GX	UMa				
(305)	5634	GC	Vir				
(306)	5676	GX	Boo				
(307)	5689	GX	Boo				
(308)	5694	GC	Hya				
(309)	5746	GX	Vir				
(310)	5846	GX	Vir				
(311)	5866	GX	Dra				
(312)	5897	GC	Lib				
(313)	5907	GX	Dra				
(314)	5982	GX	Dra				
(315)	6118	GX	Ser				
(316)	6144	GC	Sco				
(317)	6171	GC	Oph				
(318)	6207	GX	Her				
(319)	6217	GX	UMi				
(320)	6229	GC	Her				
(321)	6235	GC	Oph				
(322)	6284	GC	Oph				
(323)	6287	GC	Oph				
(324)	6293	GC	Oph				
(325)	6304	GC	Oph				
(326)	6316	GC	Oph				
(327)	6342	GC	Oph				
(328)	6355	GC	Oph				
(329)	6356	GC	Oph				
(330)	6369	PN	Oph				
(331)	6401	GC	Oph				
(332)	6426	GC	Oph				
(333)	6440	GC	Sgr				
(334)	6445	PN	Sgr				
(335)	6451	OC	Sco				
(336)	6514	BN	Sgr				

	NGC	Type^b	Con	Date Observed	Observe Again?	Telescope & Magnification	Notes
(337)	6517	GC	Oph				
(338)	6520	OC	Sgr				
(339)	6522	GC	Sgr				
(340)	6528	GC	Sgr				
(341)	6540	OC	Sgr				
(342)	6543	PN	Dra				
(343)	6544	GC	Sgr				
(344)	6553	GC	Sgr				
(345)	6568	OC	Sgr				
(346)	6569	GC	Sgr				
(347)	6583	OC	Sgr				
(348)	6624	GC	Sgr				
(349)	6629	PN	Sgr				
(350)	6633	OC	Oph				
(351)	6638	GC	Sgr				
(352)	6642	GC	Sgr				
(353)	6645	OC	Sgr				
(354)	6664	OC	Sct				
(355)	6712	GC	Sct				
(356)	6755	OC	Aql				
(357)	6756	OC	Aql				
(358)	6781	PN	Aql				
(359)	6802	OC	Vul				
(360)	6818	PN	Sgr				
(361)	6823	OC	Vul				
(362)	6826	PN	Cyg				
(363)	6830	OC	Vul				
(364)	6834	OC	Cyg				
(365)	6866	OC	Cyg				
(366)	6882	OC	Cyg				
(367)	6885	OC	Vul				
(368)	6905	PN	Del				
(369)	6910	OC	Cyg				
(370)	6934	GC	Del				

	NGC	Type[b]	Con	Date Observed	Observe Again?	Telescope & Magnification	Notes
(371)	6939	OC	Cep				
(372)	6940	OC	Vul				
(373)	6946	GX	Cep				
(374)	7000	BN	Cyg				
(375)	7006	GC	Del				
(376)	7008	PN	Cyg				
(377)	7009	PN	Aqr				
(378)	7044	OC	Cyg				
(379)	7062	OC	Del				
(380)	7086	OC	Cyg				
(381)	7128	OC	Cyg				
(382)	7142	OC	Cep				
(383)	7160	OC	Cep				
(384)	7209	OC	Lac				
(385)	7217	GX	Peg				
(386)	7243	OC	Lac				
(387)	7296	OC	Lac				
(388)	7331	GX	Peg				
(389)	7380	OC	Cep				
(390)	7448	GX	Peg				
(391)	7479	GX	Peg				
(392)	7510	OC	Cep				
(393)	7606	GX	Aqr				
(394)	7662	PN	And				
(395)	7686	OC	And				
(396)	7723	GX	Aqr				
(397)	7727	GX	Aqr				
(398)	7789	OC	Cas				
(399)	7790	OC	Cas				
(400)	7814	GX	Peg				

[a] In order of increasing NGC number.
[b] Type: GX, galaxy; BN, bright nebula; PN, planetary nebula; OC, open cluster; GC, globular cluster.

Appendix C Photo credits

In order of appearance

Cover
Stephen James O'Meara

Introduction
Stephen James O'Meara
Courtesy Larry Mitchell
Stephen James O'Meara
Courtesy Larry Mitchell
Hubble Heritage Team (STScI/AURA/NASA)
Hubble Heritage Team (STScI/AURA/NASA)
Digitized Sky Survey[1]
Stephen James O'Meara
Stephen James O'Meara
Stephen James O'Meara
Stephen James O'Meara
Digitized Sky Survey[1]
Digitized Sky Survey[1]
Digitized Sky Survey[1]
Digitized Sky Survey[1]
Digitized Sky Survey[1]

Chapters 1–12
The full-page photographs opening each season (winter, spring, summer, and fall) were taken by Stephen James O'Meara.

The full-page photographs opening each month (January through December) are from the Digitized Sky Survey[1], Northern Hemisphere, courtesy the Palomar Obervatory and NASA/AURA/STScI.

All photographs of the Herschel 400 objects with northern declinations are from the Digitized Sky Survey[1], Northern Hemisphere, courtesy the Palomar Obervatory and NASA/AURA/STScI.

All photographs of the Herschel 400 objects with southern declinations are courtesy of the UK Schmidt Telescope (copyright in which is owned by the Particle Physics and Astronomy Research Council of the UK and the Anglo-Australian Telescope Board) and the Digitized Sky Survey created by the Space Telescope Science Institute, operated by AURA, Inc, for NASA, and is reproduced here with permission from the Royal Observatory Edinburgh.

Author photograph
Donna O'Meara

Index